Hubbard Brook

RICHARD T. HOLMES AND GENE E. LIKENS

Hubbard Brook

THE STORY OF A FOREST ECOSYSTEM

To Gretchen
Partners with IES

Gene Likens
June 2016

Yale UNIVERSITY PRESS

NEW HAVEN AND LONDON

Frontispiece: The Gorge on the mainstem of Hubbard Brook
(Photo by N. L. Cleavitt)

Published with assistance from the Mary Cady Tew Memorial Fund,
and from the President's Office of Dartmouth College, with special
thanks to Carol L. Folt, and from Furthermore: a program of the
J. M. Kaplan Fund.

Yale University Press books may be purchased in quantity for
educational, business, or promotional use. For information, please
e-mail sales.press@yale.edu (U.S. office) or sales@yaleup.co.uk
(U.K. office).

Designed by Richard Hendel
Set in Utopia and Aller type by Tseng Information Systems, Inc.
Printed in China.

ISBN 978-0-300-20364-6 (cloth : alk. paper)
Library of Congress Control Number: 2015948837

A catalogue record for this book is available from the British Library.

This paper meets the requirements of ANSI/NISO Z39.48-1992
(Permanence of Paper).

10 9 8 7 6 5 4 3 2 1

Preface vii

Acknowledgments ix

Timeline: From the Glaciers to the Present xi

Prologue: Step into the Forest—Today 1

CONTENTS

PART 1. THE FOREST ECOSYSTEM AS A LABORATORY

1 Ecosystem and Ecological Studies at Hubbard Brook 9

2 The Small Watershed-Ecosystem Approach 17

PART 2. CHARACTERISTICS OF THE WATERSHED-ECOSYSTEMS

3 Physical Setting and Climate 25

4 The Forest: Past and Present 33

5 A Rich Array of Organisms and Their Interactions 49

**PART 3. UNDERSTANDING FOREST ECOSYSTEM
STRUCTURE AND FUNCTION**

6 How Is Energy Transformed? 75

7 Hydrology: Water Balance and Flux 90

8 Biogeochemistry: How Do Chemicals Flux and Cycle? 95

9 The Discovery of Acid Rain at Hubbard Brook 114

**PART 4. DISCOVERIES FROM LONG-TERM STUDIES
AND EXPERIMENTAL MANIPULATIONS**

10 The Consequences of Acid Rain and Other Air Pollutants 127

11 The Effects of Forest Harvesting and Other Disturbances:
Whole-Watershed Manipulations 138

12 How Does the Forest Ecosystem Recover After Harvesting and
Other Disturbances? 148

13 How Stream Ecosystems Are Integrated with Their Watersheds 155

14 What Causes Population Change in Forest Birds? 167

15 Scaling Up: Ecosystem Patterns and Processes Across the Valley 186

16 How Is Climate Change Affecting the Forest Ecosystem? 201

PART 5. BROADER IMPACTS AND LOOKING TO THE FUTURE

17 Reaching Out: Hubbard Brook's Influence on Environmental Policy, Management, and Education 215

18 A Look Ahead: The Forest Ecosystem in the Future 225

Epilogue: Step into the Forest—2065 233

APPENDIX 1. Scientific Units: Conversions and Abbreviations 237

APPENDIX 2. Scientific Names and Lists of Selected Organisms 239

Notes 243

Bibliography 249

Index 265

PREFACE

This small corner of New Hampshire has become the most intensely studied piece of landscape on Earth and the pioneering studies which have flowed from this idea have made the name Hubbard Brook world famous among ecologists and foresters.

— DIANE DUMANOSKI

The Hubbard Brook Ecosystem Study, now under way for more than 50 years, has been an extraordinarily productive source of ecological information, derived from the research efforts of dozens of scientists from a large number of institutions. The research, conducted mostly in the Hubbard Brook Experimental Forest and nearby Mirror Lake in north central New Hampshire, was initiated in 1963, and has developed into one of the longest-lasting and most comprehensive investigations of forest and associated aquatic ecosystems anywhere. It has also served as a model for other ecosystem studies worldwide.

Using the forest as a laboratory, Hubbard Brook scientists have combined intensive field investigations with powerful, long-term manipulative experiments and modeling at a variety of spatial scales that have been designed to answer critical questions about the structure, function, and temporal development of forest and associated aquatic ecosystems and their component parts. Awareness and understanding gained from this research undergird the long-term management of natural resources for water supply, water quality, wildlife, timber yield, and sustained forest growth in the region and beyond.

More than 1,500 peer-reviewed scientific publications from this effort have contributed new understanding about the structure, function, and change of terrestrial and aquatic ecosystems, and have been widely and frequently cited in the scientific literature. As the project celebrated its 50th anniversary in 2013, however, we perceived a need to summarize and highlight the major findings, and to make them more accessible to a wider audience, from naturalists and educated general readers to land managers, policy makers, students, and professional biologists, including ecologists who might not be familiar with the Hubbard Brook studies.

The book is organized in five parts. The first introduces the conceptual framework, the initial research plans, and the origin of the research. The second describes the forest ecosystem at Hubbard Brook, its setting in the White Mountains of New Hampshire, its history, and its biological characteristics. Part 3 highlights research findings related to the structure and functioning of these forest ecosystems. The fourth and longest section highlights some of the major discoveries from long-term studies and experimental manipulations, including the effects of atmospheric pollution (particularly acid rain) and climate change. The last section, Part 5, covers the educational, policy outreach, and other applications of the research findings, including the programs of the Hubbard Brook Research Foundation, established in 1993 as a support and policy outreach arm of the Hubbard Brook Ecosystem Study (hereafter the Study).

Although "Hubbard Brook" is the name of the river draining the Hubbard Brook Experimental Forest, the label has gained popular colloquial use as well, referring to both the place and the studies that have gone on there. We use the term "Hubbard Brook" in this broad sense—for example, "research was done at Hubbard Brook." When we refer to the river and the context is unclear, we use "mainstem Hubbard Brook," which specifically refers to the major stream that flows from the Hubbard Brook valley into the Pemigewasset River. Individual tributaries in the valley that feed into the mainstem have their own names—Falls Brook and Zig-Zag Brook, for example.

As noted, this book contains the results of research conducted not only by ourselves and our research groups but also by our many colleagues, collaborators, students, and technicians as contributors to the Study. When we use the first person plural ("we did this" or "we studied that"), we intend the usage to encompass these many individual scientists and their research teams, in the spirit of the collaborative research efforts that have been a long-standing feature of the Hubbard Brook research community. We extend thanks to our many colleagues who have been and are still conducting research at Hubbard Brook for their important contributions over the years, and acknowledge their foresight, intellectual input, and hard work that have led to the findings presented in this volume. A full roster of the Hubbard Brook scientists, as well as a comprehensive and searchable list of publications from Hubbard Brook, can be found on the Study's website (http://hubbardbrook.org).

Although this book is intended as a general synthesis of the research effort at Hubbard Brook, it has not been possible to include everything that has been achieved or discovered. We hope that the results we have discussed will engage our readers, and encourage them to look further into areas that interest them and have not been fully addressed. Finally, the writing of this book has been a cooperative effort by the two of us, and any critical omissions or errors of interpretation are ours alone.

ACKNOWLEDGMENTS

We would like to acknowledge the many institutions and foundations that have funded research at Hubbard Brook. Multiple grants from the U.S. National Science Foundation (NSF) have been the backbone of funding for the research. In more recent years, support from the Long-Term Ecological Research (LTER) program at NSF has provided core funding for a number of diverse studies, and several grants from the NSF's Long-Term Research in Environmental Biology (LTREB) program have funded long-term biogeochemical monitoring and the bird population studies. Additional support has come from The Andrew W. Mellon Foundation, the Environmental Protection Agency (EPA), the National Oceanic and Atmospheric Administration (NOAA), and from many of the researchers' home universities. We particularly acknowledge the support of Dartmouth College, Cornell University, University of Connecticut, and the Cary Institute of Ecosystem Studies in our research and professional pursuits.

The U.S. Forest Service has been a partner in the Study from the beginning. Dr. Robert S. Pierce initiated hydrological, meteorological, and soil research at Hubbard Brook in 1956, and co-founded the Hubbard Brook Ecosystem Study in 1963, with Gene E. Likens, F. Herbert Bormann, and Noye M. Johnson. His successors, Dr. Chris Eagar and Dr. Lindsay Rustad, and scientists and staff at the Northern Research Station of the U.S. Forest Service have all made important contributions to the research results and to the long-term maintenance of this research program. We especially acknowledge the importance of W. E. Sopper, H. W. Lull, C. A. Federer, J. W. Hornbeck, J. S. Eaton, C. W. Martin, and T. G. Siccama for their contributions to the development of this research program. The White Mountain National Forest is acknowledged for its maintenance of critical infrastructure, such as roads, inside the forest.

We have benefited from the input and help of many colleagues during the writing of this book. The following individuals have offered helpful comments and suggestions on various drafts: John Aber, Amey Bailey, Scott Bailey, John Battles, Emily Bernhardt, Don

Buso, Tom Butler, John Campbell, Lynn Christenson, Natalie Cleavitt, Charles Cogbill, Tim Fahey, Melany Fisk, Sarah Garlick, Scott Goetz, Peter Groffman, Steve Hamburg, Sujay Kaushal, Nina Lany, Gary Lovett, Kate Macneale, Bill Reiners, Nick Rodenhouse, Tom Sherry, David Sleeper, and Jackie Wilson. Also, special thanks to Don Buso, Natalie Cleavitt, Nina Lany, Winsor Lowe, and Kathie Weathers for contributing ideas and text for some of the chapters. Many of our colleagues, students, and field assistants offered photographs for use in the book, and those have greatly enhanced its appearance. We especially thank Maxwell Likens for suggesting the relevance of *The Lorax* by Dr. Seuss to our projections in Chapter 18.

Others have also helped in important ways: Mary Martin prepared the maps of the Hubbard Brook valley; Linda Mirabile (of RavenMark, Montpelier, Vermont) redesigned several of the original conceptual models, improving their clarity and making them more attractive; Amey Bailey provided updated graphs of climate variables and trends at Hubbard Brook; Don Buso produced a large number of graphics; Bill Nelson redrew or upgraded many of the charts and figures; Nick Rodenouse helped with the bird trend graphs; and Amy Zhang helped in the early stages of preparing the illustrations. Finally, we thank Jean Thomson Black of Yale University Press for her editorial guidance, and her associates Samantha Ostrowski and Phillip King for leading us through the many steps of the publication process. Matt Gillespie assisted with the preparation of tables and text.

The publication of this book would not have been possible without the financial support of Dartmouth College and of the Furthermore Foundation. We are greatly indebted to Carol L. Folt, former interim president of Dartmouth College, for her encouragement and support.

And, we cannot end this section without mentioning the important roles of our wives, Deborah Holmes and the late Phyllis Likens. They constantly urged us on, read and commented on numerous drafts, and helped in many ways in the preparation of the manuscript. Their encouragement and support were vital to our completion of this book. We owe them more than we can possibly relate here.

TIMELINE
From the Glaciers
to the Present

14,000 years ago	Glaciers begin to retreat from the Hubbard Brook valley.
13,000 years ago	Tundra plants such as grasses and sedges become established.
12,000 years ago	Willow, heath, and other shrubs are present.
ca. 12,000 years ago	Evidence of early human inhabitants in the region.
11,500 years ago	First tree pollen (spruce) becomes preserved in lake sediments.
1,000 years ago	Seasonally nomadic native Americans, members of the Pawtucket Confederation, are living in the area.
1770s	First Europeans settle Thornton and Peeling (Woodstock).
1823–1858	Sawmills are operating at Mirror Lake.
1852	Fifteen sawmills active in the town of Woodstock.
1850–1880s	Logging, mostly of spruce and hemlock, occurs in the Hubbard Brook valley.
1900–1920	Extensive logging of softwood trees continues; hardwoods are also being cut.
1920–1921	Hubbard Brook watershed sold to the U.S. government, and subsequently incorporated into the White Mountain National Forest.
1938	A Category 3 hurricane strikes the New England region, including the Hubbard Brook valley, on September 21.
1939–1940	Salvage logging of trees damaged by the hurricane.
1955	The Hubbard Brook Experimental Forest established by the U.S. Forest Service.
1956	Robert S. Pierce initiates hydrological research.
1963	The Hubbard Brook Ecosystem Study initiated with the support of the National Science Foundation.
1963–1964	Two years of extensive drought occur.
1965–1966	First watershed-scale manipulation undertaken with the deforestation of Watershed 2.
1969	A heavy late-summer hailstorm shreds parts of the forest canopy foliage.
1969	Comprehensive research on bird populations initiated.
1969–1971	A defoliating caterpillar, the saddled prominent, consumes about 40 percent of the canopy leaves in some parts of the forest.
1970	Second watershed-scale manipulation takes place with the commercial clear-cutting of Watershed 101.

1970, 1972, 1974	Third watershed-scale manipulation undertaken in the form of a commercial stripcut of Watershed 4.
1976	The Hubbard Brook Experimental Forest designated as an International Biosphere Reserve, under UNESCO's Man and the Biosphere Programme.
1978	The Hubbard Brook Experimental Forest joins the National Acid Deposition Program/National Trends Network (NADP/NTN) for measuring precipitation chemistry.
1983–1984	Fourth watershed-scale manipulation occurs with a whole-tree harvest of Watershed 5.
ca. 1985	Moose population begins to increase.
1986	First Long-Term Research in Environmental Biology (LTREB) grant awarded by the National Science Foundation.
1988	The Hubbard Brook Experimental Forest established as a National Science Foundation Long-Term Ecological Research (LTER) site.
1988	The Hubbard Brook Ecosystem Study becomes part of the Environmental Protection Agency–funded National Dry Deposition Network (NDDN).
1990	The Hubbard Brook Ecosystem Study becomes part of the Clean Air Status and Trends Network (CASTNet).
1993	The nonprofit Hubbard Brook Research Foundation established.
1998	A January ice storm causes extensive damage to trees and other vegetation on the south-facing slope.
1999	Fifth watershed-scale manipulation occurs with the aerial application of calcium silicate to Watershed 1 to test for acid rain effects.
2000	Moose population in the Hubbard Brook valley reaches its peak, then begins slow decline.
2006	Massive freezing of forest soils occurs during the winter.
2011	Major deluge from Tropical Storm Irene occurs in late August.
2013	Microburst hits the upper central part of the Hubbard Brook valley in early June, with widespread damage to canopy trees.
2013	The fiftieth anniversary of the founding of the Hubbard Brook Ecosystem Study celebrated.

Source: Updated and modified from Likens 1985 and Cogbill 1998

Hubbard Brook

PROLOGUE

Step into the Forest—Today

SUMMER

Step into the forest on a warm summer morning and let your senses take over. The first features you might notice are the serene quiet and earthy smells. Then, as your eyes and ears become attuned to this new environment, you notice the large sugar maple and yellow birch trees, and even a few large old American beeches that have survived the devastating beech bark disease, which hit the forest several decades ago. Tangles of hobblebush and a variety of saplings and seedlings cover the forest floor, often rather densely. The surface of the ground is quite uneven, largely the result of depressions and mounds produced when large trees were tipped over and uprooted by strong winds in the past. The babbling drainage stream, or brook, as it is called in New England, is quite audible as it makes its way downhill, cascading over boulders large and small, and over cobbles that cover the incised stream channel. Listen and you might hear frass (caterpillar feces) falling from the canopy of the forest and the song of a scarlet tanager or a red-eyed vireo as they pursue insects in the higher branches.

The large trees are spaced so that their individual canopies overlap very little, but nevertheless the overall canopy of the forest is generally closed, and most of the direct sunlight is blocked from reaching the forest floor. Here and there the sun makes its way through the canopy and produces flecks of sunlight on the understory leaves below. The forest floor is covered with dead leaves, a few just starting to fall from the dominant deciduous trees that constitute the forest, and others on the ground not yet fully decayed from previous autumns. There are some old logs and many dead branches rotting away on the ground, and occasionally a large boulder or even exposed bedrock may be visible. Cushions of moss cover logs and boulders. Numerous tree seedlings and herbaceous plants such as blue-bead lilies and ferns are common components of the herbaceous plant layer of the forest. There may be remains of the more abundant group of spring ephemeral plants, such as trout lily and red trillium, or round-leaved orchids. If you are

The Hubbard Brook forest in summer. (Photo by M. G. Betts)

lucky, you might see a red eft salamander moving along the forest floor, or even a moose. And there are, of course, the all-too-frequent hums of mosquitoes and bites from stealthy black flies lingering after their spring abundance.

AUTUMN

In autumn you notice the deciduous trees starting to shed their leaves, but before they fall, the leaves lose their green color as their chlorophyll pigments degrade, exposing the underlying orange and yellow colors of the carotenoid and the reds of the anthocyanin pigments. The fall colors, so characteristic of these northern hardwood forests, paint the slopes with brilliant oranges, reds, yellows, and browns, and then the trees begin to shed their leaves. Abscission cells at the base of the stems (petioles) attaching the leaves to twigs and branches weaken the attachment of the leaves, which eventually fall to the ground. An invisible but extremely important process is the resorption of nutrients, especially nitrogen, from the leaves back into the

The forest in autumn. (Photo by W. S. Schwenk)

permanent tissues of the tree before leaf fall. In this way, nitrogen, a vital nutrient in the forest ecosystem, is retained and ready for use next spring when the trees break dormancy to form new leaves.

By about late October, most of the leaves have fallen and formed a carpet on the forest floor, giving a distinctive smell as you walk in the forest at this time of year. The timing of leaf fall depends on temperature (and the imminent frost), wind, rain, and, of course, day length. Many of the young beech saplings, however, retain their dead leaves all winter.

You would be wise to wear a jacket or raingear during this time of year in the forest. There is relatively little variation in amount of monthly precipitation, but November is the wettest month (based on long-term data), and air temperatures are dropping. Look for the colorful mushrooms emerging from the leaf layer on the ground or growing on rotting logs and tree trunks. They can be as attractive as the spring flowers that are prized by so many human visitors to the forest. You may also notice the scratch marks of black bear on the trunks, especially of the large American beech. These

Moose browsing on understory vegetation in winter. (Photo by D. C. Buso)

were probably made by the bears in their attempts to reach the beech mast (seeds) in the upper canopy.

The brook may be noisily gurgling from the higher flows that occur at this time of year, and along with rustling leaves on the forest floor this may be the most audible sound in the forest, because most of the migratory birds have already left for their more southerly winter sites. Amphibians, such as salamanders, and mammals, such as black bears, are preparing for hibernation during the long, cold winter. Blue jays, chickadees, and chipmunks are all checking and rechecking the caches of food they have hidden for the coming winter months.

WINTER

Winter may be the quietest season in the forest at Hubbard Brook, and the dazzling white of the snow on a cloudless, crystal-clear day can be breathtaking. Temperatures are typically below freezing, and the snowpack can be quite deep, at times more than a meter, requiring you to use snowshoes or skis to move around. Maybe you have come to the forest on a snowmobile, and if so, your ears may still be ringing from the sound of its engine. Such sounds are foreign in this mature forest, far away from the bustle of the highway and human settlements below. The amount of snow under conifers like balsam fir and red spruce will normally be less than that between these trees,

Early spring in the forest, with hobblebush in flower. (Photo by S. A. Kaiser)

because some of the snow is caught by the needles and branches of these evergreen trees and sublimates (evaporates directly from solid to gaseous form) without reaching the forest floor.

If you are fortunate, you may see a moose browsing on twigs and branches of hobblebush or balsam fir. Look carefully, and you may see chickadees, a woodpecker or two, and maybe a barred owl, all winter residents of the forest. In years after a synchronous masting (heavy seed production) by the deciduous trees, there will be flocks of winter finches, such as siskins, goldfinches, and grosbeaks. When you look carefully underfoot, you may find tracks of red squirrels where they scampered between trees, or even those of a fisher, a medium-sized mammal related to weasels, which feeds on rodents, including porcupines, and even birds' eggs. Woodland jumping mice and chipmunks have hibernated for the winter, but voles and red squirrels remain active. Black bears are hibernating in the shelter of a downed tree or in some other depression, but they are very difficult to see. Salamanders, frogs, and toads have

burrowed into the unfrozen soil to hibernate. The snow cover provides an insulating blanket, so the soils do not normally freeze during the winter, although sometimes they do when midwinter thaws lead to a loss of snow cover and the lack of snow lasts for a period of days.

The brook is covered with ice and snow, but if you listen carefully, you will hear it whispering along under this insulating cover, providing habitat for aquatic invertebrates and transporting nutrients and other chemicals downstream. The stream and soil ecosystems are full of living organisms with their activities going on unobserved beneath the cover of the snowpack. Look carefully, however, and you may see insects such as springtails and stoneflies crawling on the surface of the snow.

SPRING

As you return to the forest in spring, you notice that it is alive with activity. Perhaps this is the most exciting time to step into the forest. The snowpack is melting rapidly, although some spring precipitation may fall as snow, even into May. April is the period with the most stream flow because of melting snow, water-saturated soils, and abundant precipitation, so you may hear the brooks roaring during this time of year. It is difficult to cross the brook without getting your feet wet. If you look downslope through the leafless canopy in April, you may see the ice cover breaking up and melting on Mirror Lake, an event that is occurring earlier each year, on average, because of global warming.

Larger animals have come or are just coming out of hibernation, looking for food and a mate, and are basking in the warm sunshine. Deciduous trees are starting to break dormancy from the long winter and beginning to show new leaves and flowers. Sap is flowing up from the roots, bringing dissolved sugars to the twigs, a feature of sugar maple trees that has been used for thousands of years by humans in the production of what we now call maple syrup. Ephemeral herbs, such as spring beauty, trilliums, trout lily, and violets are abundant, almost completely carpeting the forest floor in some places. All of these newly sprouted leaves evaporate water by transpiration, an important source of water loss in the forest ecosystem. By late April or early May, the first of the migratory birds such as hermit thrushes and winter wrens have returned; males are singing and defending territories, while females are beginning to build nests. Hungry blue jays and red squirrels may feast on a banquet of eggs in the nest of an early breeding bird. You may soon be acutely aware of large numbers of mosquitoes and black flies in their search for a blood meal. Also, this is a good time to keep your eye on the cavities of the big, old trees in the forest, because you might see a young owl looking back at you or maybe even a flying squirrel. And beware under your feet where you might see the translucent white or pink so-called indian pipes emerging from the forest floor where they parasitize underground fungi.

■

Sensing these numerous components of the forest, both living and nonliving, throughout the year is both wondrous and delightful to all the senses, but also somewhat overwhelming in its complexity. How do these components interact to support the many forms of life, recycle nutrients and water, and respond to disturbances such as windstorms, air pollution, or a changing climate? How do they behave as a system to interact and adapt to the distinct seasonal changes in the northern environment of this forest?

PART 1

THE FOREST ECOSYSTEM AS A LABORATORY

Winter view of the south-facing slope of the Hubbard Brook
valley after the whole-watershed manipulation experiments,
circa 1970–1971. (Photo by G. E. Likens)

1

ECOSYSTEM AND ECOLOGICAL STUDIES AT HUBBARD BROOK

Some 70 years ago, Aldo Leopold captured the essence of the dynamics and complexities of ecological systems, or ecosystems, particularly at the watershed scale, when he wrote the following description:

> Soil and water are not two organic systems, but one. . . . All land represents a downhill flow of nutrients from the hills to the sea. . . . Plants and animals suck nutrients out of the soil and air and pump them upwards through the food chains; the gravity of death spills them back into the soil and air. Mineral nutrients, between their successive trips through this circuit, tend to be washed downhill. . . . The downhill flow is carried by gravity, the uphill flow by animals. There is a deficit in uphill transport, which is met by the decomposition of rocks.

He was right about this. But how do water and nutrients move through a system, what roles do plants and animals have, how do ecosystems change over time, and how are they affected by disturbance, both natural and human-caused? To answer such questions, a large number of scientists from many different disciplines have been intensively studying the terrestrial and aquatic ecosystems at Hubbard Brook in north central New Hampshire for the past five decades.

In this book we have attempted to summarize and highlight what has been learned from this extraordinarily productive and comprehensive long-term research program. In doing so, we illustrate how understanding ecosystem and ecological processes can inform environmental policy and long-term management of natural resources for water supply, wildlife habitat, and sustained forest growth in the region and beyond. The findings presented are from the Hubbard Brook Ecosystem Study, which has been conducted largely in the Hubbard Brook Experimental Forest, which occupies most of the Hubbard Brook valley, and in nearby regions of the White Mountains in New Hampshire (fig. 1.1).

Figure 1.1. An aerial view looking west up the Hubbard Brook valley in early autumn, 1996, with several experimentally manipulated watersheds at the upper right, and the Robert S. Pierce Ecosystem Laboratory of the U.S. Forest Service and part of Mirror Lake in the foreground at left. (Photo by J. F. Franklin)

THE ECOSYSTEM AS A UNIT OF STUDY

We define an ecosystem as a spatially explicit unit that includes all of the organisms, along with all components of the abiotic environment, inside its boundaries, as well as their interrelationships and interactions.[1] As implied by this definition, an ecosystem's *structure* includes its biotic and physical-chemical components, such as vegetation, microorganisms, animals, chemicals, soil, and water, while its *functions* are such processes as plant productivity and respiration (including energy capture and growth), decomposition, and nutrient cycling. The interactions among structure and function, living and nonliving, are what make ecosystems vital and interesting, but at the same time so difficult to study.

Ecosystems also have a history and a future and continually change with time in both structure and function. They provide critical ecological services,

such as regulating local climate, affecting water quality and quantity, sequestering and storing carbon, and providing habitat for wildlife. Studying ecosystem components and their interactions as an integrated system is important for understanding not only how ecosystems function but also how they change in response to both natural and anthropogenic factors, such as climatic events, air pollution, and forest harvesting.

A systems approach to the study of ecosystems involves multidisciplinary studies of all components and their interactions, with the goal of providing a holistic understanding of how the system operates, as well as for making predictions about future changes. Such an approach provides a powerful way to conceptualize the diverse interactions among the large numbers of living and nonliving components occupying a defined unit of the landscape. This approach also helps to integrate results from different disciplines, from broad-scale air-land-water relationships, such as the effect of rainwater on soil chemistry, to very local interactions among the living components, or between living components and their nonliving environment.[2]

Because any natural ecosystem is dauntingly complex, the original idea of researchers at Hubbard Brook was to choose for study a clearly defined unit of forest drained by one stream network (watershed), and then to pose questions based on a simple metaphor: that the chemistry of stream water could be used to diagnose a watershed-ecosystem in the same way that a physician uses the chemistry found in blood or urine to diagnose the health of a human patient (Box 1.1). When something unusual is found in the chemistry of blood or urine, then the physician needs to analyze what is going on inside the body of a patient to try to find the reason for the malfunctioning of the system. In a similar way, when streamwater chemistry changes, the ecosystem scientist would need to go inside the black box of the ecosystem to examine the structure (for example, individual organisms or populations) or function (such as weathering of the substrate, decomposition, predation, or nitrogen cycling) of the ecosystem to find the cause of the change.

Moreover, it was clear from the beginning of the Study that the black box of this complicated watershed-ecosystem needed to be opened to reveal and understand its diverse internal structure. Opening the black box led to many studies of critical ecosystem components, such as the natural history, dynamics, and interactions of plant, animal, and microbial populations and communities. In addition, because humans affect the structure, function, and change of watershed-ecosystems—directly, through forest cutting, for example, or indirectly through air pollution and climate change—they also needed to be considered and evaluated as integral components of the ecosystem.

In many cases, this black box approach to studying watershed-ecosystems required long-term

Box 1.1. What Is a Watershed-Ecosystem?

In the studies at Hubbard Brook, research has focused on watershed-ecosystems. A watershed-ecosystem is both a watershed and an ecosystem, where the water drainage divide, or phreatic boundary, of a catchment area is used to define the boundary of an ecosystem. The phreatic boundaries may be difficult to determine exactly, but they are approximated by the topographic boundaries, which are more readily defined. Using these topographic boundaries, it is possible to measure the movement of water, chemicals, and animals into and out of the watershed-ecosystem. At Hubbard Brook, the entire valley from the ridge tops to the Pemigewasset River can be considered one large watershed-ecosystem. For research and experimental purposes, nine small watersheds on the upper slopes of the valley are designated as experimental watersheds (numbered W1 through W9), and these have been the sites of the most intensive studies.

The small-watershed approach has been used to great advantage in the Study in developing quantitative measures of inputs and outputs of water and chemicals for these discrete watershed-ecosystem units of the Hubbard Brook Experimental Forest and for conducting large-scale experiments involving manipulations of whole watersheds.

investigations of structure, function, and change, which enhanced the understanding and utility of the other, more focused, studies. In practice, this rather straightforward metaphor worked very well, being guided from the beginning by a rigorous, but still simple, conceptual model of the ecosystem's structure and function (fig. 1.2).

This conceptual model considered a watershed and the stream network draining it as an inseparable

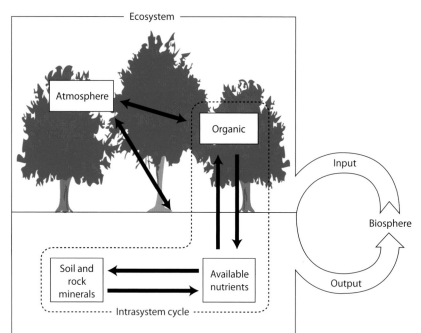

Figure 1.2. The initial conceptual model of nutrient cycling and flux relationships for a watershed-ecosystem, showing sites of accumulation and major pathways of movement of nutrients in the system. Cycling is defined as movements or exchanges inside the ecosystem boundaries, while flux refers to movements across the boundaries of the ecosystem. Chemical elements without a gaseous phase at normal biological temperatures, such as calcium, form an intrasystem cycle. Inputs and outputs for the ecosystem occur through geologic, meteorologic, and biologic vectors. (Modified from Bormann and Likens 1967; reprinted by permission of AAAS)

functional unit, which provided the basis for the small watershed-ecosystem approach used in the Study.[3] This model identified the primary ecological and biogeochemical features and interactions of the watershed-ecosystem, and was used to guide thinking and development of research questions as the Study matured. The initial questions that were addressed included whether there is a net gain or net loss of nutrients in an ecosystem over time, how the chemistry of precipitation changes as it passes through the ecosystem, and how the biota affect nutrient cycling. These are questions central to the science of biogeochemistry, which links the effects of biology, geology, chemistry, hydrology, and many other disciplines on the flux and cycling of water, nutrients, and other chemicals in an attempt to forge a more complete understanding of how complicated ecological systems function and interact, especially over time.

THE HUBBARD BROOK ECOSYSTEM STUDY

The Study was initiated in the early 1960s as a collaborative effort among several ecologists,

hydrologists, and ecosystem scientists (Box 1.2). Nine small watersheds, each of them tens of hectares in size, on the upper slopes of the valley served as the ecosystems used for experimental study. There were no precedents to guide these studies, because similar comprehensive studies of natural ecosystems did not exist. It was reasoned that a solid base of studies on nutrient-hydrological interactions should be constructed first, and subsequent expanded studies could be built on these.[4]

The official starting date for the Study was June 1, 1963, when the first samples of stream water were collected (fig. 1.3). The first precipitation sample was collected several weeks later, on July 24. The nine small watersheds were the test tubes in our laboratory, and research commenced, primarily on biogeochemistry, hydrology, and plant ecology. As the Study evolved, questions began to arise about other ecosystem components, leading to additional investigations. In 1969, comprehensive long-term studies of bird populations were initiated.[5] With time, the scope of studies and the number of cooperating scientists increased significantly, as they took up detailed

The Hubbard Brook Experimental Forest was established in 1955, and is administered by the Northern Research Station of the U.S. Forest Service, part of the Department of Agriculture. In 1963, F. Herbert Bormann (a terrestrial ecologist), Gene E. Likens (an aquatic ecologist), and Noye M. Johnson (a geologist), then professors at Dartmouth College, and Robert S. Pierce (a soil scientist), U.S. Forest Service Project Leader, initiated the Study to investigate the ecological, hydrological, and biogeochemical dynamics and interactions in watershed-ecosystems at Hubbard Brook. The interests and expertise of these scientists blended and complemented one another strongly. In the 1960s, cooperative research between Forest Service scientists and academic scientists was somewhat unusual, but at Hubbard Brook, a fruitful partnership was forged from the beginning under the leadership of Dr. Pierce, and has continued to this day.

The core funding for much of the ecological and biogeochemical research at Hubbard Brook has come from grants provided by the National Science Foundation, including through its Long Term Ecological Research (LTER) and Long Term Research in Environmental Biology (LTREB) programs.

Figure 1.3. Gene Likens collecting water sample from a stream at Hubbard Brook. (Photo from G. E. Likens)

conduct research at Hubbard Brook, on topics ranging from soil hydrology to the effects of moose on forest vegetation to the impact of global climate change. It is a very active research site.

As the research program expanded over time and new investigators became involved, the conceptual framework broadened to include more detailed considerations of ecosystem structure and function, how these are determined, and the services they provide (fig. 1.4). This new model conceptualizes how ecosystem patterns and processes might be determined or controlled by abiotic and biotic factors (state variables) and by a variety of stochastic ("random") processes. In turn, it illustrates how ecosystems provide a number of services that influence plant productivity (carbon capture), water quantity and quality, the maintenance of biodiversity, and other functions.

LONG-TERM ECOSYSTEM AND ECOLOGICAL STUDIES

Truly long-term studies extending over many decades, like those at Hubbard Brook, are particularly valuable for the unique insights they provide about ecosystem structure, function, and particularly change, but they are quite rare. To be successful, long-term research

studies of plant, animal, and microbial communities and their dynamics, and of soils and geology. The scale of studies expanded further to include patterns and processes across the entire Hubbard Brook valley and in the region. In this way, the Study brought together scientists from the U.S. Forest Service and other federal agencies with academic scientists from institutions throughout the world. Today, some 40–50 investigators from more than 20 institutions

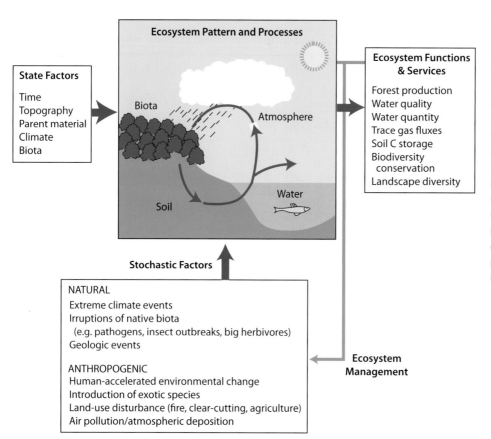

Ecosystem Pattern and Processes

State Factors

Time
Topography
Parent material
Climate
Biota

Biota

Atmosphere

Soil

Water

Stochastic Factors

NATURAL
Extreme climate events
Irruptions of native biota
  (e.g. pathogens, insect outbreaks, big herbivores)
Geologic events

ANTHROPOGENIC
Human-accelerated environmental change
Introduction of exotic species
Land-use disturbance (fire, clear-cutting, agriculture)
Air pollution/atmospheric deposition

Ecosystem Functions & Services

Forest production
Water quality
Water quantity
Trace gas fluxes
Soil C storage
Biodiversity
  conservation
Landscape diversity

Ecosystem
Management

Figure 1.4. An expanded conceptual diagram for the Hubbard Brook Ecosystem Study, showing how abiotic and biotic factors influence ecosystem patterns and processes and, ultimately, ecosystem functions and services. Stochastic factors that are a result of human activities can be influenced or mitigated by management to affect ecosystem services. (Modified from Groffman et al. 2004; reprinted by permission of Oxford University Press)

and monitoring must be driven by questions, and follow an adaptive monitoring approach, in which the development of conceptual models, questions, data collection and analysis, and data interpretation are linked in an iterative process.[6] An example of a successful long-term, place-based research program is at the Rothamsted Experimental Station (now Rothamsted Research) in Harpenden, Hertfordshire, England, which began in 1843.[7] Sustained research and monitoring at Rothamsted have been successful largely because of its tractable, well-designed questions to guide the research and monitoring, appropriate statistical designs for experiments, strong leadership, monitoring that is consistent in maintaining core programs yet nimble in supporting new or revised questions, and ability to maintain stable funding.[8] In general, this approach to long-term research and monitoring has been used at Hubbard Brook, which greatly enhances the value

and credibility of the long-term environmental measurements (Box 1.3).

THIS BOOK

It is especially difficult to summarize 50 years of research by a large number of investigators, but basically, the effort has been driven by the scientists and the ideas they have tested in this place-based, long-term study. The research at Hubbard Brook has been diverse and unusual in producing a large number of important scientific findings. We highlight these findings and how they are relevant to the long-term management of natural resources, such as water supply, water quality, wildlife, timber yield, and sustained forest health and growth, especially in the northern hardwood forest ecosystem. Indeed, practical application of the research was an important objective of the Study from the beginning. Quoting from the original proposal to the National

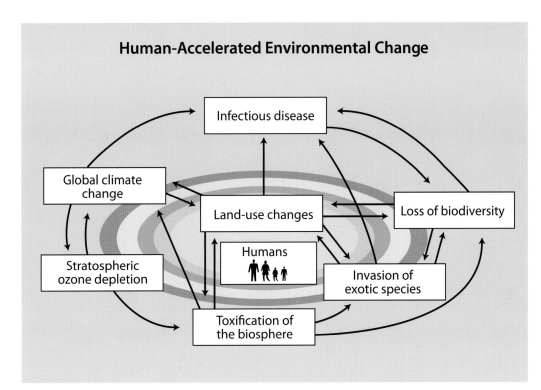

Human-Accelerated Environmental Change

Infectious disease

Global climate change

Land-use changes

Loss of biodiversity

Stratospheric ozone depletion

Humans

Invasion of exotic species

Toxification of the biosphere

Figure 1.5. The large and interactive effects of humans in accelerating major components of environmental change, with their linkages. Humans are at the center of these changes, with their feet on the accelerator. (Modified from Likens 2001)

Box 1.3. Why Is Long-Term Monitoring Valuable?

Long-term, high-quality monitoring has been a hallmark of the research program at Hubbard Brook. Without consistent sampling and careful application of analytical techniques over time, many findings reported from the Study would not have been evident or substantiated. This includes, for example, the long records of precipitation and streamwater chemistry that have been crucial for documenting the occurrence and then decline of acid rain (Chapters 9 and 10), the strong correlations between sulfur dioxide emissions and sulfate concentrations in precipitation and stream water (Chapter 9), the long-term record of masting cycles by forest trees (Chapter 5), and the decline in bird populations (Chapter 14). Knowing the patterns of environmental change over long periods allows not only for the identification of important trends but also for evaluating the impact of unusual events, such as an ice storm, a particularly cold winter, or a sudden outbreak of defoliating insects (Chapter 4). Long-term monitoring can tell us definitively whether an event such as a very cold winter is unusual or not.

Environmental monitoring at Hubbard Brook exemplifies the use of monitoring as part of a large integrated research program. Measurements have been designed to meet strict guidelines for addressing clear and important questions and for using consistent methods.[a] As new methods become available, their use is overlapped with the previous ones, allowing for careful calibration and continued integrity of the long-term record.[b]

NOTES

a. Lovett et al. 2007; Lindenmayer and Likens 2009.

b. Buso et al. 2000.

Science Foundation: "Eventually, after the basic . . . relationships of the watershed are known, it will be possible to evaluate the effects of land management practices . . . [and] provide practical guidelines for land managers."[9]

Now, faced with greatly increasing problems of human-accelerated environmental change and demands on resources (fig. 1.5), the Hubbard Brook approach to understanding complicated ecosystem function, and particularly the ecosystem response to anthropogenic disturbances, is more valuable than ever. Some of our major findings regarding changes caused by acid rain, clear-cutting of forests, alteration of major biogeochemical cycles and fluxes (of nitrogen, for example), bird decline, and climate change are directly relevant to current environmental policy and conservation concerns.

2

THE SMALL WATERSHED-ECOSYSTEM APPROACH

Watersheds (the term used in North America for a water catchment basin or discrete drainage area) have served scientists for almost 90 years as a useful unit of study for quantifying hydrologic and erosion dynamics in landscapes. Early work at the Coweeta Hydrologic Laboratory in North Carolina in quantifying hydrologic and erosion dynamics was pioneering.[1] It was at Hubbard Brook, however, that this approach was first applied comprehensively and quantitatively, and in an integrated way, to understand how water, nutrients, and organisms interact ecologically in topographically defined watershed-ecosystems. The opportunity to exploit nutrient-hydrologic-biologic interactions for greater ecological understanding of forest ecosystems has now been pursued in the watershed-ecosystems of the Hubbard Brook Experimental Forest for more than 50 years.

The fundamental components of a small forest watershed-ecosystem are a topographically defined basin drained by a stream network, the basin's soil and bedrock, and the plants, animals, and microorganisms that live within the boundaries of it. The stream network is fed by precipitation, but some of the water provided in precipitation is returned to the atmosphere through evaporation (fig. 2.1).[2] The small-watershed approach uses this unit to identify, study, experimentally manipulate, and ultimately understand the processes operating in such ecosystems. Using this approach, F. Herbert Bormann and Gene E. Likens expanded their initial model to describe the pathways and flux patterns for nutrients and other chemical elements in forested watershed-ecosystems (fig. 2.2).[3] This updated conceptual model depicted and clarified the major biogeochemical and ecological components of an ecosystem, identified their relationships, and showed cycling pathways and input and output vectors for a watershed. The model identified and directed quantitative measurements of these major pathways (what needed to be measured and where), facilitated the posing of important questions about these relationships, and was successful in stimulating thinking, developing questions, and guiding the research.

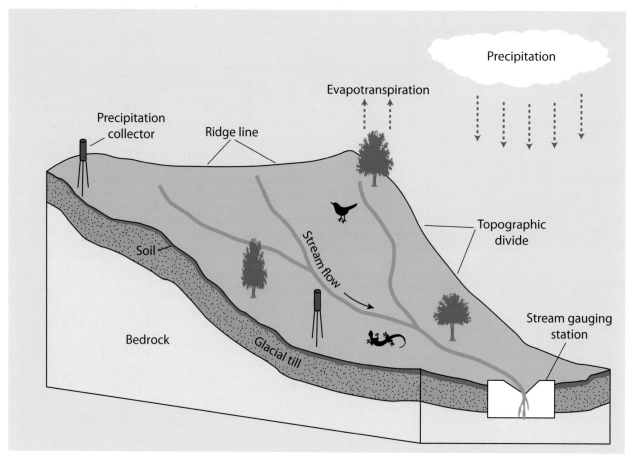

Figure 2.1. Fundamental components of a small watershed-ecosystem at Hubbard Brook, showing topographic boundaries, underlying bedrock, input from the atmosphere (via precipitation), and output (evapotranspiration and stream) pathways. Plants, animals, and microorganisms constitute the biotic components of the watershed-ecosystem. (Modified from a diagram by J. L. Campbell, Northern Research Station, U.S. Forest Service)

Chemical elements and compounds, such as calcium and nitrate, are transported into and out of the forest ecosystem by geologic, meteorologic, and biologic vectors. Geologic inputs and outputs are defined as dissolved or particulate matter that is carried into or out of the system by moving water or by colluvial action (driven by gravity). Geologic *inputs* cannot exist when the ecosystem is defined as a watertight watershed. As a matter of budgetary accounting, the generation of nutrients by chemical weathering is considered a release from existing pools inside the ecosystem rather than input across the ecosystem's boundary (Box 2.1). Meteorologic

fluxes come from gaseous materials or dissolved and particulate matter in precipitation, dust, or other windborne materials—such as when leaves and pollen are blown across the boundary of the ecosystem. Biological inputs and outputs result from movements of animals across the ecosystem's boundary, and tend to cancel one another. Meteorologic outputs from humid systems like those at Hubbard Brook tend to be very small for most elements, except for those with a prominent gaseous phase like carbon, sulfur, nitrogen, and chlorine. Thus, in accounting for chemical elements and compounds in watershed-ecosystems at Hubbard Brook, without a prominent gaseous

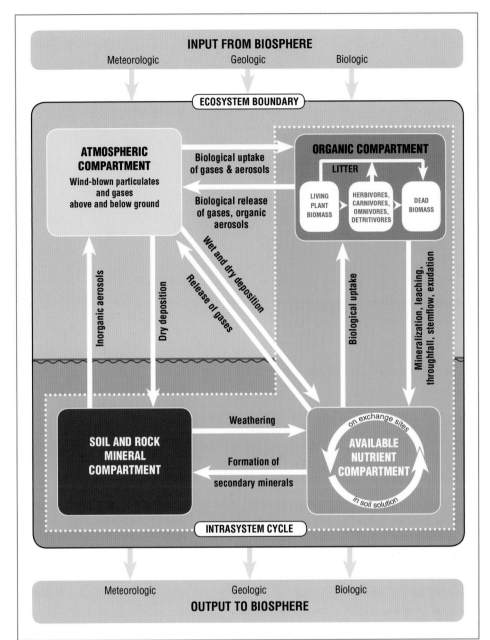

INPUT FROM BIOSPHERE

Meteorologic   Geologic   Biologic

ECOSYSTEM BOUNDARY

ATMOSPHERIC COMPARTMENT

Wind-blown particulates and gases above and below ground

Biological uptake of gases & aerosols

Biological release of gases, organic aerosols

ORGANIC COMPARTMENT

LITTER

LIVING PLANT BIOMASS   HERBIVORES, CARNIVORES, OMNIVORES, DETRITIVORES   DEAD BIOMASS

Inorganic aerosols

Dry deposition

Wet and dry deposition

Release of gases

Biological uptake

Mineralization, leaching, throughfall, stemflow, exudation

SOIL AND ROCK MINERAL COMPARTMENT

Weathering

Formation of secondary minerals

AVAILABLE NUTRIENT COMPARTMENT

on exchange sites

in soil solution

INTRASYSTEM CYCLE

Meteorologic   Geologic   Biologic

OUTPUT TO BIOSPHERE

Figure 2.2. The second-generation watershed-ecosystem conceptual model describing the fundamental components and relationships of nutrient flux and cycling within a watershed-ecosystem, such as those studied at Hubbard Brook. (Modified from Likens et al. 1977; reprinted by permission of Springer Science+Business Media)

phase, the major terms that need to be considered are reduced to meteorologic inputs in precipitation and geologic outputs in stream water.

The small watershed-ecosystem approach, which was developed at Hubbard Brook and is now used widely around the world, "provides a means of studying the interrelationships between the biota and the hydrologic cycle, various nutrient cycles, and energy flow in a single [ecological] system. . . . Our ecosystem model probably would have remained an intellectual curiosity had not we conceived of the 'small watershed technique' for measuring input-output relationships."[4] This watershed-ecosystem technique allows for quantitative measurements in

Box 2.1. What Are Ecosystem Boundaries?

Ecosystem boundaries are defined as the boundary (plane) across which substances or organisms can pass and be quantified. Ecosystem boundaries are difficult to determine even for a lake or an island, not to mention a forest. They are usually defined to meet the pragmatic needs of the investigator who wants to make quantitative measurements of the flux of water or a chemical substance or organisms. But once this definition is made, such as for a watershed in the case of the Hubbard Brook studies, quantitative measurements can be made, and can link the watershed-ecosystem to the surrounding biosphere by a system of inputs and outputs. Small watershed-ecosystems can therefore be visualized as integrated units of the landscape, where the water and chemical cycles and fluxes are linked (and integrated).[a]

NOTE

a. Bormann and Likens 1967.

Figure 2.3. A combination San Dimas flume and 90° V-notch gauging weir for measuring stream flow in Watershed 6. The weir is anchored to the bedrock, with wing-walls extending upslope to capture the water draining downhill. The ponding basin upstream of the V-notch is covered and heated during the winter to avoid ice buildup in the V-notch. The height of water (stage-height) in the notch is calibrated to amount of stream flow so only stage-height measurements done in the small gauge house are recorded. The U.S. Forest Service currently is transitioning from mechanical devices for recording stream stage height (float and pen recorders) to electronic devices for data collection. This allows real-time data transmission from the weirs directly to computers at the Robert S. Pierce Ecosystem Laboratory. (Photo by A. S. Bailey, Northern Research Station, U.S. Forest Service)

very complicated ecosystems rather than remaining only as a theoretical concept.

THE FOREST WATERSHED-ECOSYSTEMS AT HUBBARD BROOK

After careful analysis and thought, the Hubbard Brook Experimental Forest was selected in the mid-1950s by the U.S. Forest Service as a representative drainage area for studies of forest/hydrologic interactions in New England.[5] At that time, weirs were installed to measure stream flow in headwater drainages (fig. 2.3). Precipitation gauges and meteorological instruments were established simultaneously in nearby cleared areas of the forest, so researchers could begin taking measurements of the meteorology for these headwater watersheds and their hydrological and biogeochemical inputs and outputs (fig. 2.4). Also, since 1988, an EPA-funded Clean Air Status and Trends

Network (CASTNet) flux tower has been operated to estimate gaseous and atmospheric particulate inputs (those that arrive through dry deposition, rather than in precipitation).

The Hubbard Brook forest turned out to be an ideal laboratory, especially as a place to undertake studies of the interaction of hydrologic and nutrient cycles at the watershed-ecosystem scale, for a number of reasons. Among these:

- Gauging weirs and their wing walls could be attached to bedrock, so that water draining

Figure 2.4. A wet-dry precipitation collector used in the early years of the Study to collect precipitation (wet deposition) and small particles from the atmosphere during dry conditions (dry deposition). When precipitation falls on the heated sensor (the small brown rectangle at front), the lid moves to cover the open (dry deposition) bucket. When the precipitation event is finished, the water on the sensor evaporates, opening the electrical circuit, and the lid moves back in place over the wet-only bucket. This device provided relative, not quantitative, measures of dry deposition. (Photo by G. E. Likens)

downhill is directed through the weir for quantitative measurement of stream flow.

- Topographical boundaries of the watersheds chosen for experimental study could be clearly defined and identified.
- Hydrologic characteristics of the watersheds, especially those on the south-facing slope of the valley, were similar to regional watersheds, even those of much larger size.[6]
- Insignificant deep seepage of water through bedrock within discrete headwater watersheds allowed for quantitative construction of hydrologic and chemical budgets. A recent reevaluation of the watersheds on the south-facing slopes concluded that the bedrock underlying these watershed-ecosystems is relatively watertight and that deep seepage is negligible.[7]
- Adjacent watersheds had relatively homogeneous biology and geology (but

50 years have shown details to be more diverse than originally thought).

- The Hubbard Brook valley is downwind of major atmospheric pollution sources in the midwestern United States, and there are no major local air pollution sources.
- There was a sufficient number of gauged watersheds to conduct paired-watershed research, including entire watershed manipulation experiments. Because ecosystems are so complicated, any two watersheds, even adjacent ones, are not necessarily similar, and thus cannot be considered as scientific "controls," as in many laboratory experiments. Instead, adjacent nearby watersheds are referred to as "reference" watersheds, which is helpful for evaluating results from experiments, particularly over long intervals of time.[8]
- Finally, and perhaps most important, the landscape units chosen for study were small enough for quantitative studies of ecological and biogeochemical dynamics but large enough to be relevant to management questions.

APPLICATION OF THE SMALL WATERSHED-ECOSYSTEM APPROACH

The small watershed-ecosystem approach is useful because it provides a unique opportunity to learn about the workings of these complicated ecosystems, and—more important—to identify and quantify the connections these inputs and outputs provide to the rest of the biosphere. These connections give the "vital signs" for these ecosystems and are critical for developing management strategies, such as responses to inputs or outputs of critical nutrients (such as calcium or nitrogen) or pollutants (toxic metals such as lead, for example, or atmospheric acids).

The small watershed-ecosystem approach at Hubbard Brook has been highly productive for several reasons. These include providing a natural unit of the landscape suitable for quantitative study and experimentation and reducing problems with difficult-to-measure components (such as flux of water lost in deep seepage). Controlling for these problems allowed for estimating erosion, weathering, and evapotranspiration at the ecosystem level. Evapotranspiration, for example, is estimated as the difference between precipitation and streamwater output. The small-watershed approach provides a means to study interactions among hydrologic cycles, nutrient cycles, and energy flow, and allows for quantitative testing of the effects of various land-management practices or pollutants on interactions between the hydrologic and nutrient cycles.

The watershed-ecosystem model also provides the framework for visualizing and quantifying a series of living and nonliving components, such as plant and animal populations and communities, organic debris, available nutrients in the soil, primary and secondary minerals, and atmospheric gases, all linked together by food webs, nutrient cycles, and energy flow. The model was crucial for identifying the important pathways in such a complicated ecosystem.

Detailed knowledge of these components and the links involved at Hubbard Brook, all functioning at diverse scales, have led to an understanding of the interrelationships within the systems and of the ramifications of any manipulation or disturbance applied at any point in the system. These ecological and biogeochemical features have been studied at several spatial scales, ranging from the local or plot scale to small watersheds, the forest, the valley, and finally to the region. Such cross-scale studies have been instrumental in understanding how the results from Hubbard Brook studies fit into a broader geographic context. The watershed-ecosystem conceptual model can be used at any of these scales by defining the boundaries appropriately.

An additional feature on the regional scale is the inclusion of Mirror Lake, which has been investigated as part of the Study since the beginning. From a global perspective, it is exceedingly rare that a comprehensive lake ecosystem study would be embedded in and coincident with a comprehensive terrestrial ecosystem study, particularly over a long period. This juxtaposition has provided an opportunity to compare, contrast, and integrate the interactions among structure, function, and change in adjacent lake and forest ecosystems. The research on the Mirror Lake ecosystem has been described elsewhere, and will not be considered in detail here.[9]

Long-term biogeochemical and ecological research and monitoring, coupled with watershed-scale experimental manipulations to evaluate the response of ecosystems to disturbance, have been the hallmark of the Hubbard Brook studies since 1965.[10] Moreover, long-term research and monitoring at the watershed-ecosystem scale have been enriched with plot and whole-ecosystem manipulations, as well as with modeling efforts, to generate and test hypotheses concerning ecosystem response to disturbance, such as acid rain and clear-cutting. Experimental manipulations have also been used to test mechanisms at the community and population levels in the studies of the plant and animal components of the systems. Experimentation at any scale is a powerful tool in science, and it has been used widely and successfully in the Study. Validation of the small watershed-ecosystem approach as a powerful scientific technique in addressing complicated problems at the landscape scale of complexity (ecological, hydrological, biogeochemical) has been confirmed in the long-term studies at Hubbard Brook, particularly through experimental manipulation of entire watershed-ecosystems.

PART 2

CHARACTERISTICS OF THE WATERSHED-ECOSYSTEMS

The northern hardwood forest at Hubbard Brook.
(Photo by A. G. Muniz)

3

PHYSICAL SETTING AND CLIMATE

Where is Hubbard Brook and what are its characteristics? Hubbard Brook is located mostly in the townships of Woodstock and Ellsworth, Grafton County, in north central New Hampshire, about 116 km inland from the Atlantic Ocean and 210 km north of Boston. The 3,160-ha Experimental Forest occupies most of the valley, and is both part of and contiguous with the much larger White Mountain National Forest (317,478 ha). The Hubbard Brook valley is oriented in an east-west direction, with slopes predominately north- and south-facing (figs. 3.1 and 3.2). Elevations range from 222 m above sea level at Mirror Lake to 1,015 m on Mount Kineo along the southwestern rim. Mirror Lake lies near the eastern end of the valley, and much of its watershed was added to the National Forest in the 1980s.

Small streams flow downward from high on the hillsides to form the mainstem of Hubbard Brook, which drains from the valley to the east, connecting with the Pemigewasset River (the "Pemi"). The Pemi becomes the Merrimack River, and eventually discharges into the Atlantic Ocean at Newburyport, Massachusetts.

In addition to the stream system, other aquatic habitats include Mirror Lake and several small beaver ponds and associated wet areas near the western end of the valley. More than a hundred years ago, the outflow from Mirror Lake supported sawmills and a tannery. Now, a few cottages occur along about half of the lake's shoreline, and the lake is used primarily for recreation, such as swimming and fishing, as well as for research. Other human settlements, old pastures, and second-growth forests occur in the lower parts of the valley between Mirror Lake and the Pemi.

THE EXPERIMENTAL WATERSHED-ECOSYSTEMS

Nine small headwater catchments (12 to 68 ha) in the forest have been designated by the U.S. Forest Service as experimental watershed-ecosystems. All nine have gauging weirs where water flow and chemical concentrations are measured, allowing calculation of nutrient fluxes. These gauged watersheds are located along the upper slopes of the forest; six face south

Figure 3.1. Aerial view of the south-facing slope of the Hubbard Brook valley in winter, 1970, with higher peaks of the White Mountains in the background. Two experimental watersheds are evident: Watershed 4 (left), with horizontal strips of forest harvested, and Watershed 2 (right), completely deforested in 1965–1966. (Photo from J. W. Hornbeck, Northern Research Station, U.S. Forest Service)

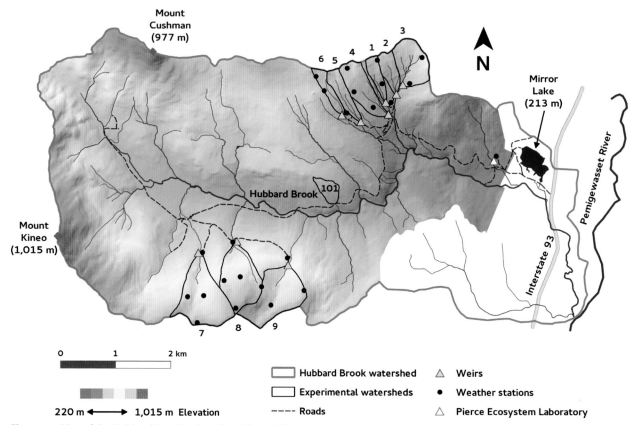

Figure 3.2. Map of the Hubbard Brook valley, from Mount Kineo and Mount Cushman along the high-elevation ridges to the Pemigewasset River. The Hubbard Brook Experimental Forest, with topographic contours shown in color relief, is drained by Hubbard Brook and its tributaries. The map shows the locations of the experimental watershed-ecosystems numbered 1 through 9, and 101, weirs, and a network of weather stations. Mirror Lake and the lower part of the Hubbard Brook watershed are shown as the uncolored area at right, outlined by a solid gray line. (Map by Mary Martin, University of New Hampshire)

at altitudes of 442–792 m, and three face north at 610–910 m. Four of these watersheds (1, 2, 4, and 5) have been used for experimental manipulations. Watershed 6 serves as the biogeochemical reference watershed—an undisturbed ecosystem, used as a baseline for comparison with the results from the experimentally treated watersheds. Watershed 3 serves as the hydrological reference for the south-facing catchments. A tenth, nongauged watershed (Watershed 101) was clear-cut in 1970 to test whether the hydrological and biogeochemical effects that had been found in other watershed experiments could be replicated after commercial harvest by clear-cutting.

Meteorological stations are distributed in and near these watersheds at relatively high density, about 1 per 13 ha in the south-facing watersheds. A narrow 10-kilometer gravel road provides access to the experimental research areas on both the south- and north-facing slopes and the western end of the valley.

TOPOGRAPHY

The terrain throughout the forest is hilly and often steep, with an average slope of 20–30%. Large granite boulders deposited by glaciers (glacial erratics) are scattered about. One of the features that one notices when walking through the forest is the unevenness of the ground. This surface constitutes what is called pit-and-mound topography, which is the result of large trees having been uprooted by high winds, leaving a mound of soil next to the pit once occupied by the root system (fig. 3.3). Over time these mounds settle, the pits partially fill with soil washed off the upturned root system by rain and melting snow, as well as with forest litter, herbaceous plants, and fallen debris, but the lumpy terrain remains for decades or possibly even centuries. This uprooting of trees mixes mineral soil from below with nutrient-rich surface layers, affecting weathering rates and biogeochemical cycling and the availability of germination sites for seeds of some forest plants, such as yellow birch. These local natural disturbances of the soil from uprooting of trees are therefore important to ecosystem nutrient cycling and forest regeneration processes.[1]

Figure 3.3. Pit-and-mound topography of the forest floor is created by large trees, such as this one, tipping over in heavy winds, leaving a pit where root systems were and forming an adjacent mound of soil. This sugar maple was toppled during a microburst in early June 2013. (Photo by R. T. Holmes)

The forest was logged extensively in the late nineteenth and early twentieth centuries, but except for some salvage logging after the 1938 hurricane and the experimental harvesting between the mid-1960s and the 1980s, the vegetation has remained relatively undisturbed since about 1920. The presence of pit-and-mound topography and the lack of stone walls indicate that most of the forest was never cleared for cultivation or grazing, unlike the area around Mirror Lake, the Pemigewasset valley, and other nearby low-lying parts of the region.[2]

GEOLOGY AND SOILS

A continental ice sheet covered all of New England until it began to recede between 23,000 and 28,000 years ago. By about 14,000 years ago, the ice front had receded to the vicinity of Hubbard Brook. As the ice melted, debris from the glacier was deposited on the bedrock, providing the stony till from which today's soil has developed. At present, the till lying

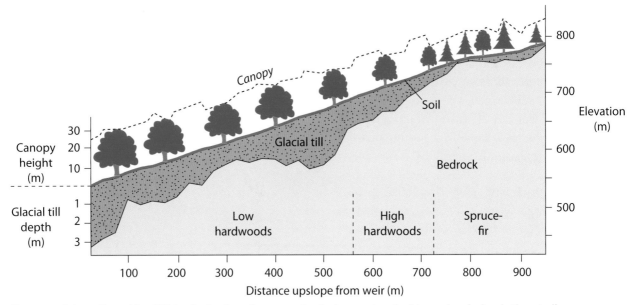

Figure 3.4. Schematic profile of Watershed 6, the reference watershed-ecosystem for biogeochemical and other studies. Depth of glacial till and tree canopy heights decrease with elevation. The upper layer of till is more porous and is underlain by a discontinuous compacted, impermeable layer (pan). (Modified from Bormann et al. 1970)

Figure 3.5. Soil profile at Hubbard Brook. The grayish-white layer below the dark organic (humus) forest floor constitutes a bleached and chemically leached horizon, and identifies the soil type as a spodosol. The whitish layer grades into a red-orange layer of accumulated organic matter, clays, iron, and aluminum oxides. Below this is the subsoil, which varies from partially weathered to unweathered mineral components. (Photo by S. P. Hamburg)

over bedrock in the valley ranges from zero to several meters in thickness, varying with topography and elevation (fig. 3.4). The bedrock is highly metamorphosed sedimentary and granitic rock. In the area of the experimental watersheds, this bedrock is relatively watertight, and losses of water by deep seepage are considered negligible.[3]

Over time, weathering of the glacial till brought about by water draining through the system, together with the cool continental climate and the development of a vegetative cover that added humus and organic acids, resulted in the formation of what is called a spodosol soil type. Spodosols are well-drained, acidic, and relatively infertile soils that occur under forests in cool, moist climates. This soil type is characterized by distinctive layers (fig. 3.5) that have developed over centuries as water percolates through the overlying organic material (the 3- to 15-cm black layer of humus), carrying leached materials, such as dissolved organic matter and minerals (for example, aluminum and iron), into the underlying mineral soil. In some areas of the forest, the subsoil develops into a hardpan of compacted till.

The well-drained nature of the soil (coarse-grained, with high porosity) and the presence of underlying bedrock or hardpan result in fast runoff of water. Most precipitation infiltrates the soil, and there is very little, if any, overland flow, except after very heavy rains. Soils typically remain unfrozen during the winter, because of the insulating effect of a deep snow cover and a thick organic layer consisting mostly of plant matter in varying stages of decay.

STREAM ECOSYSTEMS

The surface network of streams in the watersheds has a classic branching or dendritic pattern, originating as small first-order streams on upper slopes near the ridge line, progressing to second- to fourth-order streams as branches join, and culminating in the mainstem of Hubbard Brook, a fifth-order stream in the lower reaches of the valley (fig. 3.6). First-, second-, and third-order streams, which form the headwaters of this network, drain water from below the ridgelines and upper slopes to approximately the mid-elevation of the forest where the gauging weirs are located. These smaller streams have flows ranging from zero during occasional summer droughts to hundreds of liters per second during snowmelt or storm events. Stream flow usually increases noticeably during autumn, when transpiration drops as deciduous trees lose their leaves.

From June through October, headwater streams in the experimental watersheds are heavily shaded by the overarching forest canopy, and do not vary in temperature by more than a degree or two Celsius during a day. Annual water temperature ranges from just above 0°C when the streams are covered with ice and snow to about an average of 18°C in late summer, similar to the soil temperatures at that time.[4] The maximum temperature of water in the streams is about 21°C and occurs just before leaf-out during the spring, when incoming solar radiation is unimpeded by the forest canopy. Stream flow is lowest in the summer, and peaks in April coincident with snowmelt.

The mainstem of Hubbard Brook originates as an outlet from a beaver pond in the west end of the valley,

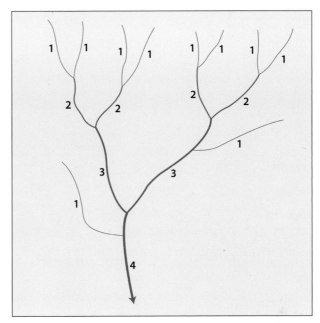

Figure 3.6. Strahler-Horton hierarchical scheme for describing the stream network in a drainage area. In this system of classification, two unbranched first-order streams join to form a second-order stream, two second-order streams join to form a third-order stream, and so on. (G. E. Likens, unpublished)

and is about 14 km long, with 16 major tributaries. In all, the valley includes a total of about 100 km of perennial streams (flowing year round) and 25 km of ephemeral streams (flowing only at snowmelt or after heavy rains).

Stream waters at Hubbard Brook typically contain low concentrations of dissolved elements and nutrients, such as calcium, potassium, and magnesium, but they are normally supersaturated with dissolved oxygen. They are acidic, with average pH of less than 5.5 in south-draining headwater streams. Stream water in Watershed 9 on the north-facing slope is the most acidic (mean pH of about 4.4), and is typically brown stained with dissolved organic matter.

Stream channels make up only about 1–2% of the land area, but are nonetheless important in many ecosystem processes (fig. 3.7). They transport water and dissolved and particulate nutrients out of the watershed-ecosystems. They are also sites where

Figure 3.7. The mainstem of Hubbard Brook, a fifth-order stream, during moderate flow conditions. (Photo by D. C. Buso)

nutrients are transformed, stored, or released, thus altering the characteristics and dynamics of chemical and organic matter outputs from the interconnected terrestrial ecosystems of which they are an integral part. The biotic components of the stream, from algae to salamanders, are diverse and dynamic. As such, streams function as ecosystems in their own right.

Water in the headwater streams usually contains low concentrations of suspended materials, except during storm flows. About 33 kg/ha of particulate matter (inorganic plus organic), for example, is transported out of Watershed 6 by stream flow each year.[5] Watersheds with mixed land use patterns typically have much higher output of particular matter (sediment yield). The Susquehanna River in the mid-Atlantic region, for example, with about 88% forest cover in its watershed, has an annual sediment yield about five times greater than that of Watershed 6 at Hubbard Brook. Agricultural watersheds have much greater sediment losses.[6]

Branches and other plant debris often become trapped in streams behind logs spanning (damming) the stream channel. These debris dams play key roles in many physical, chemical, and biological processes in stream ecosystems. They are hot spots for the anaerobic microbial transformation of inorganic nitrogen into gaseous form, for example, which leads to loss of nitrogen gas from the system, a process termed denitrification.[7]

CLIMATE
Short, cool summers and long, cold winters characterize the climate at Hubbard Brook. Notable features of the climate include distinct seasons, high changeability of the weather (hourly, daily, weekly), a wide range in both daily and annual air temperatures,

and relatively equal distribution of precipitation throughout the year. On average, precipitation falls every third day at Hubbard Brook. In summer, there are often thunderstorms accompanied by lightning. In early spring and late autumn, cold rains and ice storms, accompanied by strong winds, periodically damage vegetation. Winds are predominately from the north or northwest, bringing air masses (and pollutants) from the middle of North America. Major storms in winter often come from the northeast (the infamous nor'easters).

High-quality measurements of meteorological and hydrologic variables made by the U.S. Forest Service have been a hallmark of the Hubbard Brook studies since the 1950s. Variables being measured include solar radiation, air and soil temperatures, precipitation amount, humidity, wind direction and speed, snow depth and water content, snow cover, and stream flow.

Solar radiation is the major energy source for plants of the forest, and drives many ecosystem processes. It has been measured at the meteorological station near the Forest Service headquarters at Hubbard Brook since 1958. Monthly values for this period average 4 to 5 Megajoules (MJ) per square meter in winter to about 20 MJ in summer (fig. 3.8). On a really bright clear day in summer, the amount can reach nearly 30 MJ, while on a cloudy day it might be as low as 3 MJ.

Air temperatures at Hubbard Brook vary dramatically through the year. In winter, long periods with temperatures between –12°C and –18°C are common. Summers are short and cool, with maximum daily temperatures averaging 23.8°C in July, whereas minimum daily temperatures average 13.5°C. Based on measurements made since the mid-1950s, the highest air temperature recorded for the forest is 37°C, near the R. S. Pierce Ecosystem Laboratory on three dates (June 18, 1994, September 9, 2002, and July 22, 2011). The record low is –37°C on the north-facing slope, on January 15, 1965.[8]

Average dates for the last frost of spring and the first of autumn are May 11 and October 3, respectively, giving an average frost-free season of 146 days.[9] In

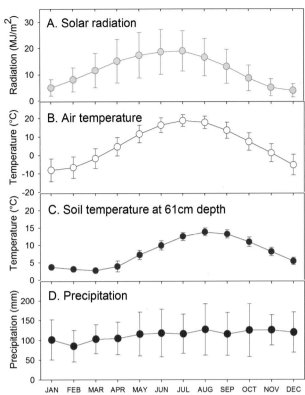

Figure 3.8. Within-year variation in solar radiation (A), air temperature (B), soil temperature at 61 cm depth (C), and precipitation amount (D) at Hubbard Brook. (Adapted from Bailey et al. 2016)

practice, the growing season for trees extends from about mid-May, when leaves have flushed out, until mid-September, when leaves begin to fall. These means and dates have been gradually changing, and the growing season is lengthening as the climate has become warmer in recent years.

When snow falls early in the winter and a snow cover is maintained through the winter season, forest soils remain unfrozen, with soil temperatures averaging just above freezing. When snow cover is thin, develops late in the autumn, or thaws during the winter (which is becoming more frequent), soil frost can develop, which affects many ecosystem processes, such as nutrient cycling, as well as soil microorganisms and burrowing animals such as salamanders. Soils warm rapidly in April with increasing solar radiation, and without shade from still leafless trees. Maximum soil temperatures are reached in July and August, averaging about 16°C, which is slightly lower than the daily mean air temperature (18°C) at that time.

Yearly precipitation at Hubbard Brook averages 143.4 cm, with about 14 cm falling per month, distributed equally through the year. About one-third to one-half of annual precipitation is snow, which typically persists as snowpack from December through early to mid-April. Melting of the snowpack during the winter sometimes occurs, and these occasions have become more frequent in recent years.

4

THE FOREST
Past and Present

What are the characteristics of the forest at Hubbard Brook, and how have they developed over time? The predominant vegetation is characteristic of what is called the northern hardwood forest, a forest type distributed in a band spanning the border between the United States and Canada from the Great Lakes to the Maritime Provinces, and south to the upper slopes of the Appalachian Mountains. In New Hampshire, this forest complex is found from about 200 to 850 m above sea level. Sugar maple, American beech, and yellow birch are the most abundant trees, with red spruce, balsam fir, and paper birch becoming more common at higher elevations and eastern hemlock in the lower parts of the valley (fig. 4.1). These northern hardwood species are replaced around Mirror Lake and near the Pemigewasset River by oaks and pines that are characteristic of lower elevations along the river valleys and more southerly latitudes.

THE HUBBARD BROOK VALLEY
AFTER THE GLACIERS

The forests in the valley, however, have not always been like this. Until about 14,000 years ago, an extensive ice sheet covered most of New England. As the glaciers retreated, they left behind a rough, barren surface of glacial till, essentially ground-up rock ranging in size from large boulders to fine clay particles, but mostly sand and silt, on top of bedrock. Sediment deposits in Mirror Lake provide evidence for when plant communities and eventually forests came to exist. These sediments contain pollen grains, fragments of plants, larger fossils, and inorganic materials that blew or washed into the lake from the surrounding watershed after the glaciers retreated. By examining cores of sediment taken from Mirror Lake and other lakes in the region (fig. 4.2), and then separating the cores layer by layer and dating these layers, it has been possible to identify the relative frequency of occurrence of plant species at different times during the past ten thousand years. The organic remains, especially pollen grains, found at each level of the core (oldest ones at the bottom, most

Figure 4.1. The northern hardwood forest at higher elevations, with a mix of deciduous and coniferous tree species and a well-developed shrub layer consisting mostly of hobblebush and ferns. (Photo by A. G. Muniz)

recent near the top) in each layer were dated using a radiocarbon isotope technique.[1]

Analyses of pollen and other fossilized plant material in these cores allowed for reconstruction of the history of the plant communities as they colonized these barren postglacial surfaces (fig. 4.3). The first plants to become established in the vicinity of Mirror Lake about 14,000 years ago were ones characteristic today of tundra communities in alpine or far northern areas, such as sedges, grasses, and some herbaceous species (Table 4.1). Some of these plants housed nitrogen-fixing bacteria capable of taking nitrogen

Figure 4.2. Drilling a core from the sediments of Mirror Lake, circa 1973. (Photo by G. E. Likens)

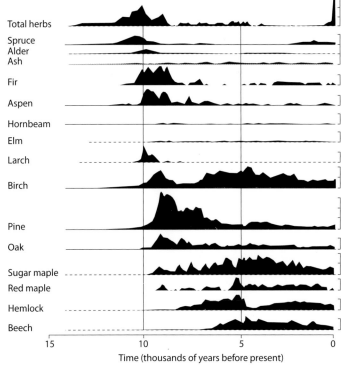

Figure 4.3. Fossil plant pollen in sediment cores taken from Mirror Lake provides information on the changes in vegetation in the Hubbard Brook region since the time of the glaciers. Pollen accumulation rates for herbs, fir, aspen, larch, and maple are magnified ten times to make the changes over time more visible. Each segment of the brackets on the right side represents 10,000 pollen grains per cm² per year deposited in the sediments. (Modified from Davis 1985; reprinted by permission of Springer Science+Business Media)

from the air and converting it into plant matter. Precipitation brought in additional nitrogen. These sources of nitrogen, along with nutrients released from the rocks (such as calcium, magnesium, potassium, and phosphorus), allowed the first pioneer plants to live in the poor soil conditions that were left after the glaciers retreated. Nitrogen is particularly critical, because it is an important component of proteins that control the metabolic processes required for plant growth, but was largely absent in the early soil. One of the nitrogen-fixing pioneer plants, mountain avens (*Dryas integrifolia*), is still extant today in subarctic and arctic regions, although it no longer occurs in New England. Its nearest population to Hubbard Brook today is in the Gaspé Peninsula of southern Quebec.[2]

When these pioneer plants died and decomposed, nutrients and organic matter were added to the system. The nutrients became sources of essential elements for other plants to use, and the organic matter increased the water-holding capacity of the soil. Plant roots and surface organic matter stabilized the soil and prevented erosion. As soil fertility grew, other plants suited to these more favorable conditions invaded, including shrubby ones such as willow, alder, heather, and juniper.

The first evidence for trees in the pollen record appeared about 2,000 years after the glaciers retreated. Spruce pollen and needles were found in the sediments of Mirror Lake as early as 11,300 years ago. By 10,000 years ago, spruce was widespread in the White Mountain region at all elevations, and then it declined, probably because of the warming climate. Other tree species increased at this time, typically invading from the south. These included alder (another nitrogen-fixer), birch, aspen, balsam fir, and cedars.[3]

By about 9,000 years ago, the main forest cover in the area surrounding Mirror Lake consisted of a mix of conifers, aspen, and the first oaks and white pine. This mix was followed by the migration northward of sugar maple, hemlock, and beech. The relative abundance of these tree species changed over time, as judged from

Table 4.1. Chronology of Forest Development in the Hubbard Brook Region
Time span following retreat of the glacier, as indicated by preserved plant remains (especially pollen) in sediment cores taken from Mirror Lake. BP = Before the present (1985).

Years BP	Sequence of events since glaciation
14,000	Glaciers melting, followed by first silty sediments
13,000	Tundra plants (sedges, grasses and some herbs; also nitrogen-fixing pioneer species such as *Dryas*)
12,000	Shrubs evident, including willow, heath, and juniper
12–11,000	First tree pollen (spruce)
10,000	Nitrogen fixers such as alder present; soils developing and retaining nutrients and organic matter
10–9,000	Spruce declining; other species such as alder, birch, balsam fir, and cedar/juniper invading from the south
9,000	Arrival of oak and white pine, some birch and maple
8–9,000	Small amount of charcoal in the core indicates that fire was present, but not extensive
8,000	Sugar maple appears and remains common to the present
7,000	Eastern hemlock common, then declines abruptly about 4,800 BP
6,000–300	Oak, pine, and hemlock decline; forest becomes dominated by red spruce and American beech, with some birch, maple, and hemlock mainly at middle to lower elevations (300–700 m above sea level)

Source: Davis 1985

the quantity of pollen in the sediment cores. Hemlock, for example, declined suddenly about 4,800 years ago, perhaps because of an insect outbreak or a disease that specifically affected hemlocks.[4]

Only small amounts of charcoal were found in the sediment cores, suggesting that fire, either natural or used by early native Americans, did not play a major role in the development of these forests. By the nineteenth century, pollen of many grasses and weeds appeared in the sediment cores, a development that reflected farming practice employed by early European settlers.[5]

There are several important messages for understanding the development of the Hubbard Brook forest, based on these analyses of the Mirror Lake sediment cores.

- The vegetation and plant communities in the Hubbard Brook region have changed dramatically multiple times over the past 14,000 years.
- Once forests became established, their composition was dynamic, with tree species arriving, then increasing, decreasing, or fluctuating in abundance. These changes involved plant species contending with changes in climate, the arrival of invading and sometimes competitively superior species, changing soil characteristics over time, and the occurrence of disturbances of various types, including outbreaks of pests and diseases and pollutants from humans.
- Nitrogen-fixing plant species were abundant only during the early development of these forests, thousands of years ago. Few if any occur today in the south-facing watersheds, and they probably have not been present there for at least 1,000 years. All nitrogen added recently, therefore, has come from the atmosphere.
- The pollen record indicates that periods of stability in tree species composition over the past 10,000 years have been relatively rare, and when they did occur, they lasted no more than one or two thousand years, long enough for only a few generations of the long-lived canopy trees.

The forest community at Hubbard Brook has thus been one of frequent change, and many different factors have determined and influenced the plant species present and the nature of these forest ecosystems over time. Therefore, the forest as we see it now is just a snapshot of a process unfolding over timescales much longer than human memory or written history.

THE FOREST AT THE TIME OF EUROPEAN SETTLEMENT

Visitors to Hubbard Brook often ask: What was the forest like when the first European settlers arrived (~1600–1700s), that is, the primeval forest? Data from several sources address this question. First, the pollen record from Mirror Lake sediments shows that the most abundant tree species in the presettlement forest (circa 1500–1700) were American beech, sugar maple, hemlock, and spruce, a mix very similar to that occurring today (except for the paucity of birches in the presettlement forest).[6]

Another way of determining what the forest was like 300–400 years ago is to examine the relative abundance of species used as witness trees by early land surveyors to mark boundaries of townships or other property transactions (Box 4.1). Of 34 such trees located in the historical records from the Hubbard Brook valley, 35% were red spruce and 38% were American beech.[7]

Another analysis using more than 2,500 witness tree records and forest surveys from a larger region in central New Hampshire, including Hubbard Brook, confirmed these results. In this more extensive study, red spruce and American beech were the most common species, and this result held across the elevations. Also, maples and birches (especially paper birch) were much less frequently represented then than in present-day forests. The relative abundances of other tree species, such as balsam fir, hemlock, white pine, and oaks, in the earlier period were similar to those today.[8]

From the witness tree records, we can surmise that the tree composition of the Hubbard Brook forest in the presettlement period (circa 1500–1750) was different from the one we see today, with almost ten

Box 4.1. What Is a Witness Tree?

Land surveyors in the early settlement period often recorded the species of trees used for marking boundaries when they were laying out townships or describing property transactions. These witness trees were generally large, and assuming the surveyors consistently chose the nearest tree to the surveyed point of a certain size without regard to species, their relative abundance in this historical record provides an index to the canopy tree species composition of the forest at that time. Use of witness trees to reconstruct forest history also assumes that the surveyors made accurate determinations of the species they used as witness trees.

times more red spruce, especially at middle and high elevations, and with American beech as the dominant hardwood species and an intermediate amount of birch and relatively limited number of sugar maple.[9] Furthermore, judging from forests in The Bowl, a Research Natural Area in the White Mountains about 28 km east of Hubbard Brook and one of the few remnant old-growth stands of northern hardwoods in New Hampshire, the presettlement forest in the region probably contained more larger trees, more frequent openings in the canopy because of tree falls, and a greater mix of differently aged trees. Partly because of the more frequent openings or gaps (created mostly by fallen large old trees), the understory would probably have been denser because of more light reaching the lower strata. These are largely conjectures, but they give a general picture of what the forest may have been like a few centuries ago.

This history of the forest underpins what we find today, and provides an important perspective for our 50-year study of forest ecosystem dynamics.

HUMAN INFLUENCES: PREHISTORY TO 1955

Although humans have been present in central New England for at least 10,000 years, they appear to have had little impact on the forests in the White Mountains region until the past 300–350 years. Before European settlers arrived, most native Americans hunted and fished, gathered nuts and fruits, and practiced some subsistence farming, growing maize, squash, and other crops. No direct evidence of their presence in the Hubbard Brook valley exists, but these people were often seminomadic, moving their living places several times each year, depending on the season, local abundances of game, and interactions with other tribes.[10] Although even the earliest inhabitants of the region used fire in their daily activities, fire does not seem to have been a major force affecting forests in the Hubbard Brook region, as judged by the sediment records from Mirror Lake and from the lack of charcoal in hundreds of soil pits dug by researchers at Hubbard Brook.[11] Small quantities of charcoal found in the deeper sediments of Mirror Lake dating back to 11,000 years ago probably came mostly from cooking fires.[12] The lack of fire in northern hardwood forests, in the past and today, has led them to be referred to as asbestos forests, meaning they are very fire resistant except during periods after tree harvests when there are large amounts of logging debris (slash) left on the ground.[13]

The arrival of Europeans, however, changed this story dramatically (Table 4.2). The first to settle near Hubbard Brook arrived in the 1770s, clearing small patches of forest on the lower slopes of the valley, near but not in the flood plain of the Pemigewasset River, planting crops, grazing small numbers of cattle and sheep, and harvesting cordwood for cooking fires and heating. As more settlers arrived and their families and activities grew, their impact on the land intensified. This happened largely after 1830, when there was a dramatic increase in the importance of wool production to local farmers and consequently a need for more grazing lands.[14] More trees were cut for boards and other building materials (mainly pine and spruce) and bark (particularly hemlock, which contained tannins used to cure leather). By the mid-1800s, several commercial enterprises were operating in the vicinity of Mirror Lake and the lower parts of the valley, including sawmills, gristmills, bobbin mills, and tanneries. As the demand for wood products increased following the Civil War, people began to cut trees in the surrounding hills, including the middle and upper parts of the Hubbard Brook valley. At this time, they harvested spruce for boards and hemlock from which tannins were extracted from bark. The logs were either used locally or, starting in the 1850s, transported south by means of log drives on the Pemigewasset and Merrimack Rivers. This first phase of logging in the Hubbard Brook region slowed in the late 1800s, coinciding with the arrival of the railroads that gave loggers access to other forests across the region.[15]

In the early 1900s, however, logging intensified again. A large steam-operated sawmill was built along Hubbard Brook in West Thornton in 1907, and logging activities in the valley expanded rapidly thereafter (fig. 4.4). Based on sawmill records, within a year or two loggers were removing an estimated 11 million board feet of timber each year from the valley.[16] At this time there were four logging camps in the forest, where harvesting was directed toward merchantable spruce, along with some balsam fir and hemlock (fig. 4.5). By about 1915, hardwoods were being cut for bobbins and pulp, with at least 3 million board feet removed by 1920. An overall estimate is that by 1920 logging had removed 166 million board feet of logs, which represented most of the merchantable timber from the forest.[17] Merchantable timber in this case refers to those trees large enough and of a species of high enough value to be worth the time and effort to extract. The forest was not clear-cut in these early years, but selectively logged.

The logging practices of the nineteenth and early twentieth centuries were very different from those used today. They were more selective, with a limited number of species and sizes chosen for harvest. Trees were cut with handsaws and axes, and the logs were dragged out by horses or oxen. This was usually done in the winter because it was easier to pull the logs

Table 4.2. Chronology of Human Impact on the Hubbard Brook Valley

Time span following the arrival of Europeans (mid-1700s) to the establishment of the Hubbard Brook Experimental Forest

1760–1840s	**Small-farm period**
1760s	Thornton and Peeling (Woodstock) first settled by Europeans
ca. 1770	First settlement (small farm) in lower Hubbard Brook valley
1786	Three families with small subsistence farms near Pemigewasset River. Forest cleared for houses and pastures or fields. Softwood cut for lumber, hardwood for heating, hemlock for bark (tannins); sugar maples favored for syrup production. Some cordwood cut, but the forest in the middle and upper parts of valley still largely intact.
1810s	First sawmill on Mirror Lake producing rough-sawn boards. Also a grist mill.
1848–1883	**Small industry and farming period**
	Sawmills (lumber, shingles) and tannery (for leather) on Mirror Lake. Logging of valley started in 1840s, mostly for spruce and some hemlock. Hardwood cut for cordwood (heating), hemlock for leather production (tannins), and spruce for lumber. Most logs floated south on Pemigewasset River to the Merrimack River. Railway arrived in 1883.
1883–1906	**Tourist and industry period**
	Logging declined, in spite of the railway for transport. Local pulp mills still active. Tourist industry brought many people to the White Mountain region in summer.
1906–1920	**Logging industry period**
1907	Large steam-operated sawmill operating next to Mirror Lake. Contract called for all merchantable (defined as over 6 inches, or 15 cm, in diameter) spruce, balsam fir, and hemlock to be cut.
1915	Increased harvest of hardwoods, removing at least 3 million board feet per year.
1920	Logging ended. Most large spruce gone from the valley; hardwoods severely thinned.
1920–	**Public-ownership period**
1920	Land in middle and upper parts of the valley purchased by U.S. government
1921	Land transferred to the U.S. Forest Service (USFS)
1927	Hubbard Brook State Game Refuge established by the New Hampshire Fish and Game Commission, mainly for management of white-tailed deer; 1,000 ha on north slope near Mount Cushman surrounded by a single strand of wire. Hunting was prohibited.
1938	Game Refuge abandoned.
1939	Post-hurricane salvage logging.
1955	Designated by USFS as the Hubbard Brook Experimental Forest.
1956	Installation of first gauging stations on several small headwater watersheds.

Sources: Likens 1972, 1985b; Cogbill 1989

across the snow, which had the incidental effect of limiting the amount of disturbance of the ground. As a result, logging had a much gentler impact on the land than modern methods that often involve removing all trees, using large machinery.

In 1920, after having removed much of the larger and more economically valuable (merchantable) timber, the logging companies sold their forest landholdings in the valley to the federal government, which at the time was authorized by the Weeks Act of 1911 to establish national forests east of the Mississippi River with the aim of protecting the watersheds of navigable rivers. Numerous conservation organizations, but especially the Society for the Protection of New Hampshire Forests, were instrumental in this process. In 1921, the land

Figure 4.4. Sawmill near Mirror Lake at the base of the Hubbard Brook valley in the early 1900s. (Photo from www.hubbardbrook.org)

Figure 4.5. Logging in the forest at Hubbard Brook in the early 1900s. (Photo from www.hubbardbrook.org)

containing Hubbard Brook and surrounding areas was transferred to the U.S. Forest Service and incorporated into the adjacent White Mountain National Forest.[18]

The forest at this time was well stocked with trees that were growing rapidly in response to the cutting over the previous half century. Except for a short-term wildlife management project and salvage logging following the 1938 hurricane, there was little human activity in the forest after 1921 until the Forest Service established the Hubbard Brook Experimental Forest in 1955 and research began.

EXTREME WEATHER EVENTS, DEFOLIATOR OUTBREAKS, AND PATHOGENS

Although logging activities in the valley during the late 1800s and early 1900s have probably been the most conspicuous perturbation to the forest ecosystem in recent times, other events have occurred. Besides damaging windstorms, these include ice and hail storms, droughts, insect outbreaks, diseases, and pathogens (Table 4.3). Such events have had variable but often important impacts on the structure and function of the forest ecosystem. Another less conspicuous but extremely important perturbation during this time was acid rain (Chapter 9).

Extreme weather events. The hurricane of 1938 remains one of the deadliest and costliest hurricanes ever to reach New England. It occurred in late September, and resulted in major uprooting and breakage of trees throughout the forest, but especially on southeast-facing slopes and among older overstory hardwoods. Damage was patchy, with heavily hit areas experiencing 70% reduction in canopy cover, while other places remained almost untouched. Overall, the average reduction in canopy cover in the forest was estimated at about 20%. In the three years following the hurricane, an estimated 4.2 million board feet of timber, primarily hardwoods, was removed from the valley as part of a salvage operation.[19]

The effect of this hurricane was still evident 30 to 50 years later. When the Study began in the early to mid-1960s, pin cherries, which sprouted from buried seeds in the more heavily damaged stands following

Table 4.3. Severe Weather and Other Disturbance Events at Hubbard Brook, Since 1938

1938	Category 3 hurricane on September 21
1939–1942	Salvage logging following 1938 hurricane
1963–1965	Major drought
1969	Heavy hailstorm on August 25
1969–1971	Outbreak of a defoliating caterpillar, the saddled prominent
1977	Beech bark disease first detected; spreads widely thereafter
1998	Ice storm in early January
	Heavy hailstorm on August 24
2001–2003	Drought conditions
2006	Massive freezing of forest soils during winter because of light snow cover
2011	Tropical storm Irene on August 28 causes heavy rains
2013	Microburst in early June damages forest in upper central part of valley

the hurricane, were evident in most areas of the forest. These were present through the mid-1970s, but most had died and fallen by the late 1980s.[20] In the years following the hurricane, tree growth (as measured by increase in diameter of the boles) increased sharply, averaging more than sevenfold greater than previous years.[21] Nearly 50 years later, the trees had largely recovered, masking the patchy effect of the hurricane damage across the forest.[22]

A major weather-induced disturbance that affected the forest was an ice storm that occurred in early January 1998, with the ice remaining on the trees for several days. This storm damaged forests across all of northern New England and southeastern Canada, with more than 427,000 ha of forest affected in New Hampshire alone.[23] At Hubbard Brook, most of the damage occurred at higher elevations, mainly on the south-facing slope between 600 and 800 m above sea level.[24] Ice from a few millimeters thick to as much as 20 mm coated branches and boles, pruning twigs, causing major breakage of branches, snapping boles, and in some areas uprooting whole trees (fig. 4.6). Deciduous trees were the most damaged, with crown loss estimated at between 10% and 70%. Within the zone of ice accumulation, the ice storm did not affect conifers much. Among the dominant hardwood species, mainly American beech was affected (73%), followed by yellow birch (25%) and sugar maple (2%). Large beech trees suffered the most damage, primarily because many were infected with beech bark disease and many were already in poor health.[25] Smaller trees, and especially paper birch, were affected by permanent bending of stems rather than by breakage. This effect was very noticeable at mid- to high elevations on Watershed 4.

The storm resulted in the reduction of leaf surface area of about one-third in the growing season following the storm, again with greater loss at higher elevations.[26] The leaf surface area recovered to pre-storm levels by the third growing season (2001).[27] A longer-term effect was a change in vertical distribution of foliage, related to increased regeneration by ice-damaged (and diseased) American beech trees. These reductions in leaf surface and change in foliage distribution reduced energy fixation through photosynthesis in the forest ecosystem for one to several years following the storm. This disturbance also increased the loss of nitrogen and other nutrients in stream water.

Soil freezing can also affect forest processes. In most winters, deep snow covers the ground and insulates the forest floor from freezing. But when snow is late to fall or is not deep and temperatures are exceptionally cold, the ground can freeze. Such major soil freezing events have occurred four to five times at Hubbard Brook over the past 50 years, sometimes causing water to freeze in the stream channels and flow over the surrounding land (fig. 4.7). More significantly, soil frost inhibits biological activity, which normally continues in the unfrozen forest floor under the snow. Freezing conditions can cause the death of fine roots, soil organisms, including many of those actively involved in decomposition, and even overwintering amphibians.[28] Soil freezing events can also influence soil chemistry, leading to increased loss of nitrogen and other nutrients in stream water.[29]

Defoliator outbreaks. Of the hundreds of species

Figure 4.6. An ice storm in early January 1998 caused severe damage to the forest, especially at upper elevations on the south-facing slope. (Photos by J. Pett-Ridge)

of moths (Lepidoptera) that live in the forest, only one, the saddled prominent, has been recorded as reaching outbreak levels at Hubbard Brook (fig. 4.8). This occurred between 1969 and 1971, with caterpillars reaching peak numbers in 1970. In that year, the caterpillars were estimated to have consumed about 44% of the canopy leaves on the south-facing slope, with heavier defoliation in some localized areas.[30] At the Bartlett Experimental Forest, about 35 kilometers northeast of Hubbard Brook, large areas of forest were totally defoliated (fig. 4.9). In most years, when caterpillar numbers are at endemic (non-outbreak) levels, the average rate of leaf consumption is about 4–6% of the leaf tissue present.[31]

Figure 4.7. A frozen stream on March 25, 2006, during an unusually cold period when the soil froze to a depth of tens of centimeters. The lack of snow cover on this date resulted from an early snowmelt. (Photo by D. C. Buso)

The effect of this defoliation on the forest ecosystem in 1970 was to reduce photosynthesis for that season by reducing the amount of leaf tissue in the canopy, and to increase the rate of transfer of nutrients from leaves to the soil as caterpillars consumed the leaves and produced frass that fell to the ground in midsummer.[32] This event corresponded to increased nitrogen concentrations in stream water, which in fact reached the highest annual levels yet recorded.[33] This high nitrogen loss is thought to be due both to increased cycling of nitrogen during this caterpillar irruption and to the occurrence of soil frost in the intervening winters.[34]

Pests and pathogens. Another group of organisms that influence ecosystem structure and function in the forest are exotic (non-native) insect pests and pathogens that cause disease, mainly of forest trees.[35] These can have either short- or long-term effects on tree health and mortality, resulting in shifts in tree species composition, and changes in such processes as water use, nutrient uptake and storage, tree growth and productivity, and food web dynamics (fig. 4.10).

At Hubbard Brook, beech bark disease is caused by two fungal pathogens in the genus *Neonectria,* one of which is native to North America and the other apparently was introduced from Europe.[36] These pathogens are spread by an introduced insect, the woolly beech scale, *Cryptococcus fagisuga.* The scale

Figure 4.8. Larva of the saddled prominent moth, a major consumer of canopy foliage at Hubbard Brook in some years. The most severe defoliation by this species during the 50-year Study occurred from the summer of 1969 to 1971. (Photo by W. S. Schwenk)

Figure 4.9. Frank Sturges at the Bartlett Experimental Forest near Hubbard Brook on August 6, 1971, the year the Bartlett forest was extensively defoliated by saddled prominent caterpillars. The more normal condition of these forests in mid-summer is seen in Figure 4.1. (Photo by R. T. Holmes)

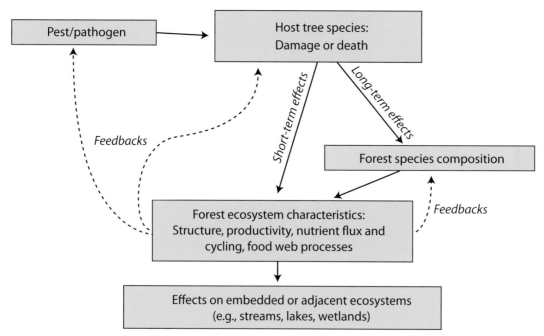

Figure 4.10. Potential effects and interactions of pests and pathogens (disease) on forest ecosystem processes. (Modified from Lovett et al. 2006; reprinted by permission of Oxford University Press)

Figure 4.11. Diggings of a pileated woodpecker in an American beech dying from beech bark disease. Dead or dying beech trees are invaded by beetle larvae and other wood boring insects, which provide food for woodpeckers. (Photo by N. L. Cleavitt)

years. These dead trees gradually shed bark and limbs, and eventually are blown over or broken off during strong winds.

Deaths of these large beech trees have had numerous effects on the forest ecosystem at Hubbard Brook. One has been to create gaps in the forest canopy, letting more light into the forest. Increased light affects photosynthetic and transpiration rates, growth and distribution of plants on the forest floor, germination rates of seeds and tree regeneration, as well as the vertical distribution of foliage. When these mature beech trees become stressed from disease, their root systems send up numerous sprouts or suckers. These root sprouts at first form dense stands or even thickets of young beech saplings in the lower strata of the forest. Within a few years these dense stands make walking through the forest difficult, leading researchers at Hubbard Brook to coin the term "beech hell." Beech hell at Hubbard Brook was most evident in the 1980s and 1990s, but by the 2000s, many of these saplings had died, either from competitive effects in these dense stands or from shading by the canopy overhead. The remaining saplings have grown taller and now form a dense foliage layer below the forest canopy. The fact that beech hell no longer is evident suggests that the greatest impact of beech bark disease may have passed, although a keen observer will note that most mid- and large-sized beech trees still show signs of heavy infection. In fact, a survey conducted in 2010–2012 indicates that 80% of the American beech trees on Watershed 6 have symptoms of beech bark disease.[37]

These changes in canopy and foliage structure brought about by beech bark disease have affected animals in the forest, too. Canopy structure strongly influences bird foraging patterns and bird community structure, even affecting the presence or absence of particular species. Likewise, a reduction in the number of large beech trees may reduce the quantity of beech seeds produced, leading to a decrease in such animals as bear, rodents, and winter finches that depend on these masting events. Because beech is an important codominant tree of these forests and

insect was accidentally introduced into Nova Scotia around 1890 on young trees imported from Europe. Since then, the disease has spread throughout eastern North America. At Hubbard Brook, it was first noticed in the mid-1970s, although it may have been present at low levels over the previous decade. Large beech trees are particularly susceptible, and many have died from the disease during the past 30–40 years (fig. 4.11).

The fungal infection causes cankers to form on the bark, which gradually spread, slowing the tree growth and eventually killing the tree (fig. 4.12). This process may occur over a period of a decade or more. Large, mature-sized dead standing beech trees have been a relatively common sight in the forest for the past 20

Figure 4.12. Diseased American beech bole (left), compared with a relatively healthy, unaffected bole (right). The pathogen is a fungus, which is spread by the woolly beech scale, an alien insect species. Scars in the bark of the tree at right are claw marks from a black bear. (Photos by R. T. Holmes, N. L. Cleavitt)

is being severely affected by this pathogen, many changes in the forest ecosystems may occur in the coming decades. These changes continue to be a focus of ongoing investigations.

THE FOREST TODAY

The forest at Hubbard Brook today carries a legacy of all of the events and processes discussed above. It can best be characterized as an uneven-aged, unmanaged, second-growth northern hardwood forest. Most trees range in age from 60 to 120 years, with a few older ones that were too small to cut, or escaped cutting because of poor form, during the harvest of the early twentieth century (fig. 4.13). In the forest as a whole, seven tree species account for 90% of the tree basal area: yellow birch is the most common (9.0 m² per hectare), followed by sugar maple (4.9), red spruce (3.5), American beech (3.3), red maple (2.1), paper birch (2.1), balsam fir (1.5), and all other species (2.8).[38] These species vary in relative abundance in different parts of the valley, leading to the recognition of distinguishable tree associations or communities. Other less common but frequently encountered tree species are white ash and eastern hemlock.[39]

On the south-facing slope and especially on Watershed 6 where much of the research has been conducted, sugar maple, American beech, and yellow birch are codominants (equally common), with scattered individuals or clumps of red spruce, balsam fir, red maple, and eastern hemlock.[40] The spruce

Figure 4.13. The oldest known tree in the Hubbard Brook valley, a 425-year-old red spruce on Mount Kineo. (Photo by N. L. Cleavitt)

and fir increase in frequency at the higher elevations, especially on ridges and rocky divides. Paper birch, a mid-successional species, was once relatively common on the higher slopes but has been in decline over the last decade or two. Hemlock, although not present on Watershed 6, is most common and sometimes forms dense stands in the cooler drainages along the streams, on the north-facing slope, and in the lower portions of the forest.

The structure of the forest canopy varies with the mix of tree species present and with elevation, aspect, and slope. Canopy height decreases with elevation, averaging about 25–30 m at lower to mid-elevations, with scattered emergent trees, mostly white ash, reaching to 40–45 m. Near the ridgelines, trees rarely exceed 10–15 m in height. The vertical foliage profile over most of the forest basically consists of two layers, the canopy and the understory. The understory contains the seedlings and saplings of sugar maple and especially American beech, one widespread and abundant shrub, hobblebush, and a rich but patchy herb flora. The density, distribution, and species composition of these plants in the shrub layer vary both spatially and temporally, depending on soils, slope, moisture availability, and past disturbance history. Both the distribution and local density of hobblebush, for example, are affected by the occurrence and size of gaps in the canopy that result from falling dead or storm-damaged trees, which let in more light to the understory. Hobblebush seems to be negatively affected by the presence of dense stands of beech saplings resulting from the effects of beech bark disease. The distribution of hobblebush in the forest has, in fact, changed dramatically over the past 50 years, becoming less dense or disappearing from some areas while increasing in others.

The number and condition of dead trees also affect the forest's structure. In 2001, for example, on one 10-ha mid-elevation study area near the base of Watershed 6, out of 5,000 tree stems (boles) greater than 10 cm in diameter, 630 were dead (14.4%). Of these 630 dead stems, 155 were standing dead (with most branches still present), while the remaining 475 were snags (dead trees without major branches, often broken off aboveground by strong winds).[41] Moreover, throughout the forest the ground is littered with dead wood from fallen tree boles, branches, and twigs.

The forest at Hubbard Brook is thus dynamic, and constantly undergoing change. Not only have these changes occurred over the past 10,000 years, but also in the 50-year time frame of our Study. These changes are likely to continue and perhaps accelerate with ongoing changes in climate warming, the invasion of new pests and pathogens, and other natural and anthropogenic disturbances.

5

A RICH ARRAY OF ORGANISMS AND THEIR INTERACTIONS

Natural history is the observation and description of the natural world, with the study of organisms and their linkages to the environment being central. . . . Direct knowledge of organisms—what they are, where they live, why they behave the way they do, how they die— remains vital to science and society.

— JOSHUA J. TEWKSBURY ET AL. 2014: 300

The plants, animals, and microorganisms of an ecosystem constitute its internal living parts that, in effect, make the system work. Thus, to study and assess the functioning of the forest ecosystem at Hubbard Brook, it has been important to learn which species are present in the forest, how many there are, where they occur, what they do, and how they interact with one another. In this chapter, we introduce and briefly survey what we know about the taxonomic (species) and ecological diversity of the major groups of organisms at Hubbard Brook, along with pertinent life history traits and characteristics.[1]

Some of the organisms at Hubbard Brook have been relatively well inventoried, including the liverworts, mosses, vascular plants, select insect groups, and the vertebrates (Table 5.1, Appendix 2). The diversity and taxonomic identities of others, such as the lichens, fungi, many of the invertebrates (spiders, centipedes and millipedes, and most insects), and the soil-dwelling microorganisms (bacteria, protozoa, algae, and most fungi), are at best incompletely known. Notably missing at Hubbard Brook are poisonous or truly noxious organisms, such as are often found in forests farther to the south (Box 5.1).

PLANTS

As of 2015, 258 species of vascular plants have been recorded in the Hubbard Brook valley. These include 27 species of trees, 38 shrubs, and 193 herbs (including ferns). There are no lianas (woody vines) in the forest, and the only epiphytic plants (nonparasitic plants growing on the surface of other plants) are some mosses and lichens. This contrasts with forests farther south and especially those in tropical regions, which contain many epiphytes, such as orchids, bromeliads, and ferns, as well as many lianas.

Trees. The forest at Hubbard Brook consists mostly of deciduous tree species, with a mix of conifers (fig. 5.1). Based on forest-wide surveys, the seven most common trees in the valley, in descending order, are yellow birch, sugar maple, red spruce, American beech, red maple, paper birch, and balsam fir. Other relatively common but locally distributed species

Table 5.1. Taxonomic Diversity of the Better-Studied Organisms in the Hubbard Brook Valley

	Number of species
Fungi	212[a]
Lichens	75[a]
Nonvascular plants	
Liverworts	40
Mosses	148
Vascular plants (by family)	
Clubmosses and allies	9
Ferns	16
Sedges	19
Rushes	4
Lilies	12
Orchids	9
Grasses	12
Asters	33
Honeysuckles	7
Heaths	9
Violets	3
Conifers	5
Maples	4
Birches	5
Beech/oak	2
Ash	2
Cherries	4
Other plants	103
(Total vascular = 258)	
Animals	
Insects	
Ground beetles	190+[b]
Moths, butterflies	151[c]
Mayflies	21
Stoneflies	42
Caddisflies	51
Black flies	4
Snails and slugs	17
Fish	13
Amphibians	14
Reptiles	7
Birds	127
Mammals	31

Source: Updated from Holmes and Likens 1999
[a]Preliminary data from C. Wood and N. Cleavitt, unpublished data. Fungal surveys based on observed aboveground fruiting bodies.
[b]From Fiumara 2006.
[c]From Stange et al. 2011.

Box 5.1. Noxious Organisms

Hubbard Brook lacks poison ivy, poisonous snakes, and other truly noxious organisms that are harmful to humans, making it a relatively benign place to work and visit. There are, however, a few insects that can be quite unpleasant, such as pesky black flies in late May and June, seemingly omnipresent mosquitoes from June through August, and biting deer flies in midsummer.

include eastern hemlock, white ash, quaking aspen, big-tooth aspen, and two subcanopy species, striped maple and mountain maple. In general, spruce, fir, and paper birch become more common with increasing elevation, whereas hemlock occurs most commonly at lower elevations, especially along stream drainages.[2]

Yellow birch, sugar maple, and American beech dominate the northern hardwood forest on Watershed 6 and other parts of the south-facing slope where the most intensive research has been conducted. These three species have both similarities and differences in their life history traits (Table 5.2). All three produce small inconspicuous flowers in early spring, and their pollen is spread mostly by wind, although bees often visit the flowers of sugar maple and may play a role in pollen dispersal. Seeds of all three species ripen in the autumn. Those not consumed by insects, squirrels, or birds fall to the ground in late autumn or winter, and germinate the following spring.

Seeds are not produced every year, but in episodic events, usually every two to three years, and often synchronously among these three species (fig. 5.2). This is called masting, and why it happens is not fully understood.[3] Producing many seeds all at once may satiate seed predators before all the seeds have been consumed, allowing some to escape predation, germinate, and grow. Doing this at two-to-three year intervals may starve seed consumers

Figure 5.1. The northern hardwood forest at Hubbard Brook in autumn. The canopy is dominated by deciduous species such as sugar maple, yellow birch, and American beech, with some conifers. (Photo by D. C. Buso)

in the intervening years or at least prevent them from maintaining or building up large populations.[4] An alternative possibility is that synchronous flowering may increase the efficiency of pollination, especially for wind-pollinated species such as the trees at Hubbard Brook.[5] Whatever its selective advantage, masting has many consequences in the forest ecosystem, from influencing the success of tree seedling survival to affecting the survival and reproductive success of animals that live in the forest.

Table 5.2. Life History Traits and Patterns of Reproduction by the Codominant Northern Hardwood Tree Species at Hubbard Brook

	Average seed mass (mg)	Relative shade tolerance	Seed dispersal distance (m)	Masting frequency (yrs)	Regeneration patterns and requirements
American beech	286	+++	< 10	2–3	Closed-canopy root sprouting
Sugar maple	65	+++	< 100	2–4	Closed canopy
Yellow birch	10	+	> 400	2–3	Canopy gaps, roughed-up soil

Source: Adapted from Schwarz et al. 2003

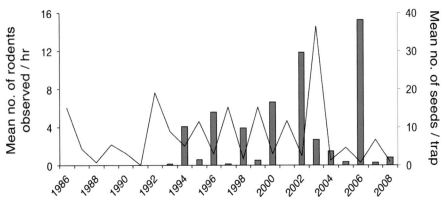

Figure 5.2. Mast production by two dominant hardwood tree species, sugar maple and American beech, occurs at two- to three-year intervals (bars). High seed production leads to increases in the numbers of large, diurnally active rodents (red squirrels and eastern chipmunks, numbers combined), which peak in the summer following the masting events (line). (Modified from Holmes 2011; reprinted by permission from Elsevier)

The three codominant tree species differ in other aspects of their reproductive strategies, including seed size and dispersal, shade tolerance, and regeneration requirements. American beech seeds are produced in fruits containing two or three large seeds, which fall to the ground beneath the parent tree and are dispersed a few meters at most. Sugar maple fruits consist of two seed wings, or samaras, each typically containing a medium-sized seed. The winged seeds of sugar maples may be transported by wind up to 100 meters away from the tree. The relatively large seeds of beech and maple are further redistributed and moved short distances by chipmunks, squirrels, and mice that cache food. In years after masting at Hubbard Brook, American beech seedlings resulting from seed germination reach densities as high as 52 per square meter (fig. 5.3).[6] Both American beech and sugar maple seeds are shade tolerant, and germinate under the forest canopy. Once germinated, however, these tree seedlings are subject to heavy predation, disease, and mechanical damage from falling trees and winter conditions, which quickly reduces their survival. Because of masting, new tree seedlings are produced in pulses or cohorts at two- to three-year intervals, which can be tracked as they grow and die (Box 5.2).[7]

American beech, besides having large seeds that germinate well under shady conditions, also reproduces asexually by sprouting or suckering from its roots. Such vegetative reproduction frequently

Figure 5.3. Recently germinated seedlings of sugar maple (left) and American beech (right) occur at high densities on the forest floor in springs after years of high mast production. (Photos by N. L. Cleavitt)

Box 5.2. What Determines Tree Seedling Survival?

A study of the survival and causes of death for a natural cohort of sugar maple seedlings across 22 sites in the forest revealed that more than 70% died within the first growing season, and only 3.4% were still alive after a period of 7 years. Significant differences in survival, however, were observed among sites, which differed in elevation, seedling characteristics, and the prevalence of damage agents. The main causes of seedling deaths were fungal infection (*Rhizoctonia* spp.) and caterpillar herbivory. Other principal causes of mortality, in order of importance, were winter injury, mechanical damage, and rodent (mainly red-backed vole) tunneling. The impacts of all damage agents varied significantly in severity among years. Regeneration of sugar maple thus depends significantly on the site where the seed starts to germinate and the impact of various natural enemies and physical damage.[a]

NOTE

a. Cleavitt et al. 2014.

occurs following disturbance or other environmental stresses, such as beech bark disease.

In contrast with sugar maple and American beech, yellow birch produces a large number of very tiny seeds that are dispersed long distances by wind.

For germination, the small birch seeds require bare mineral soil, often in disturbed ground or sites where the ground surface has been scarified by logging activities. Because of these special germination requirements, yellow birch trees are often found

Figure 5.4. Yellow birches often grow on tops of mounds or straddle large boulders. This positioning results from the specific requirements of their seeds for disturbed or disrupted soil surfaces for germination. Such sites are created when trees topple during windstorms or when other kinds of disturbances, such as forest harvesting, alter the forest floor. (Photo by R. T. Holmes)

growing on thin soils, including the tops of soil mounds resulting from trees toppled by strong winds, in rocky areas, or even straddling large boulders (fig. 5.4). As a consequence, few if any yellow birch seedlings or saplings are found in the understory of intact older-aged forests, and the persistence of this species in the forest landscape over the long term depends on periodic large-scale disturbances.[8]

Shrubs. The most common and widely distributed plant in the lower strata of the forest is hobblebush, which is a tall (1–3 m), sprawling shrub that often grows in dense patches (fig. 5.5). These tangles of

Figure 5.5. A patch of hobblebush, the most common understory shrub in the forest at Hubbard Brook. (Photo by S. A. Kaiser)

low-growing stems make walking through the forest difficult at times (one possible origin for its name). Moose extensively browse the leaves and especially twigs of hobblebush, primarily during the winter. Hobblebush also provides nesting sites for one of the most common birds species, the black-throated blue warbler.

Other shrubs including raspberry (several species), elderberry, honeysuckle, blueberry, willow, and alder also occur in the forest, but these tend to be widely scattered or restricted to special situations. Raspberry, for example, occurs mostly in disturbed places, such as large gaps caused by falling trees, sites recently logged, or even in clearings around weather stations.

In contrast, shrubby willows and alders are found only along streams in the lower part of the valley and in localized wetlands, such as beaver ponds.

Plants of the forest floor. The ground layer of the northern hardwood forest is home to a diverse set of low-growing herbaceous plants, as well as tree and shrub seedlings.[9] In early spring, the ground is covered by a variety of herbs, including trout lilies, red trillium, spring beauty, and yellow violets (fig. 5.6). These species sprout from underground tubers or bulbs, and begin to grow as soon as the snow has melted and before the overhead trees have leafed out, taking advantage of sunlight and the warming of the forest soils by solar radiation. They grow quickly and

Figure 5.6. Spring ephemeral herbs of the forest floor: top, painted trillium, marsh violet; middle, spring beauty, trout lily; bottom, yellow violet, red trillium. These vernal specialists are perennial plants that appear in early spring, grow, produce flowers and fruit, and then their aboveground parts die by late May to early June as the forest canopy closes overhead. (Photos by W. S. Schwenk, except painted trillium by N. L. Cleavitt)

produce flowers in May, and then senesce and their aboveground parts die by early June; they are referred to as spring ephemerals.

Even though photosynthetically active only for a brief period, these herbaceous plants of the forest floor are important in ecosystem nutrient cycling. They take up nutrients such as potassium and nitrogen from the soil at a time when streams are at high flow and such nutrients could otherwise be lost from the system through runoff. In this way they serve as a short-term sink or vernal dam, incorporating nutrients into plant biomass and conserving the nutrient capital of the ecosystem.[10]

Other herbaceous plants on the forest floor that are more capable of growing under the shady forest canopy begin to appear in late May and reach their full size in late June through July (fig. 5.7). These include clubmosses, ferns (several species), sorrels, lilies, and several species of orchids, including pink lady slippers and the large round-leaved orchid. These herbs provide a nectar source for many small insects. Their foliage is eaten by snowshoe hare, deer, and moose, as well as by caterpillars and slugs, while their fruits and seeds provide food for a variety of insects, rodents, and birds.

Another interesting plant on the forest floor is indian pipe, which is a locally common plant at middle elevations, especially where American beech is common. It is a 10- to 20-cm-tall flowering plant that lacks chlorophyll; instead of photosynthesizing, it gets its energy (nutrients and carbohydrates) by tapping into underground mycorrhizal fungi, which are themselves in a mutualistic relationship with forest trees. Unlike the fungi, however, indian pipe only takes in materials and gives nothing back, making it a parasite on the fungi and indirectly on the forest trees.[11] This rather special arrangement makes it a mycoheterotroph (an organism that obtains its energy from a fungus).

The flora of the Hubbard Brook valley also includes other herbaceous species that have specialized habitat requirements, such as particular light conditions, soil types, and moisture levels. Several species of sedges and rushes, for instance, occur only in boggy areas or seeps, which are scattered about the valley. Several weedy or invasive species, such as some of the grasses, asters, and goldenrods, are limited to roadsides, forest edges, and other disturbed sites.[12]

FUNGI, MICROARTHROPODS, AND OTHER SMALL ORGANISMS

Fungi and other small organisms are important components in nearly all terrestrial ecosystems, where they account for most of the decomposition of dead organic matter. These organisms are extremely diverse in forest soils, as indicated by a recent study in an Alaskan forest that found, using genetic methods, at least 6 million fungal "species" living in the soil layers there.[13]

Fungi are important in plant ecology as both helpers and pathogens. The familiar mushrooms and shelf fungi are the fruiting bodies of fungi that live in decaying wood and other organic matter, and serve to break down plant tissues, aiding decomposition and releasing nutrients for uptake by plants. Preliminary surveys on transects on the south-facing slope of the forest in 2003–2005 yielded approximately 212 species of these "macro" fungi (mushrooms) living on the surface of the ground, and in dead wood.[14]

Other fungi, such as the microscopic mycorrhizae in the soil, are much more numerous. These organisms are notoriously difficult to separate taxonomically, although preliminary studies using genetic (DNA) approaches indicate they are extraordinarily diverse in Hubbard Brook soils.[15] These fungi are critically important ecologically, forming mutualistic relationships with forest trees, in which both the tree and the fungi benefit. They live in the soil, attached to fine roots, or even within the roots of the major tree species. These mycorrhizae obtain the carbohydrates they require from the roots and in return facilitate the uptake of moisture and nutrients, such as nitrogen and phosphorus, by the plants. At Hubbard Brook, the early formation of mycorrhizal fungal associations within seedling roots have been shown to enhance the survival of sugar maple seedlings.[16]

Not all fungi have a positive effect on their host species, however. In 2007 at Hubbard Brook, a pathogenic fungus in the genus *Rhizoctonia* was found to be the main source of mortality for sugar maple seedlings, and in 2012 another unidentified pathogenic fungus caused high mortality in the round-leaved orchid population.[17]

Many other types of soil-dwelling organisms are also abundant throughout the upper layers of the soil. Soil nematodes are common throughout the forest soils, with the highest abundances recorded at mid- to high-elevation sites where soils are generally wetter (about 18 per 50 grams of soil) compared with lower elevation sites (about 2 per 50 grams of soil).[18] Well over 1,000 bacterial taxa, based on genetic methods of determination, have been detected in Hubbard Brook soils, and this probably represents only a fraction of the total numbers there.[19] Nitrogen-fixing bacteria seem to be scarce in the forest. The only plant species with these symbiotic bacteria that we are aware of in the valley are two species of alder, which have limited distributions along the lower mainstem of Hubbard Brook and in some beaver-maintained wetlands in the upper parts of the valley.

Other small organisms live in the upper layers of the forest soil, the most abundant being springtails (insects in the order Collembola) and mites (Arachnids in the subclass Acari). These tiny (< 1 mm) arthropods promote decomposition and nutrient cycling in the ecosystem by consuming subterranean fungi and other microorganisms and by fragmenting leaves and woody matter fallen to the ground. Most of these arthropods (70%) occur in the upper organic layer consisting primarily of undecomposed litter, and 20% are found in the dark layer of well-decomposed humus deeper in the soil, with the remaining 10% in the underlying mineral soil. The abundance of these microarthropods is impressive, ranging from 100,000 to 250,000 per square meter, and vary with elevation and plant community type. On average, about 150,000 of these detritus-consuming organisms live beneath the surface of every square meter of the forest floor![20]

MACROINVERTEBRATES

Larger invertebrates in the forest include species that feed on living plant tissues (herbivores), many that break down dead plant material (detritivores) and live mostly in the forest floor and in the soil, and yet others that prey on other animals, both above- and belowground. In addition, there are parasitic and even hyperparasitic (parasitic on another parasite) hymenoptera, mostly wasps, as well as numerous invertebrate predators, including spiders, centipedes, and certain beetles. Taxonomic identification of these invertebrates at Hubbard Brook has been limited to a few key groups, but even these groups are incompletely known.

Moths and butterflies. One insect group that has received considerable attention at Hubbard Brook is the Lepidoptera (moths and butterflies). Thus far, at least 151 species in 15 families have been identified from the forest, but many more undoubtedly occur.[21] Moths make up most of these species, which exhibit an impressively diverse set of life history traits and feeding patterns (Box 5.3). The most frequently encountered Lepidoptera are the owlet moths (Noctuidae, 47 species identified to date) and the geometer, or inchworm, moths (Geometridae, 38 species). Other well-represented and conspicuous moth families are the prominents (Notodontidae, 15 species), hawk or sphinx moths (Sphingidae, 8 species), tiger and lichen moths (Arctiidae), and the large saturnid moths (Saturniidae, 4 species), including the impressive and common luna and polyphemus moths.

In contrast to the moths, only six species of

(opposite) Figure 5.7. Summer herbs of the forest floor: top, hay-scented fern, rose-twisted stalk; middle, pink wood sorrel, blue-bead lily; bottom, running pine (clubmoss), indian pipe. Unlike the spring ephemerals, these species are more shade tolerant and reach maximum size and reproduce in mid- to late summer. (Photos by W. S. Schwenk, except clubmoss by S. A. Kaiser and indian pipe by N. L. Cleavitt)

Box 5.3. A Diverse Array of Moths

There are many different species of moths in the forest, and in addition, the ones that are present exhibit a wide variety of life history patterns and larval feeding requirements. Some species, for instance, overwinter as eggs, and others as larvae or as pupae. Species also differ in when during the year their adults emerge and lay eggs. For some species this happens in early spring, others in early, middle, or late summer, and some can be found flying about the forest as late as November.

Feeding habits are equally if not more diverse. Although the larvae of most moth species consume leaves of the deciduous trees, some caterpillars feed exclusively on one specific plant species, whereas others are polyphagous and feed on the foliage of two or more host plants. But other feeding specialties occur. Caterpillars of at least seven species that occur in the Hubbard Brook forest are specialists on conifers (mostly feeding on needles), four on herbaceous plants in the understory, and four on mosses and lichens. Larvae of the American idia moth (*Idia americalis*) feed on dead leaves and fungi, and thus participate in the detrital food web. Larvae of the little white lichen moth (*Clemensia albata*) feed on lichens, cyanobacteria (formerly called blue-green algae), and mosses growing on tree bark. Not all caterpillars are leaf-chewers.[a]

NOTE

a. Stange et al. 2011; N. Lany, personal communication.

Two common small moths in the Hubbard Brook forest: the American idia moth (left), and the little white lichen moth. (Photos by D. Small)

butterflies have been recorded, representing two families, the Paplioniidae (swallowtails) and the Nymphaliidae (viceroy, white admiral, and mourning cloak). The most common of these is the Canadian tiger swallowtail, usually seen flying about in sunny spots in the forest understory.

Larval Lepidoptera, or caterpillars, mostly of the owlet, geometer, and notodontid families, are the most important herbivores in the forest (fig. 5.8). They also serve as food for most of the songbirds nesting in the forest as well as for other animals, and largely for that reason they have been studied in some detail since the

Figure 5.8. Leaf-consuming Lepidoptera larvae, mostly moths, are the major herbivores in the forest ecosystem. Examples shown here represent six different families of Lepidoptera: top, Notodontidae, Saturniidae; middle, Noctuidae, Geometridae; bottom, Erebidae, Lasiocampidae. (Photos by M. P. Ayres and B. Marlin, top; W. S. Schwenk and M. P. Ayres, middle; and M. P. Ayres and G. Hume, bottom)

mid-1980s (fig. 5.9). Caterpillar abundance fluctuated widely (more than sixtyfold) between 1981 and 2012 (fig. 5.10). In the late 1960s and early 1970s, before quantitative measurements of caterpillar abundances were made, an even greater irruption or outbreak occurred. In this case, a species of notodontid moth, the saddled prominent, began to increase in the summer of 1968 and reached peak abundance in 1970

before declining in 1971 and 1972. The causes of such fluctuations remain largely unknown.

Stream invertebrates. Most invertebrates in Hubbard Brook streams live on or in the sediments, and as a group they are referred to as the macrobenthos. They include larval mayflies (Ephemeroptera), stoneflies (Plecoptera), caddisflies (Trichoptera), black flies (Diptera: Simuliidae), crane

Figure 5.9. Jennifer Ma, a field research assistant, measuring the length of a saddled prominent caterpillar. Larval lengths allow for calculations of the biomass of these important herbivores in the forest ecosystem. (Photo by A. G. Muniz)

flies (Diptera: Tipulidae), and midges (Diptera: Chironomidae). Abundant water striders (Hemiptera) live on the water surface of pools. Detailed studies in the 1970s found that the aquatic invertebrate fauna in headwater streams include more than 75 taxa, including beetles (Coleoptera), springtails (Collembola), segmented worms (Oligochaeta), flat worms (Platyhelminthes), round worms (Nematoda), water fleas (Copepoda, Cladocera), and mites (Hydracarina), in addition to those mentioned previously.[22]

Many of these aquatic invertebrates are detritivores, organisms that break down and consume dead organic matter (fig. 5.11). They differ, however, in how they process this material, and constitute three functional trophic groups. Some are shredders (most crane flies and many stoneflies and caddisflies) that feed on large organic particles. Others are grazers and scrapers that feed on algae and bacteria on the surface of dead leaves or twigs and inorganic sediments (caddisflies, midges, some mayflies). And yet others are collectors and gatherers that feed on fine organic particles or filter particles from the water (some mayflies, black flies, midges). Collectively these invertebrates reduce the size of organic particles by fragmentation and maceration, which further facilitates the consumption of this organic matter by other invertebrates

and especially microbes. Other invertebrates are carnivorous predators (for example, *Rhyacophilia* sp., a caddisfly; *Hexatoma* sp., a crane fly; *Dytiscus* sp., a beetle; some stoneflies).

By consuming and utilizing organic matter to reproduce and maintain themselves, the stream invertebrates play an important role in the retention and processing of nutrients that would otherwise quickly flow downstream and thus out of the local stream ecosystem. These aquatic invertebrates make up a significant part of the heterotrophic food web of these aquatic ecosystems, and in turn, they are consumed by amphibians, fish, and birds, thereby expanding and interconnecting the food webs of stream and terrestrial ecosystems.

Snails and slugs. At least 17 species of snails have been recorded in the forest. Overall density in a study conducted just west of Watershed 6 averaged 7.6 snails per square meter of forest floor (all species combined).[23] In spite of this seemingly high diversity and abundance, most visitors to the forest rarely see these gastropods. Most are very small (less than 5 mm long), and live in moist areas on the forest floor, usually under fallen logs and dead wood, where they feed on decaying organic matter. Snails, which themselves rely on calcium for shell formation and for reproduction and growth, appear to have declined in recent years, coincident with the occurrence of acid rain and depletion of calcium, but it is not known whether this is a cause-and-effect relationship.

Slugs also occur in the forest, although the number of species here is unknown. They are larger than snails (up to 30 mm or more), but not as abundant (average of 1.8 per square meter).[24] Slugs are found on the ground surface more often than the smaller snail species, and are voracious feeders on herbaceous herbs and tree seedlings. Both snails and slugs are most evident on the surface of the forest litter and on trunks of trees on wet nights. Small snails are preyed upon by salamanders that occupy the same microenvironments, and by birds, especially as a source of calcium for eggshell production. Garter snakes and probably raccoons eat the slugs.

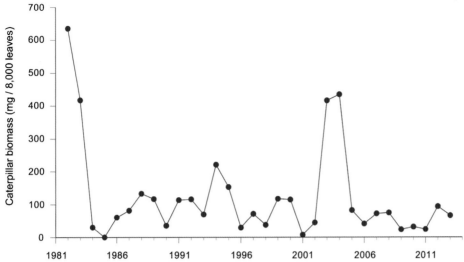

Figure 5.10. Year-to-year patterns in the abundance of caterpillars on understory foliage of American beech and sugar maple in the summers of 1986 to 2013. Researchers count free-living larvae seen during systematic visual searches of 8,000 leaves in the lower strata of the forest (trees 1–3 m tall), and measure larval body lengths. Body length is related to body mass, which can then be multiplied by the numbers counted to yield an estimate of total caterpillar biomass. (Modified and extended from Rodenhouse and Holmes 1992, and Reynolds et al. 2007; reprinted, in part, by permission of the Canadian Science Publishing, © 2007)

Figure 5.11. Larva of a caddisfly (*Lepidostoma* sp.), with its case made of bark fragments and small stones. Caddisfly larvae scrape algae and bacteria from stream substrates and shred leaves and other organic matter falling into streams, which makes them important components of the detrital food web. (Photo by R. G. Henricks)

Other invertebrates. Many other invertebrates also inhabit the forest, but have been less well studied. These include many different kinds of beetles (Coleoptera), true bugs (Hemiptera), ants, bees, wasps and sawflies (Hymenoptera), dragonflies and damselflies (Odonata), true flies such as crane flies, black flies (Diptera), and scorpionflies (Mecoptera). Spiders (Araneae) are numerous throughout the forest, and are important predators on smaller invertebrates. The webs these spiders make are also used extensively by birds in constructing their nests. Pseudoscorpions (pseudoscorpionida), centipedes (Chilopoda), and millipedes (Diplopoda) occur in the forest litter, and earthworms have recently become established (Box 5.4).

VERTEBRATES

Fish. Although 13 species of fish have been reported from the valley, most are confined to Mirror Lake and the lower parts of the mainstem of Hubbard Brook between the lake and the Pemigewasset River. Only three species of fish—brook trout, blacknose dace, and slimy sculpin—are found in the stream system within the forest. These occur mostly in the mainstem of Hubbard Brook and the lower portions of the tributaries that feed into it.[25]

Amphibians. Amphibians are common but infrequently encountered residents of the forest at Hubbard Brook (fig. 5.12, Appendix 2). They are found in the forest litter, under rocks or logs, and in the streams, seeps, and wetlands throughout the valley. Most are seen only during or after rainy periods, and even then mostly at night. Seven species of frogs and toads and six salamanders have been reported from the forest.

Among frogs, the wood frog and the American toad are the two most commonly encountered on the forest floor, whereas green frogs and spring peepers occur along streams and in boggy and other moist locations. All of these species seek standing water for egg laying, including Mirror Lake, beaver ponds in the upper parts of the valley, or in the case of the wood frog,

Box 5.4. Earthworms: Aliens in the Forest?

Some types of organisms that might be expected to occur at Hubbard Brook are absent or very rare. Earthworms, for example, are scarce and hard to find, and have been recorded only in recent years.[a] A survey in 2013–2014 found none in hundreds of samples taken from the forest floor in Watershed 6, but small numbers in samples from Watershed 4.[b]

Earthworms, in fact, are not found naturally in most parts of the northern United States or Canada, having been eliminated by the glaciers tens of thousands of years ago.[c] Most earthworms present today in gardens, pastures, and farmlands close to Hubbard Brook and in other parts of New England are descendants of those that arrived with European settlers, and others have been introduced from Asia. Although considered beneficial in agricultural systems, these aliens have invaded forests near human settlements in the Northeast, where they are speeding up and altering the decomposition of forest litter, which negatively affects nutrient availability, seed germination, and other forest floor processes.[d] If earthworms should increase in abundance at Hubbard Brook, which appears likely, we might expect changes to occur in litter depth, nutrient cycling patterns, plant regeneration, and other properties of the ecosystem in the years ahead.

NOTES

a. T. J. Fahey, N. L. Cleavitt, and G. E. Likens personal communications.

b. C. E. Johnson, personal communication.

c. Edwards and Bohlen 1996.

d. Bohlen et al. 2004; Huang et al. 2010.

Figure 5.12. Four common frogs and toads at Hubbard Brook: top, American toad, green frog; bottom, wood frog, spring peeper. (Photos by N. L. Cleavitt and D. Gochfeld, top; N. L. Cleavitt and W. S. Schwenk, bottom)

small temporary (ephemeral or vernal) pools that are widely scattered about the valley.

Salamanders in the forest are active mainly at night, especially during rainy weather, when they forage for insects, snails, centipedes, spiders, and other invertebrates. Three species—dusky, two-lined, and spring salamanders—breed exclusively in streambeds and forage either in streams or in leaf litter within a few meters of the streams. Spring salamanders can range up to 10 m from the streams in search of food.[26]

In contrast to these stream-dwelling species, the red-backed salamander is widely distributed in the litter of the forest floor, often under rotting logs. This species lays its eggs in moist leaf litter, in burrows, or under rotting logs, and females tend their egg clusters and defend them aggressively against predators. The density of this salamander species in the early 1970s averaged $0.26/m^2$, and their biomass per square meter exceeded that of all bird species combined at that time.[27] Censuses conducted between 1995 and 2006 using the same procedures, however, found that their numbers had dropped to about $0.075/m^2$, a decline of about 74% in 25 to 30 years.[28] The cause of this decline is not known, although it might be related

Figure 5.13. The eft is the juvenile stage of the red-spotted (eastern) newt and lives mostly in or on the litter of the forest floor. (Photo by N. L. Cleavitt)

Figure 5.14. The eastern garter snake is the only common reptile in the Hubbard Brook forest. (Photo by T. M. Jones)

to the effects of acid rain or to changes in moisture levels brought about by climate change. Further investigations are clearly needed.

Adults of the red-spotted newt live in Mirror Lake and in ponds in the upper parts of the valley. The bright orange-red juveniles of this salamander species, called efts, leave the ponds in late summer and spend the next 3–6 years living terrestrially before returning to water to breed (fig. 5.13). At Hubbard Brook, efts are frequently seen moving across the forest floor, especially during and after rainy periods. Casual observers to the forest often think the adult newts and the efts are different species because they are so different in appearance and they occur in different habitats.

Reptiles. Compared with more southerly regions of the United States, the reptile fauna of the Hubbard Brook valley is depauperate. There are no lizards, and only a few records of turtles, mainly near Mirror Lake. Only one snake, the eastern garter snake, is found within the forest (fig. 5.14). It is relatively common and widely distributed from low elevations to the ridge tops, and is an important predator on forest floor invertebrates and small mammals. There is no evidence from Hubbard Brook that the garter snake

preys on bird eggs or nestlings, as happens in other regions of North America.

Birds. Birds are conspicuous components of the forest at Hubbard Brook, and have been the subject of intensive study for more than 45 years. A total of 127 species has been recorded in the valley, with about half of these being vagrants or migrants that pass through during spring or autumn.[29] The bird community is thus dynamic, with dozens of species coming and going on different schedules and spending different amounts of time in the forest each year. Moreover, not all species are widespread or common, some having specific habitat requirements that limit their distributions to certain parts of the valley, such as the wetlands around beaver ponds or the conifer-dominated high-elevation ridges.

The numbers and species of birds occurring in the forest change dramatically from season to season. Eleven species live there year-round (permanent residents, such as woodpeckers, blue jays); seven are often (but not always) present just in autumn and through much of the winter (winter visitors, mostly nomadic seed-eating finches); another seven regularly pass through in spring and fall but spend little time there (transient migrants); and about 50 truly migratory species arrive in spring, breed, and

Figure 5.15. Six representative Neotropical migrant songbird species that breed in the northern hardwood forest at Hubbard Brook: top, ovenbird, Canada warbler; middle, red-eyed vireo, blackburnian warbler; bottom, scarlet tanager, rose-breasted grosbeak. Neotropical migrants are species that breed at temperate or higher latitudes, and then migrate to winter in the New World tropics, from southern Florida to South America. They comprise the majority of birds breeding at Hubbard Brook. (Photos of ovenbird and blackburnian warbler by J. Klaves; others reproduced by permission of Powdermill Avian Research Center, Carnegie Museum of Natural History)

then depart before autumn (summer residents). These summer residents spend only 2–3 months in the ecosystem, after which they move south to wintering areas either in the mid-Atlantic region or the southern United States (short-distance migrants) or to the Caribbean islands, Central America, or South America (long-distance or Neotropical migrants). These Neotropical migrants constitute the majority of the birds breeding in the forest (fig. 5.15, Appendix 2).[30]

The comings and goings of these bird species over the course of a year result in strong seasonal change in the numbers and biomass (collective weights) of birds within the forest ecosystem. Bird numbers and biomass are at a peak during the summer, when both the migratory and resident species are breeding, and reach their lowest values in midwinter when only residents and a variable number of winter visitors are present.[31]

This seasonal pattern largely reflects changes in the availability of food resources, with most of the birds exploiting the flush of food, mainly insects and other arthropods, from mid-May through the summer, a period in which most nesting and production of young takes place. Species present in winter are ones that are able to live in cold climates and exploit the few specialized resources available at that time, such as the small insects and spiders that live under or on tree bark, or seeds of the major tree species when available. The latter resource is used by winter visitors, mostly seed-eating finches that arrive in autumn and spend variable lengths of time in the forest. The presence and abundance of these seed-eating species vary strongly from year to year, depending on which tree species are masting and how plentiful the mast is. In years when large quantities of mast are produced by sugar maple and especially American beech, for example, the major seed-eating bird species in the forest in winter are the large-beaked grosbeaks and crossbills, whereas when tiny birch seeds are available, the smaller finches, such as the siskin and goldfinch, are the species present in highest numbers.[32]

Each bird species has its own way of exploiting the forest environment, including specific habitat preference, foraging behavior, and nest site preference.[33] Species differ greatly in where and how they obtain their food, and therefore in their trophic roles in ecosystem food webs. Four basic foraging groups, or guilds, can be recognized among the species breeding in the forest, based on types of foods taken: insectivores (34 species), herbivores (2 species), sap or nectar feeders (2 species), and raptorial birds of prey (5 species).[34]

Within each of these guilds, species differ in where they search for food and the methods they use to capture their prey. Among the insectivores, for instance, there are species, such as the nuthatches, brown creeper, and black and white warbler, that specialize in finding insects on or under tree bark, and others such as most warblers, vireos, and the scarlet tanager that move along the outer branches and twigs searching the foliage.

Among those that primarily search for prey on foliage, some search the undersurfaces of leaves and then fly up and hover near the leaf surface to pluck the insect from the lower side; others move along the branches, and glean sitting prey from nearby leaves,

often from the top surfaces (Box 5.5). Therefore, depending on its physical and perceptual abilities and preferences, each species has its own way of finding and exploiting food resources in this system, which usually leads to different prey being taken.[35]

Other major foraging strategies include sap and nectar feeding (hummingbirds, sapsuckers; Box 5.6), ground feeding for invertebrates (thrushes, juncos, ovenbirds), and plant feeding, largely for seeds, buds, and twigs (ruffed grouse and wild turkey). Although turkeys were probably present at Hubbard Brook a century or two ago, they were absent from the forest at the start of the Study. They then appeared in the late 1990s to early 2000s, and are gradually spreading throughout the valley. The first confirmed breeding record was in the lower end of the valley in 2004. Turkeys and grouse forage mostly in the litter of the forest floor for insects and seeds, or feed on buds, twigs, and fruits from plants in the lower strata of the forest.

Five raptor species occur at Hubbard Brook, all at relatively low densities. Broad-winged and red-tailed hawks and the barred and saw-whet owls usually feed on rodents, such as mice, voles, and squirrels. Sharp-shinned hawks prey extensively on small birds, and have been documented taking young birds from nests.

Mammals. Thirty-one species of mammals occur regularly in the Hubbard Brook valley (fig. 5.16, Appendix 2). Most are nocturnal, occur at low densities, and are rarely seen by the casual observer. They range in size from large (moose) to small (shrews, bats), and in foraging habits from species that browse on woody shrubs (deer) to those that feed mostly on herbs or seeds (mice, squirrels), insect feeders (bats, moles, shrews), and a variety of carnivores (fisher, weasels, coyote, fox, bobcat). Several rodent species even forage on subterranean fungi and thus are part of the detrital food web.

The most visible mammals in the forest are the diurnally active eastern chipmunks and northern red squirrels. Both are abundant and widely distributed in the forest. Red squirrels, often but not exclusively

Box 5.5. Tree Species Preferences by Foraging Insectivorous Birds: Implications for Forest Management

Differences in architecture among tree species, including leaf size and shape, leaf position in relation to nearby branches, and length of petiole (leaf stalk), either facilitate or constrain how birds locate and capture their insect prey. Sugar maple leaves, for example (a), have long petioles and leaves that are positioned upward relative to the branch. This allows birds to search under leaf surfaces, where most insects occur, as they move along the branches. In yellow birch (b), the leaves tend to hang down below branch level, and under-leaf insects must be sighted from below and captured by birds in upward hovering flights. Bird species differ in their physical and perceptual abilities in ways that affect how they search for and capture prey among the foliage of these different tree species.[a] The abundance and types of insect prey also differ among tree species.[b] Therefore, the tree species composition of a forest affects which bird species can successfully forage and exist there. This behavior has implications for forest management in that selective cutting or removal of particular tree species can influence the occurrence or persistence of particular bird species, a finding important for sustaining forest bird diversity.[c]

NOTES

a. Holmes and Robinson 1981.

b. Holmes and Schultz 1988.

c. Robinson and Holmes 1984; Holmes 2011.

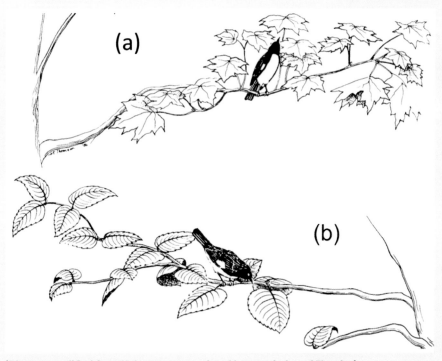

(Diagram modified from Holmes 2011; reprinted by permission of Elsevier)

Box 5.6. Commensalism Between a Woodpecker and a Hummingbird

The sap/nectar foraging guild at Hubbard Brook involves a commensalistic relationship between the yellow-bellied sapsucker, a migratory species of woodpecker, and the ruby-throated hummingbird. The sapsucker drills many small holes in trees, such as maples and birches, and sap oozes from these. The sapsucker then revisits these sap wells routinely, feeding on the sap and the insects attracted to it. Hummingbirds also visit these feeding sites, even following the sapsucker between trees, possibly learning the location of the sites.

(Photo by B. Parsons)

The hummingbird also feeds on both the sap and the attendant insects. Because there are few nectar-producing flowers in this northern temperate forest, the hummingbird depends heavily on this food resource. In fact, it is unlikely that the species would even occur in this forest without these sap wells. Although sapsuckers may displace hummingbirds occasionally from the feeding holes, the two species seem to coexist. This relationship is an example of commensalism, in which one species (the hummingbird) benefits without seriously affecting the other (the sapsucker).

associated with the more coniferous sections of the forest, are active year-round and forage for seeds, fruits, and occasionally insects, both on the ground and in the trees. Chipmunks, which hibernate during the winter, seek similar kinds of food primarily from the forest floor during the snowfree season, but also climb into shrubs and the lower parts of the canopy for fruits, seeds, and other foods. Both have been documented taking eggs and nestlings of songbirds breeding in the forest, and are among the major predators at songbird nests in this forest. Their numbers generally reach peak abundance in the spring and summer following a high seed production (masting) the previous autumn (see fig. 5.2), which results in high levels of predation on songbird nests

in that season at a time when seeds are becoming scarce.[36]

Other small mammals include several species of shrew, a mole, flying squirrels, and several small rodents. Bats also occur, but have not yet been studied. Of the shrews, the large (25–30 g) short-tailed shrew is the most common, and is regularly seen scurrying around in the forest litter in its search for insects. Flying squirrels are nocturnally active, hiding out during the day in old woodpecker holes and large tree cavities. The two most common small rodents are deer mice and red-backed voles, each of which ranges in density from 3 to about 12 per hectare, varying with the availability of mast.[37] The deer mouse is semi-arboreal, often climbing into the shrubs and

Figure 5.16. Some representative mammals of the Hubbard Brook forest: top, black bear, fisher, North American porcupine; middle, coyote, red fox; bottom, northern red squirrel, eastern chipmunk, snowshoe hare, and boreal red-backed vole. (Photos by N. L. Rodenhouse, T. Murray, A. Muniz, top; N. L. Rodenhouse, middle; S. Kaiser, N. L. Rodenhouse, D. G. E. Robertson, C. Conrod, bottom)

trees. It is usually active at night, and feeds on insects, herbaceous plants, and especially seeds. In contrast, the vole is active both day and night, and is confined to the litter of the forest floor where it searches for herbs, insects, seeds, and especially underground fungi. Both deer mice and voles remain active under the snow during the winter. The third common species of small rodent is the woodland jumping mouse, which hibernates during the winter, entering torpor in mid-autumn and not emerging until May. Jumping mice are patchily distributed in the forest, and seem most numerous where dense low vegetation occurs near wet areas or open water such as pools and streams.[38] These jumping mice feed on herbaceous vegetation, seeds, insects, and especially fungi, including mycorrhizae.

Among the large herbivorous mammals, moose were rare in the Hubbard Brook valley before the mid-1980s. By 2000, about 12–16 and possibly as many as 20 individuals were estimated to be in the forest during the winter, when their browsing of vegetation, especially hobblebush and understory balsam fir, was pronounced, as were their rubbing posts, which they use to rid themselves of winter ticks. Moose numbers started to decline by about 2005, and by 2012, only 2 to 6 individuals could be found in the valley during the winter.[39] This decline was probably due to various causes, including exceptionally warm winters, overbrowsing of winter forage, winter tick infestation, and a debilitating brain worm, all of which speak to problems of overpopulation and ecosystem change.[40]

By comparison, deer were common in the forest between the 1960s and the early 1980s, but declined in the 1980s, and particularly after the ice storm of January 1998.[41] Recently, however, deer have begun to increase again as moose numbers have dropped. Deer and moose do not seem to coexist at Hubbard Brook, perhaps because of a meningeal (brain) parasitic worm that does not seem to seriously impair deer but can be lethal to moose where the two species' ranges overlap.[42]

Two other herbivorous mammals live in the forest, the snowshoe hare and the North American porcupine. Both are widely distributed, but at low densities, and are consequently only infrequently seen. The presence and commonness of the hare is made most evident by its tracks in winter.

The order Carnivora is well represented at Hubbard Brook, with two species of canids (coyote, red fox), one cat (bobcat), five mustelids (fisher, mink, American marten, long-tailed weasel, short-tailed weasel), one skunk (striped), one procyonid (raccoon), and one bear (black bear). Most of these species are nocturnal or occur at low densities, and are rarely encountered. Tracks of fisher, coyote, and red fox in winter, however, show that these species are actually relatively common. Weasels are usually observed only in years when small rodents are abundant, which often follows masting events. The presence of the American marten, which is considered a relatively rare species in New Hampshire, has been confirmed by motion-activated wildlife cameras and in videotapes showing them depredating bird nests.

■

This brief overview of the living, or biotic, components of the watershed-ecosystems at Hubbard Brook highlights the organisms that are integral to the functioning of the ecosystem, and are also tightly integrated with its abiotic components. At the same time, these organisms also provide great beauty and exhibit interesting behavior, which contribute to making the northern hardwood ecosystems so appealing.

PART 3

UNDERSTANDING FOREST ECOSYSTEM STRUCTURE
AND FUNCTION

Leaf litter swept into a stream by a
spring rain. (Photo by S. A. Kaiser)

6

HOW IS ENERGY TRANSFORMED?

Without energy transfers, . . . there could be no life and no ecological systems.

— E. P. ODUM 1959

Life in the forest depends on energy from the sun captured by green plants in the process of photosynthesis. This energy is used by plants for respiration and growth. Living plant matter is eaten by herbivores, which in turn are eaten by a variety of carnivores—together these make up the grazing, or "green," food web. Plant matter not eaten by herbivores eventually dies and is broken down and consumed, along with dead animal remains, by a host of decomposer organisms and their predators in what is termed the detrital, or "brown," food web. Because energy is lost in each step of this process and dissipated as heat, ecosystems depend on a continual input of energy from solar radiation to maintain their structure and function.[1]

How do these energy transformations actually take place in a forested ecosystem, and how can these steps be measured? What are the major links and pathways by which energy is transferred from the sun through the system? How much plant biomass exists in the ecosystem, and how efficient are plants in converting energy into biomass? How do organisms partition and control the movement of energy through this complex system? Why isn't the forest knee deep in dead leaves and wood? These questions have been addressed as part of the long-term research at Hubbard Brook, using (1) field observations that identify the links and interactions among species and their environment; (2) measurements of rates of photosynthesis; (3) quantification of the numbers, biomass, and energy consumption by key components of the biota, which lead to estimates of growth and productivity (Box 6.1); and (4) measurements of energy flow from the physical environment (that is, input of solar radiation) through links in the food webs, which help to elucidate how energy is transformed in this forest ecosystem.

Understanding feeding relationships and energy flux is basic to understanding how ecosystems function and how they are maintained and change over time. It also allows for an evaluation of how organisms interact, how they reproduce and survive in this environment, and how they perform important roles in ecosystem processes.

Box 6.1 What Are Biomass and Productivity?

Biomass in an ecosystem context is defined as the standing crop or collective mass (weight) of organic matter, expressed on a unit area basis. It could be given as the biomass of sugar maples, for example, or of northern red squirrels, per hectare of forest. Or it can be the amount of dead organic matter on the forest floor per square meter. In practice, biomass of living organisms is calculated as the number of individuals of a population or species per unit area multiplied by the known weights of those individuals, with values expressed in grams, kilograms, or kilocalories (energy content) per unit of area. For comparative purposes here, these values have been converted to grams of carbon (C) per square meter (g C/m²). Carbon is often used as the unit of biomass, because it is the major structural element common to all organisms, allowing for more direct comparisons of biomass among components, such as living and dead plant biomass. Carbon constitutes about 50% of the dry weight of organic matter.

Productivity refers to the rate at which biomass is produced in an ecosystem. It is usually given in units of mass per unit area per unit time (grams per square meter per year), or in equivalent units of energy (kilocalories per square meter per year). Productivity by plants (which synthesize organic biomass by capturing energy from the sun) is referred to as primary productivity, whereas that of animals (converting the chemical energy of food into their own biomass) is secondary productivity. Plants are also referred to as autotrophs, as they manufacture their own food, while animals, fungi, and other microorganisms are heterotrophs, deriving their nutrition from already formed organic compounds such as those made by plants.

Measurements of biomass and productivity are essential not only for assessing energy relationships in ecosystems, but also for understanding the causes and consequences of environmental change, including those brought about by a changing climate or atmospheric pollution.

TROPHIC RELATIONSHIPS AND FOOD WEBS

The food web of the northern hardwood ecosystem at Hubbard Brook provides a framework for visualizing relationships and interactions among organisms and their roles in energy transfer and storage (fig. 6.1). In the grazing food web, living plant tissues such as leaves, flowers (and subsequently fruits or seeds), and to some extent wood are resources sustaining a variety of herbivores (primary consumers), ranging from many different kinds of caterpillars (Lepidoptera larvae) and other plant-feeding insects to wood-boring beetles to rodents to large herbivores such as deer and moose. Most primary consumers are in turn food for animals at higher trophic levels, including birds, amphibians, rodents, and other small mammals, which support even higher trophic links involving avian predators such as sharp-shinned hawks and barred owls, and mammalian predators such as weasels, fishers, coyotes, and red foxes.

The detrital or "brown" food web, in contrast, is based on dead organic matter, mostly plant material such as standing dead wood, dead leaves and roots, and woody debris, but also whatever animal remains reach the forest floor. This dead organic matter is broken down by bacteria, fungi (fig. 6.2), and a variety of invertebrates, such as snails, millipedes, larval crane flies, and beetles. Centipedes, beetles, spiders, amphibians, shrews, and some ground-foraging birds prey on these detritus feeders. Earthworms are not, at present, an important part of the forest ecosystem at Hubbard Brook, because native species became extinct during the last ice age and invasive ones have only recently been detected.

In the streams at Hubbard Brook, the energy base

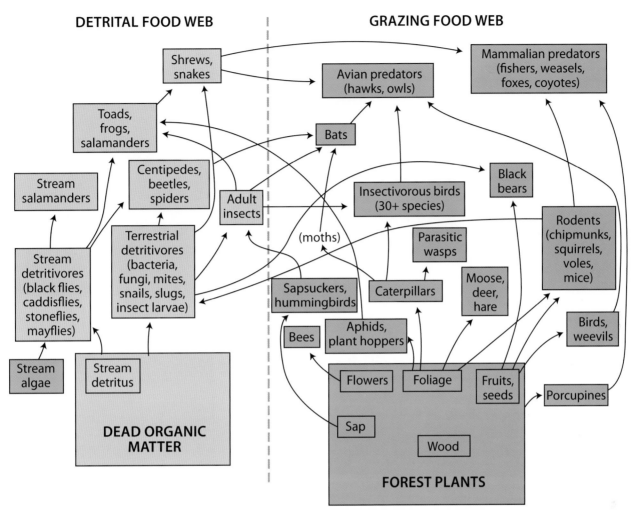

DETRITAL FOOD WEB

GRAZING FOOD WEB

Figure 6.1. The major feeding relationships among the organisms of the northern hardwood forest ecosystem at Hubbard Brook. The food web is divided into two main divisions—the detrital web (left of dashed line) and the grazing web (right). The number of links or connections within the community, and especially between the two webs, promotes stability and increases the ability of the system to respond to disturbances or perturbations. Not all known or potential connections are shown in the diagram. (R. T. Holmes and G. E. Likens, unpublished)

is derived almost entirely from leaves, branches, and wood falling or washing in from the surrounding forest and from dissolved organic matter. This dead organic matter supports a large variety of detritus-feeding organisms, such as microbes, larval black flies, mayflies, stoneflies and caddisflies, and their predators. A very small quantity of energy in the streams comes from photosynthetic mosses and algae. These primary producers are relatively scarce in headwater streams, and, in fact, algae have only

appeared there in the past two decades. The reason for this recent increase in stream algae is not fully understood, but it may be a long-term effect of changing canopy structure (light availability) as the forest develops over time or chemical dynamics resulting from acid rain.

Although lower links in these two major food webs are relatively distinct, considerable crossover occurs at the higher trophic levels, which adds to the complexity and ability of the system to be resilient to disturbance.

Figure 6.2. Many different types of fungi occur in the forest at Hubbard Brook. These are just a few of the different types of fruiting bodies, or mushrooms, that can be found: top, *Boletus, Scutellinia,* and *Craterellus;* center, *Mycena;* bottom, *Coltricia, Gomphus,* and *Russula.* (Photos by N. L. Cleavitt, except center, J. Klaves, and bottom row center, S. Kaiser)

Some organisms, for example, fit neatly into discrete roles: most caterpillars are strictly herbivores that consume plants, mostly leaves, and salamanders are predators on insects and other invertebrates. Other organisms, however, are omnivorous. Hummingbirds in the Hubbard Brook forest, for instance, feed both on sap from sapsucker holes as well as on small insects attracted to the oozing sap. Among the rodents, boreal red-backed voles and woodland jumping mice feed on leaves of herbaceous plants, seeds, and fruits, but also consume large quantities of subterranean fungi, and thus participate in both the grazing and detrital food webs. Similarly, black bears are classic omnivores, being primary consumers when they feed on herbs, roots, fruits, berries, and nuts and secondary consumers when they take beetle larvae, birds' eggs, and even the occasional fawn. In addition, since bears are known to scavenge dead carcasses of deer and other animals, they also participate in the detrital food web. In these ways, organisms interact with one another through food webs, with energy transfer being the integrating process.

PLANT BIOMASS AND PRODUCTIVITY

Plants, as the main energy fixers in the forest, are at the core of food webs, capturing the sun's energy and providing resources to herbivores and their predators, while dead organic matter forms the base of the detrital food web. Plants are also important in concentrating nutrients into tissues such as leaves and fruits that animals use for food, as well as providing shelter and places for animals to live. How much plant material (biomass) is present as living versus dead organic matter, and at what rate is it produced? Is the rate of production and turnover of plant matter in this forest ecosystem changing over time, and how are those patterns affected by climate change and other environmental events? Answering such questions requires measuring the amount of biomass in living plants, both above- and belowground, as well as the amount of dead plant material in the forest, and how these change over time. Gathering such information for a forest ecosystem is not a trivial process (Box 6.2).

Forest biomass. In the late 1990s, an estimated 29,678 g of carbon per square meter existed in the forested ecosystem (Watershed 6) at Hubbard Brook (fig. 6.3).[2] Of this, the largest proportion (58%) consisted of organic detritus, largely from dead plant material; 40% was bound in living plant biomass (above- and belowground); and 1.8% was detritus in the streams. Less than 0.2% of organic matter in the system was found in the biomass of heterotrophs, mostly microbes (96%). Of the total carbon in the detrital pool, 92% was located in the mineral soil, whereas the remainder consisted of dead trees, branches, and other coarse woody debris and humus on the forest floor. Biomass of the aboveground

living plants was about three and a half times greater than that in the root systems. Of the living biomass aboveground, most was found in the bole wood (63%), and branches and twigs (29%), with smaller amounts in bark (6%), foliage (2%), and all of the animals (< 0.1%).

Net primary productivity. To understand forest ecosystem dynamics, it is useful to know whether biomass is accumulating over time. How much energy is channeled into the production of new tissue, and how is this changing over time? Answering such questions for a forest ecosystem requires knowing the annual growth rates of perennial tissues like wood in the boles and branches of the living trees and the production of annual tissues like leaves, flowers, and fruits, and then correcting for dying trees. The result is an estimate of net primary production—that is, the amount of energy going into plant growth exclusive of that used by the plants in their own respiration or lost to the system.

At Hubbard Brook, the incremental growth of woody tissues was estimated from analyses of harvested trees and the resulting equations that related plant biomass to plant dimensions. Starting in 1991, researchers began placing aluminum tags on all trees greater than 10 cm DBH on Watershed 6 and on other plots across the forest (fig. 6.4). These individually tagged trees have subsequently been remeasured at 2- to 5-year intervals, providing data on the incremental growth of tree boles (wood). The amount of annually produced tissues was directly measured by the quantities of leaves, twigs, seeds, and fruits falling from the canopy into litter traps, while the biomass of herbaceous plants, shrubs, saplings, and smaller trees was determined by harvesting these plants at the end of the growing season. Taken together, these measurements allow for calculations of the rate of change of living plant biomass (net primary production) for the forest ecosystem over time.

Between 1956 and 1965, the aboveground net primary productivity—the annual increment—of the forest ecosystem on Watershed 6 averaged 420 g of carbon/m². For the period 1996–1998, the rate of

Box 6.2. How Do You Measure the Biomass of a Tree? Of an Entire Forest?

Measuring the biomass of living vegetation in a forested ecosystem requires many steps. For trees, the numbers and sizes (calculated from their diameters at breast height, or DBH) have been measured through field surveys. These data are then converted to biomass, using equations derived empirically that relate tree dimensions (including mass) to DBH.[a] To develop these equations, trees had to be felled, and their component parts (bole, branches, twigs, leaves) weighed and measured. Root systems were also excavated, in some cases with the "encouragement" of dynamite. Subsamples (discs or "cookies") of boles and branches were measured for determining dry weights and for analysis of their chemical (nutrient) contents. Other measures recorded were radial growth increments (tree rings), branch length and diameter, bark thickness and surface area, number of twigs, and weights of living wood, dead wood, and current twigs with leaves and fruits. This is a very laborious process.

These data were then analyzed to obtain statistical relationships between the measured DBH of the trees to their total mass, bole volume, bark surface area, and other dimensions, which yield estimates of biomass. These estimates were subsequently verified independently by direct weighing of whole trees removed during the harvest of trees on Watershed 5 in the early 1980s.[b]

Biomass of understory plants was determined by harvesting whole plants from small plots, excavating underground root systems, and hand-sorting fine roots from soil pits and soil cores. Finally, organic matter in the soil was sampled at different depths in the soil pits, and the carbon content was chemically determined. All samples were dried and weighed, yielding estimates of biomass in dry matter per unit area.

An online calculator for biomass on Watershed 6 is available on the following interactive website: www.hubbardbrook.org/w6_tour/biomass-stop/stop-7.htm.

NOTES

a. Whittaker et al. 1974; Siccama et al. 2007.

b. Siccama et al. 2007.

Soil pit. Root biomass and organic content of soil are determined quantitatively from samples taken from soil pits of known dimensions. (Photo from C. E. Johnson)

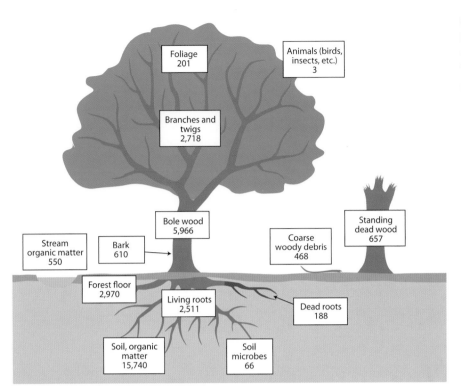

Figure 6.3. Distribution of biomass among the components of the forest ecosystem on Watershed 6. All units are grams per square meter. (Adapted from Fahey et al. 2010)

Foliage 201

Animals (birds, insects, etc.) 3

Branches and twigs 2,718

Bole wood 5,966

Standing dead wood 657

Stream organic matter 550

Bark 610

Coarse woody debris 468

Forest floor 2,970

Living roots 2,511

Dead roots 188

Soil, organic matter 15,740

Soil microbes 66

Figure 6.4. Christa Anderson measuring tree diameter at 1.4 m aboveground. Aluminum tags allow for repeated measurements of individual trees, which provide important data on tree growth rates and mortality over time. (Photo by N. van Doorn)

production was found to be only 354 g carbon/m²/yr, a lower value than had been predicted from forest growth models.[3] The net accumulation of tree biomass in the forest therefore seems to be slowing. This appears to be due to at least two factors, an increase in tree deaths and a slowing of tree growth (accumulation of new biomass). Both of these may be related to environmental stresses, such as acid rain affecting calcium availability and mobilization of toxic aluminum in the soil.[4] This hypothesis is currently being tested at Hubbard Brook through a whole watershed-ecosystem manipulation experiment in Watershed 1.

Heterotroph biomass and productivity. Heterotrophs are organisms that ingest and use organic matter produced by the autotrophic ("self-feeding," or

Box 6.3. Estimating Biomass and Energy Flow Through a Salamander Population

Even though rarely seen, the red-backed salamander is probably the most common vertebrate species in the Hubbard Brook forest. It lives terrestrially year-round, away from streams and typically under fallen logs and in the leaf litter of the forest floor. For estimating its energy use, the first step was to determine how many individuals of this species were present, and how much biomass (weight) per unit area they represented. To do this, counts were made on 25 × 25 m plots systematically located across one 36.1-hectare watershed (Watershed 4). Animals were located within these plots by thoroughly searching the ground layer, which included turning all logs and rocks and raking deep accumulations of litter and by visual counting of those encountered along fixed transects on rainy nights when these animals are active aboveground. This effort resulted in an estimate of approximately 95,000 red-backed salamanders on this watershed, or an average of 2,632 per hectare of forest. With the live (wet) weight of salamanders averaging 0.63 g (equal to about one quarter of a U.S. penny), this yielded an estimate of 1,650 g of salamander biomass per hectare.[a]

Combining information on density and biomass with data from laboratory studies on how much food these salamanders ingest (measured in energy units, kcal), how much of that energy they actually assimilate, and how much energy they use for respiration (metabolism) and growth, an energy budget can be calculated for this salamander population.

To maintain itself, the salamander population in this forest ecosystem is estimated to ingest about 10,000 kcal worth of food per square meter per year. This represents a tiny fraction (about 0.022%) of the net annual primary production by plants in this ecosystem. Of the amount of energy ingested, about 8,000 kcal are

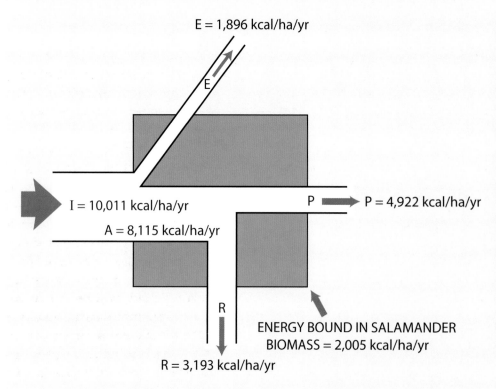

E = 1,896 kcal/ha/yr

E

I = 10,011 kcal/ha/yr

A = 8,115 kcal/ha/yr

P ⟶ P = 4,922 kcal/ha/yr

R

ENERGY BOUND IN SALAMANDER
BIOMASS = 2,005 kcal/ha/yr

R = 3,193 kcal/ha/yr

Summary of measurements and estimates of energy ingested (I), egested (E), assimilated (A), respired (R), and being channeled into production (P) of new tissue for the red-backed salamander population in the early 1970s. Biomass energy, represented by the shaded portion of the diagram, is the energy equivalent of salamander biomass in one hectare of forest. (Modified from Burton and Likens 1975a, b)

actually assimilated by the salamander population. Of the amount assimilated, about 40%, or 3,000 kcal, is used in respiration, and 60%, or almost 5,000 kcal, goes into production of new biomass, mainly as growth and reproduction. This production amount is large compared with warm-blooded vertebrates in the ecosystem, such as birds and mammals, which use about 98% of assimilated energy for respiration, with only 2% going into production. Red-backed salamanders are therefore more efficient at converting ingested energy into new biomass.[b]

Red-backed salamander, the most abundant vertebrate species in the forest at Hubbard Brook. (Photo by R. K. Lawton)

NOTES

a. Burton and Likens 1975a, b.

b. Burton and Likens 1975a, b; Holmes and Sturges 1975.

photosynthesizing) plants. The heterotrophs are the animals in the forest, along with fungi and the many other microorganisms that live mostly in the forest floor and the underlying soil. To assess the role of these heterotrophs in energy flow, their numbers, biomass, energy use, and productivity per unit area must be determined. At Hubbard Brook, for the leaf-eating caterpillars, salamanders, small mammals, and birds, this involved determining, for each group, the number of individuals per unit area, making measurements of live and dry weights, and using laboratory-based estimates of the amount of energy consumed and channeled into the production of new animal tissue. Finally, for soil organisms, indirect assays were used to estimate biomass and energy use (metabolism).[5]

Combining data for all of these major groups of heterotrophs showed that they constitute less than 2% of the living biomass of this forest ecosystem, most of which was accounted for by microorganisms. This result is not too surprising when considering that most of the larger animals, particularly vertebrates, have relatively low densities and biomass per square meter, and thus contribute very little to total biomass in the system. The red-backed salamander, for example, the most abundant vertebrate species in the forest, was found to occur at an average density of about 2,632 individuals per hectare, or 0.26 per square meter (Box 6.3).[6] A similar calculation for birds reveals that the maximum number was present in mid-July, at the peak of the breeding season: there were 322 individuals per 10 ha (all species combined), or only 0.003 individuals per square meter.[7] Rather surprisingly, the biomass of this one species of salamander is 1.9 times greater than that of the entire bird community at its mid-July peak, based on dry weight per unit area.[8] Except perhaps when moose are present in high densities, animals do not contribute much to the total biomass of these forest ecosystems.

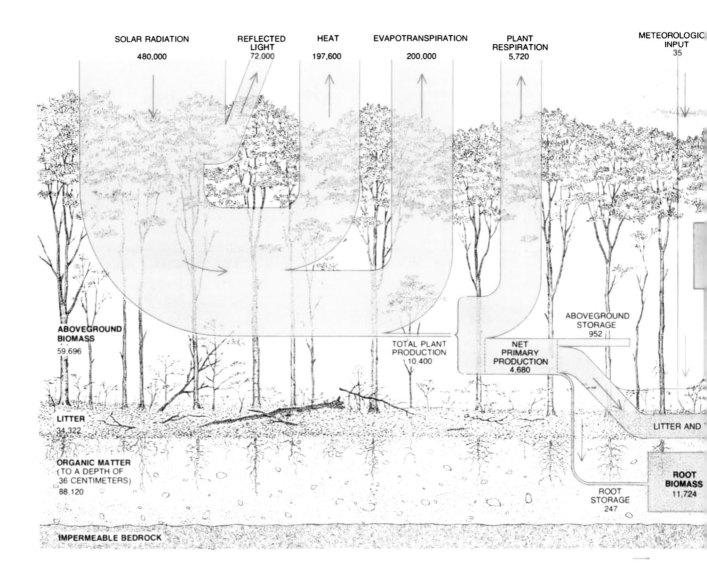

SOLAR RADIATION 480,000	REFLECTED LIGHT 72,000	HEAT 197,600	EVAPOTRANSPIRATION 200,000	PLANT RESPIRATION 5,720	METEOROLOGIC INPUT 35

ABOVEGROUND BIOMASS
59,696

TOTAL PLANT PRODUCTION 10,400

NET PRIMARY PRODUCTION 4,680

ABOVEGROUND STORAGE 952

LITTER
34,322

LITTER AND

ORGANIC MATTER (TO A DEPTH OF 36 CENTIMETERS) 88,120

ROOT STORAGE 247

ROOT BIOMASS 11,724

IMPERMEABLE BEDROCK

ENERGY TRANSFER THROUGH THE NORTHERN HARDWOOD ECOSYSTEM

Combining data on the biomass of the major groups of organisms, estimates of energy used by those organisms, and measures of solar energy and meteorological inputs and losses provides the basis for constructing an energy budget for the entire forested watershed-ecosystem (fig. 6.5). In the description of the ecosystem energy budget that follows, energy is expressed in units of kilocalories per square meter per year (kcal/m²/yr).

The energy budget for this forest ecosystem is complex because of the many components, and requires close scrutiny to appreciate and understand all of these relationships. In the section below, we summarize the major findings from the analysis of the energy flow patterns of this northern hardwood ecosystem at Hubbard Brook.[9]

How much solar energy is available and used by the plants?

- Total solar radiation reaching the ecosystem is estimated as 1,254,000 kcal per square meter per year, of which 480,000 arrives during the

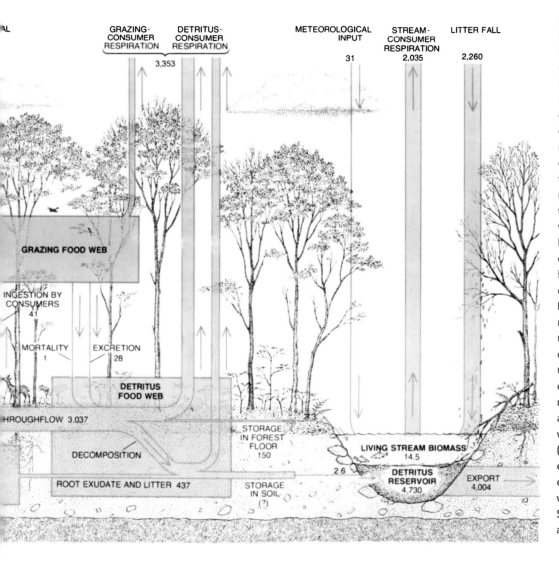

growing season, June to September, when plants are actively photosynthesizing.

- Plants fix, in chemical form, an estimated 10,400 kcal/m² of energy annually, which represents about 2% of the incoming solar radiation during the growing season. This is the energy available to plants and heterotrophs in the ecosystem.

- The remaining 98% of incident radiation is absorbed as heat by the ground or other components of the ecosystem, reflected back into the atmosphere from the ground and vegetation, or used for the important process of transpiration.

- A small but measurable amount of energy also enters the ecosystem via wind or as organic matter dissolved in rain and snow. Precipitation falling on the watershed contains an average of about 1 mg of total organic carbon per liter of rain and snow, equivalent to 35 kcal per square meter per year. Although small, this meteorological input is about half that of the energy contained in organic matter leaving the ecosystem in stream water, and is readily available to microorganisms in the ecosystem.

How is solar energy transferred and used through the food webs?

- Plants use 55% of the energy they fix for their own self-maintenance, mainly for respiration (5,720 of the 10,400 kcal/m²/yr fixed by plants). The remaining 45% (4,680 kcal) goes into the production of new plant biomass through growth or reproduction (annual net production).
- About 22% (952 kcal/m²/yr) of annual net production is stored in aboveground plant biomass (growth of wood, leaves, etc.), whereas 6% (247 kcal) is invested in growth of roots underground.
- On an annual basis, 3,037 kcal/m²/yr (or 72%) fall to the ground in the form of dead trunks, branches, and leaves, forming the energy base for the detrital food web.
- Consumption of plants by herbivores in the forest is usually low but varies greatly because the numbers of caterpillars fluctuate. The highest herbivory levels recorded were during a population outbreak of saddled prominent caterpillars between 1969 and 1971, when an estimated 44% of the leaf tissue was consumed. The caterpillars ate so many tree leaves in these years that in July patches of the forest looked like winter. Annual variation in caterpillar numbers affects the rate and amount of energy fixed in the ecosystem, as well as influencing food resource availability for animals at higher trophic levels, such as predatory insects and insectivorous birds.
- The average annual amount of energy entering the detrital pathway is estimated at about 3,503 kcal/m²/year. Most of this consists of fallen leaves, branches, and boles (83%) and root death (12%), plus smaller amounts from such sources as litter falling from shrubs, saplings, and herbs, organic matter dissolved in precipitation or washed from foliage (throughfall), fecal material,

deaths of animals, and organic material exuded by roots. More recently, studies have shown that organic compounds exuded by roots account for 10% to 15% of the annual net productivity in the forest ecosystem.[10]

- Energy used in respiration by heterotrophs is estimated to be 3,553 kcal/m²/yr, and most of this is respiration by organisms that feed on detritus, or dead organic matter.
- Energy used for growth and reproduction by animals in the grazing food web is very small relative to that of other ecosystem components (see below).

How do streams affect the energy budget?

Streams transport water along with particulate and dissolved organic and inorganic materials through and out of the system, in this way affecting the energy budget. The stream draining the reference watershed-ecosystem (Watershed 6) is small, heavily shaded by the overhead canopy of trees, and in the early 1970s when these energy budgets were calculated lacked algae or other aquatic plant life, except mosses in some places (algal blooms have occurred in recent years).

- Aquatic plants contribute only about 10 kcal of energy per m² of streambed per year.
- Most energy input to the stream (4,730 kcal/m²/yr) arrives as litterfall from the surrounding forest.
- Energy output from the stream is in the form of heat produced by the respiration of the stream organisms (2,035 kcal/m²/yr) and by the loss of organic matter in stream water leaving the ecosystem (4,004 kcal/m²/yr). Such movements downstream are strongly affected by flooding that occurs during and following heavy rains.
- Dead plant material in the stream often catches and builds up behind boulders or logs in the channel, forming dams of organic debris. These debris dams are important structures providing shelter and feeding places for stream organisms,

and they are "hotspots" for biogeochemical reactions, such as denitrification. Organisms living there account for only a very small part (14.5 kcal/m² or 0.3%) of organic matter in the streams.

What are the roles of animals and microorganisms in moving energy through the grazing and detrital food webs?

Energy use by animals and microorganisms varies with the type of organism, its abundance, and the efficiency with which it utilizes energy (fig. 6.6). Many animals at the top of the food web—such as moose, songbirds, and the ubiquitous chipmunks—are the more conspicuous and arguably the more charismatic parts of the forest ecosystem. In contrast, the microbial and microarthropod components are largely invisible, but they carry out a greater share of energy transfer. The more important results from our analyses of energy transfer by these groups are as follows.

- The amount of energy moving into and through animal populations is relatively small compared with that utilized and transformed by plants.
- The living plant tissue, the base of the grazing food chain, supports a large diversity of herbivorous insects and mammals along with their predators. These herbivores mainly consume tissues that are produced annually by plants, such as flowers, fruits, seeds, and leaves.
- Lepidoptera larvae (caterpillars) are the most important consumers of foliage, but they assimilate relatively little of the energy they ingest (14%). The unassimilated plant material passes through the insect's gut and falls to the forest floor as frass (caterpillar feces), contributing to nutrient cycling, especially in years when caterpillar numbers are high.
- About 40% of the energy assimilated by caterpillars is channeled into their growth, whereas 60% goes to their respiration. In contrast, mammals such as shrews and birds, which are warm-blooded, use about 98% of

their energy for respiration and only 2% for growth or reproduction, while salamanders channel 60% of the energy they assimilate into the production of new tissue, and 40% is used in their respiration.

- Few animals consume live woody tissue. The exceptions are some burrowing insect larvae (mostly beetles) and the porcupine. This result contrasts with more southerly forests that have more wood-consuming heterotrophs, such as termites, ants, and bark beetles.
- The dead tree boles, branches, and other plant matter that fall to the ground form a large organic pool that supports a diverse and large number of detritivores, including bacteria, fungi, numerous invertebrates, and their predators. The detrital food web remains a relatively less well studied part of the forest ecosystem at Hubbard Brook.
- The adult forms of the detritus-feeding larvae of many insects (for example, flies, beetles, and stoneflies) in the forest floor and in the streams provide important food for higher-level predators, such as salamanders, frogs, birds, and small mammals.
- Microorganisms, fungi, and other decomposers at the base of the detrital food web shred and process dead organic material, which releases nutrients for reuse by the plants for further growth. Although their biomass is small relative to that of plants in the system, their impact is huge in terms of breaking down organic matter and supporting a myriad of other species in the detrital food web.
- Similarly, even though herbivorous animals consume only a small fraction of the tree leaves, fruits, and seeds present in most years, their consumption affects rates of energy fixation (by removing photosynthetic tissue), plant reproductive success (by consuming seeds and seedlings), and even plant structure and form (for example, through browsing effects by snowshoe hares, moose, and deer). Also,

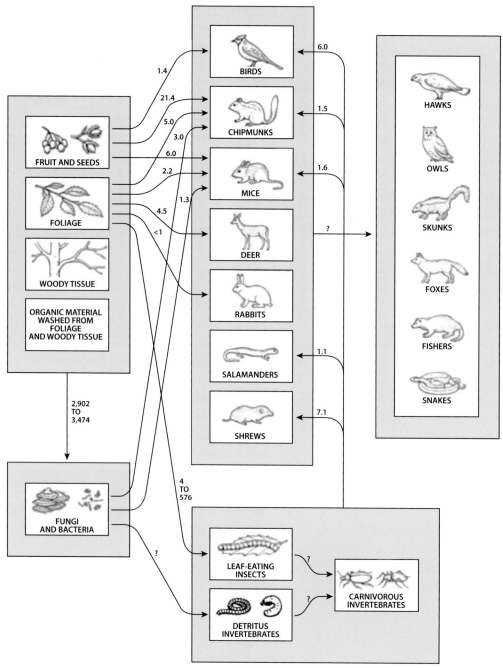

Figure 6.6. Energy flow through the grazing and detrital food webs at Hubbard Brook as measured in the early 1970s. Units are kilocalories per square meter per year (kcal/m²/yr). A relatively small amount of energy supports the large diversity of herbivorous insects, mammals, and birds and their predators in the grazing food web compared with the large amount entering the detrital food web, which is consumed by fungi, bacteria, and some invertebrates. (From Gosz et al. 1978; reproduced with permission © 1978, Scientific American, Inc., all rights reserved)

consumption of plants by herbivores speeds up the flow of partially digested materials to the detrital pool, especially during caterpillar outbreaks that result in heavy frass production.

- Some birds and mammals are potentially important dispersers of seeds, either by caching them in places where some eventually germinate, or by dropping seeds at some distance from the source plant after they have passed through digestive tracts. A prime example of the latter at Hubbard Brook is the dispersal of pin cherry seeds by birds following forest disturbance. Pin cherry is a key species involved in regeneration of the forest following major disturbances.

This quantitative description of energy flow through a forest ecosystem highlights the importance of green plants in producing organic matter that supports both the grazing and the detrital food webs. One of the surprising discoveries, however, was that the amount of dead organic biomass was actually greater than the biomass of living plants. Also, most of this dead organic material was located in the soil, where it supports many detritivores, including bacteria, fungi, and invertebrate animals such as millipedes and their predators. Some of these predators, such as birds, are usually associated with the grazing food web. To date, many details of the relationships and patterns of energy flow within the detrital food web remain unstudied, providing a challenge for future work.

We need to emphasize that most of the biomass and energy flow measurements for the Hubbard Brook forest presented here were obtained in the late 1960s and early 1970s, and therefore the numerical values cited may not be the most accurate for the forest that exists today (2015). Since the early 1980s, the forest has no longer been accumulating biomass, so the ratios of living to dead organic matter are probably different now from those obtained in the 1960s and 1970s. And the atmosphere now contains a much larger concentration of carbon dioxide that has been emitted from global anthropogenic activity, which likely affects photosynthetic rates. Nevertheless, the patterns and relationships discussed here and shown in figures 6.5 and 6.6 illustrate the intricacies of energy transfer and the interconnectedness of the diverse components of forest ecosystems.

Other changes have occurred in the forest over the course of the 50-year Study. Algal growth has increased in the streams, affecting rates of primary production. Bird and salamander populations now are about one-third of what they were in the early 1970s. And moose, which were rarely seen in the valley before the mid-1980s, increased during the 1990s and early 2000s, but now seem to be in decline. Challenges for researchers at Hubbard Brook will be to understand why these changes occurred and to predict trends for the future. Understanding these changes in physical and biological factors, and including them in analyses and models of energy dynamics in this system will be central to the future management of northern hardwood landscapes.

7

HYDROLOGY
Water Balance and Flux

Water and nutrients are the lifeblood of a forest ecosystem. From the beginning of the Study, the intent was to measure all of the inputs and outputs of water and chemical nutrients in gauged watersheds and to construct quantitative budgets for these forested watershed-ecosystems. Water is critical to this effort because water and nutrient flows and budgets are intricately interconnected and linked in watershed-ecosystems. The overall goals of the Study, which have evolved over time, have been to understand how these systems work, to gain insights about the interconnections between water and nutrient dynamics, to understand how these dynamics were connected to, and affected by, adjacent ecosystems, and how water and nutrient budgets responded to disturbances, especially forest harvesting.

The water cycle. In its basic form, the water (hydrologic) cycle at Hubbard Brook is composed of inputs of water in precipitation (rain and snow), and outputs of liquid water in stream flow and of water vapor in evaporation and transpiration (evapotranspiration). Quantitative measurement of these components is critical for budgetary (mass balance) analyses calculated for the Study.

Precipitation and stream flow. In general, water is abundant at Hubbard Brook, with precipitation occurring, on average, about three times a week throughout the year. Nevertheless, there is significant interannual variability, with precipitation in the wettest years (1973–1974 and 1995–1996) being about two times higher than the driest year (1964–1965) during the period 1956–2012. The drought during 1963–1965 was one of the driest periods in the northeastern United States in about 200 years.

The amount of precipitation falling on the south-facing experimental watersheds has been measured by the U.S. Forest Service in a fixed network of standard U.S. Weather Bureau gauges sited in cleared areas of the forest since 1956.[1] On average, 143.4 cm (56 inches), ± 20 cm, of water has fallen on Watershed 6 each year during 1963–2009 (Table 7.1). Of this amount 65% becomes stream flow, and 35% is lost through evapotranspiration. About one-third to one-half of

Table 7.1. Average Annual Hydrologic Budget for Watershed 6, 1963–2009

	cm/unit area ± SD	Percentage of input
Input (precipitation)	143.4 ± 20.0	100
Output (stream flow)	92.6 ± 20.6	64.6
Output (evaporation plus transpiration)[a]	50.8 ± 5.0	35.4

Source: Likens 2013; reprinted by permission of Springer Science+Business Media.

[a] Calculated (input minus output)

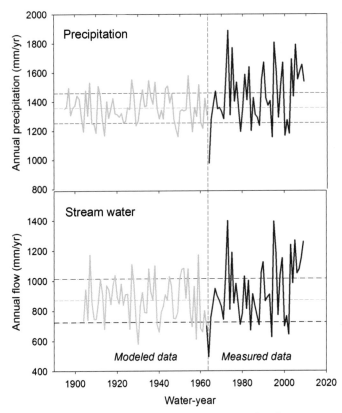

Figure 7.1. Average annual precipitation and stream flow in Watershed 6 from 1895 (precipitation) and 1904 (stream flow) to 2012. Dashed lines show the long-term average value ± 1 standard deviation. Data prior to 1963 were derived from a statistical relationship with longer precipitation and stream flow records from nearby monitoring stations. A water-year extends from June 1 to May 31. (Precipitation data from Likens 2013; reprinted by permission of Springer Science+Business Media)

yearly precipitation is snow, which typically persists as a snowpack from December through most of April. Occasional thaws occur during the winter, and these midwinter thaws have become more frequent in recent years.

Based on a regression (statistical relationship) of measured precipitation data for Watershed 6 and nearby Bethlehem, New Hampshire, and for streamflow data at Watershed 6 and the Pemigewasset River in nearby Plymouth, New Hampshire, where records extend back in time to 1895 and 1904, respectively, it was possible to estimate annual precipitation and stream flow records at Hubbard Brook back to the beginning of the twentieth century. Using this historical perspective, it is clear that precipitation at Hubbard Brook has increased significantly since 1964, and especially since 1985 (fig. 7.1). Other insights from this long-term record are that there has been more year-to-year variability since about 1964 and that the wettest and driest years during this long period occurred during the 50-year period of our Study.[2]

On average, February is the driest month and August the wettest in terms of precipitation, but overall there is little difference between monthly values. In contrast, about 47% of annual stream flow occurs during the months of March, April, and May, with 23% on average occurring in April alone. During these snowmelt periods, especially when accompanied by intense rainstorms, stream flow can

reach hundreds of cubic meters per day (fig. 7.2). Summer months (July, August, and September) usually have low stream flow, except during storms, and in dry summers with high rates of water loss by transpiration, stream flow can approach zero at the gauging weirs (fig. 7.3). Stream flow can increase noticeably and quickly during summer storms, and especially in autumn when transpiration water loss is reduced as deciduous leaves fall. Because of the high infiltration rates of water by soils at Hubbard Brook, precipitation is rapidly converted to stream flow in these watershed-ecosystems.

Figure 7.2. A headwater stream in winter. (Photo by G. E. Likens)

There is a tight connection between amount of annual precipitation and stream flow, where increased precipitation results in increased stream flow (fig. 7.4). Because of the relatively watertight bedrock in the headwaters of south-draining experimental watersheds at Hubbard Brook, evapotranspiration can be calculated with appreciable confidence from the difference between annual precipitation and annual stream flow. Both precipitation and stream flow increased in amount from 1963 to 2011, whereas evapotranspiration remained relatively constant (fig. 7.5).

MEASURING EVAPOTRANSPIRATION USING THE WATERSHED-ECOSYSTEM APPROACH

Evapotranspiration (evaporation plus transpiration) is an energy-demanding process in which liquid water is converted to vapor at all surfaces in the ecosystem. Transpiration occurs when plants transport water from the soil through the plant until it is lost by evaporation through stomata (pores) in the leaf's surface. Using the small-watershed approach, it is possible to calculate the amount of evapotranspiration by subtracting the quantity of water leaving the system at the base of the watershed in stream flow from the input to the watershed in precipitation, the difference being the amount of water lost as evapotranspiration.

Figure 7.3. Tammy Wooster inspecting a small headwater stream on the north-facing slope in summer. (Photo by D. C. Buso)

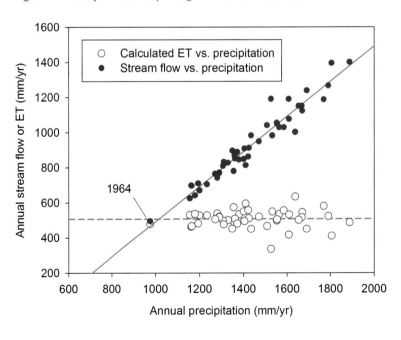

Figure 7.4. Relationship between annual precipitation and annual stream flow and annual evapotranspiration (ET) for Watershed 6 from 1963 to 2013. Annual evapotranspiration is calculated as the difference between annual precipitation and annual stream flow. Stream flow and precipitation are highly correlated at Hubbard Brook (0.99); 1964 was a year of extreme drought. (Modified from Likens 2013; reprinted with permission of Springer Science+Business Media)

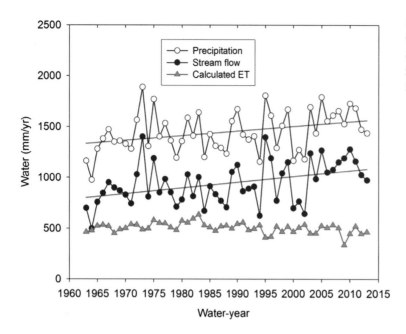

Figure 7.5. Average annual precipitation, stream flow, and evapotranspiration for Watershed 6 during 1963–2013. (Modified from Likens 2013; reprinted with permission of Springer Science+Business Media)

Transpiration on an annual basis at Hubbard Brook represents about 70% of evapotranspiration.[3] Evapotranspiration annually removes a volume equivalent to a layer of water about 51 centimeters deep over the entire watershed. For reference, 50 centimeters per year of evapotranspiration would be equivalent to 500 liters per square meter per year, or 5,000,000 liters per hectare per year, or 66,000,000 liters lost annually from the entire watershed. Over the 50 years of measurement, evapotranspiration has been relatively constant on an annual basis. The energy required to transport this much water is estimated at 288,990 kcal per square meter per year, which represents 42% of the solar radiation absorbed by the plants in the ecosystem.

Transpiration is an extremely important process for plants, because it moves water and accompanying nutrients from the soil up into the plant, helps to maintain pressure within the plant cells, and serves as a cooling mechanism for the plant when temperatures become very high. Transpiration also provides important ecosystem services by removing water without eroding the landscape and by serving as a nutrient conservation mechanism, as the water vapor does not carry away essential nutrients the way liquid water in stream flow does, with its dissolved solute load.[4]

Hydrology is central to biogeochemical activity because of the interrelationship of moving water and nutrient dynamics.[5] Precipitation and stream flow are major vectors of nutrient flux in the forest. Understanding these interactions is critical for evaluating nutrient dynamics within the forested ecosystems, and has been an important focus of the Study since 1963.

8

BIOGEOCHEMISTRY
How Do Chemicals Flux and Cycle?

The complex interactions among biology, geology, chemistry, and hydrology elucidated by the study of biogeochemistry provide the foundation for all life and for the functions in the watershed-ecosystems at Hubbard Brook. Chemical nutrients (for example, carbon, nitrogen, calcium, and sulfur) and water form the basis of all living organisms. How these chemical elements are deposited from the atmosphere, released from soil and rocks, incorporated and used by the various components of the ecosystem, cycled by and through these components, and lost in stream water or returned to the atmosphere have been central questions of the Study since 1963. The flux (across ecosystem boundaries) and cycling (within ecosystem boundaries) of these nutrients strongly affect the growth of organisms and the functioning of the forest ecosystem. Our studies have shown that the environment is changing, however, and watershed-ecosystems and their biogeochemistry are not the same as they were 50 years ago.

Learning how and why ecosystem features interact and change with time can provide insights for the management of natural ecosystems. Does the cycle of one nutrient, such as hydrogen, affect another, such as calcium? How and when is a critical nutrient, like nitrogen, retained or lost from the ecosystem? Is chloride a conservative ion (negligible storage and exchange) within the ecosystem? What policy debates are informed by the long-term information on biogeochemical flux and cycling from Hubbard Brook? For example, how effective has federal legislation been in reducing atmospheric sulfur pollution from the atmosphere? Are the effects of acid rain declining or getting worse? Detailed knowledge about the cycling and flux of water and chemical elements helps to answer such questions, and provides a basis for understanding how the watershed-ecosystem is functioning and how the system can be managed to maintain the resilience of interconnected forest and aquatic ecosystems.

As the central theme of the research efforts at Hubbard Brook from the beginning, the record of biogeochemical measurements is the longest of

Figure 8.1. Bulk (wet plus dry) deposition collector at Hubbard Brook. The loop in the tubing between funnel and reservoir (when half full of water) and water bottle on the downside of the reservoir eliminates evaporation of the sample in the reservoir. (Photo by G. E. Likens)

questions, and more, is the intent of this chapter, and will be pursued further in Chapters 9–13. The long-term biogeochemical studies at Hubbard Brook have been synthesized and detailed in a recently published book, which will be drawn upon extensively here.[1]

HOW HAVE PRECIPITATION AND STREAMWATER CHEMISTRY CHANGED DURING THE PAST 50 YEARS?

When research began at Hubbard Brook in 1963, we did not know what the chemistry of precipitation and stream water had been previously. Recent studies, however, suggest that both precipitation and stream water were relatively nutrient poor and mildly acidic (pH 5.1 and 6.0, respectively) before the Industrial Revolution (referred to as pre–Industrial Revolution, or PIR), around 1860, then increased to much higher concentrations during the 1970s, and now are becoming much more dilute again.[2] So there have been marked changes in the chemistry of both precipitation and stream water between PIR and the time when the Study began, and then subsequent to the 1970s. Increased air pollution and its biogeochemical effects on the receiving ecosystems were the primary causes of the increase in nutrients and increasing acidity of precipitation and stream water starting in the mid-1950s (figs. 8.2 and 8.3).

Because large changes (trends) in the chemistry of both precipitation and stream water occurred during the fifty years of the Study, averages for such a long period can be misleading. Therefore, to characterize the changes in chemistry during this long-term study, we have calculated the average concentrations for precipitation and stream water over two 10-year periods near the beginning and end of the 50-year Study, from 1964 to 1973, and 2000 to 2009.

Primarily because of the influence of acid rain since the mid-1950s, the chemistry of precipitation falling at Hubbard Brook was dominated by hydrogen ion and sulfate. Hydrogen ion averaged 71% of total cations over the decade of 1964–1973, and 60% for 2000–2009 (see Box 8.1, Table 8.1). Sulfate comprised an average of 63% of total anions for 1964–1973, and

any in our long-term Study. The first streamwater samples for chemical analysis were collected on June 1, 1963, just above the gauging stations on several headwater streams on south-facing watersheds. The first precipitation sample was collected on July 24, 1963. These measurements have continued without interruption ever since, making this continuous and comprehensive record of precipitation and streamwater chemistry the longest in the world (fig. 8.1).

How can the chemistry of precipitation and stream water at Hubbard Brook be characterized, and how has this chemistry changed? What are the causes of these changes, and how have they affected the ecology of the watershed-ecosystems? Answering these

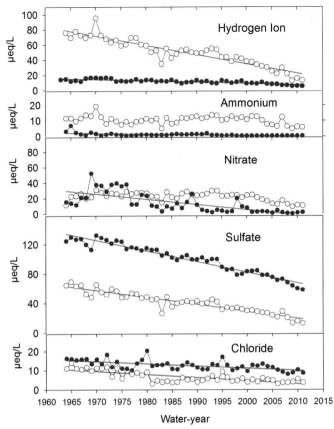

Figure 8.2. Long-term, annual volume-weighted, mean concentrations of hydrogen ion, ammonium, nitrate, sulfate, and chloride in precipitation (○) and stream water (●) for Watershed 6 during 1963–2013. Lines are shown when linear regressions are statistically significant. (Modified from Likens 2013; reprinted with permission of Springer Science+Business Media)

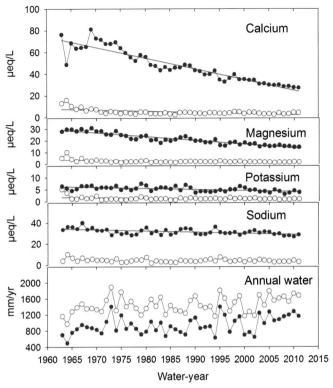

Figure 8.3. Long-term, annual volume-weighted, mean concentrations of calcium, magnesium, potassium, and sodium in precipitation (○) and stream water (●) for Watershed 6 between 1963 and 2013. Lines are shown when linear regressions are statistically significant. (Modified from Likens 2013; reprinted with permission of Springer Science+Business Media)

53% in 2000–2009. Because of weathering reactions, hydrogen ions can be exchanged with, or replaced by, other cations, such as calcium and magnesium, on mineral surfaces in the soil, releasing those ions into solution where they can be lost from the ecosystem in stream water. The chemical content of stream water is dominated by calcium and sulfate: calcium was 36% of total cations for 1964–1973 and 27% for 2000–2010, and sulfate was 70% of total anions for 1964–1973 and 69% for 2000–2010 (Table 8.2). Other elements, including some that are toxic to living organisms, such as lead and mercury, also enter the system via precipitation,

are cycled through the ecosystem, and may be lost in stream water.

WHAT IS THE SIGNIFICANCE OF TOXIC METALS AND OZONE IN THE ECOSYSTEM?

Lead. When researchers first sampled for toxic metals in precipitation at Hubbard Brook during the 1970s, they found that lead, copper, cadmium, mercury, and zinc were present in surprisingly large amounts in precipitation, even for this relatively remote location.[3] In fact, the concentrations of lead in precipitation at Hubbard Brook during the early 1970s often exceeded the United States drinking water standard of 15 μg/liter, but by 1990 concentrations of lead in precipitation and stream water were low,

Box 8.1 What Are Cations and Anions?

Cations are positively charged ions in solution, such as calcium (Ca^{2+}), sodium (Na^+), and ammonium (NH_4^+). In contrast, anions are negatively charged ions in solution, such as nitrite (NO_3^-), sulfate (SO_4^{2-}), and chloride (Cl^-).

Table 8.1. Precipitation Chemistry for Watershed 6 for Two Decades, 1964–1973 and 2000–2009

Values are volume-weighted mean concentrations. Percentages give values of total cations or total anions. Precipitation samples are collected in continuously open collectors. ANC is acid-neutralizing capacity. DOC is dissolved organic carbon.

Substance	1964–1973 Concentration (μeq/L)	2000–2009 Concentration (μeq/L)
Water (mm/ha/yr)	1,388	1,488
Calcium	7.5 (7.3%)	3.5 (7.6%)
Magnesium	3.5 (3.4%)	1.5 (3.3%)
Potassium	1.5 (1.5%)	1.0 (2.2%)
Sodium	5.1 (5.0%)	3.5 (7.6%)
Ammonium	12.1 (11.8%)	8.9 (19.3%)
Aluminum[a]	< 0.01 (0%)	< 0.01 (0%)
Hydrogen ion	73.1 (71.2%)	27.6 (60.0%)
Sulfate	59.3 (63.0%)	24.8 (54.1%)
Nitrate	23.8 (25.3%)	17.2 (36.0%)
Chloride	10.7 (11.4%)	4.4 (9.6%)
Phosphate[b]	0.3 (0.3%)	0.1 (0.22%)
ANC (bicarbonate)[c]	0.0 (0%)	0.0 (0%)
Cation sum	102.7	46.0
Anion sum	94.1	45.8
DOC[d]	91.6 μmol/L	68.4 μmol/L
Dissolved silica[a]	< 0.01=mg/L	< 0.01=mg/L
pH	4.14	4.56

Source: Likens 2013; reprinted by permission of Springer Science+Business Media.

[a] Al and Si found in trace amounts in clean precipitation

[b] Phosphate based on 1971–1973

[c] Bicarbonate = 0 at pH < 5

[d] Dissolved organic carbon (DOC) based on 1976–1977 (Likens et al. 1983)

and then very low after 1990. Much of the lead in precipitation had come from the use of leaded gasoline in high-temperature combustion engines, such as those in automobiles. Levels of lead then declined dramatically in precipitation and in other components of the ecosystem during the late 1970s and 1980s, after the Environmental Protection Agency began to phase out the use of lead as an additive in gasoline in the early 1970s (fig. 8.4). The amount of lead allowed in regular-grade gasoline was reduced from 4.0 grams/gallon to 0.5 grams/gallon in 1985 and 0.1 grams/gallon in 1986. The Clean Air Act Amendments of 1990 mandated the elimination of lead from all United States motor fuels as of January 1, 1996.

The input of lead in precipitation in 1985–1987 at Hubbard Brook was 11 times lower than in 1975-1977, and the uptake of lead by plants was 33 times less. Concentrations of lead in stream water were very low (less than 7 μg/liter) during this period, indicating that this toxic metal tended to accumulate in the organic components of the forest ecosystem. Some 97% of lead input in precipitation was retained in the ecosystem.[4]

Mercury. The methylated form of mercury released by industrial processes is highly toxic to organisms and can be found in all components of the ecosystem. It is often magnified in concentration by the biota through the food web. The atmosphere has become the primary source of mercury for forest ecosystems, and significant amounts of mercury are added to ecosystems even in areas such as Hubbard Brook that are relatively remote from industrial centers. Increasing amounts of mercury have been released into the atmosphere by emissions from coal and waste combustion during recent decades.[5] Recently, however, mercury emissions have declined, largely because municipal and medical waste incinerators have closed. As a result, wet deposition of mercury from the atmosphere in the northeastern United States declined by 1.7% per year during 1998–2005.[6] Relatively little research has been done on the ecological effects of mercury at Hubbard Brook, but hot spots of deposition and ecosystem mercury accumulation have

Table 8.2. Streamwater Chemistry for Watershed 6 for Two Decades, 1964–1973 and 2000–2009

Values are volume-weighted mean concentrations. Percentages give values of total cations or total anions. Samples are collected immediately above the gauging weir. ANC is acid-neutralizing capacity. DOC is dissolved organic carbon.

Substance	1964–1973 Concentration (µeq/L)	2000–2009 Concentration (µeq/L)
Water (mm/ha/yr)	882	1,006
Calcium	67.8 (36.0%)	30.8 (27.1%)
Magnesium	28.0 (14.9%)	15.8 (13.9%)
Potassium	5.8 (3.1%)	4.2 (3.7%)
Sodium	33.9 (18.0%)	29.8 (26.2%)
Ammonium	2.1 (1.1%)	0.3 (0.26%)
Aluminum[e]	35.7 (19.0%)	25.0 (22.0%)
Hydrogen ion	15.0 (8.0%)	7.9 (6.9%)
Sulfate	124.6 (69.5%)	74.9 (68.8%)
Nitrate	28.7 (16.0%)	2.6 (2.4%)
Chloride	15.0 (8.4%)	10.7 (9.8%)
Phosphate[b]	0.1 (0.06%)	0.0 (0%)
ANC (bicarbonate)[a]	0.0 (0%)	8.0 (7.3%)
DOC[c]	10.8 (6.0%)	12.6 (11.6%)
Cation sum	188.3	113.7
Anion sum	179.2	108.9
DOC[c, d]	149.9 = µmol/L	174.9 = µmol/L
Dissolved silica	4.0 = mg/L	1.0 = mg/L
pH	4.82	5.10

Source: Likens 2013; reprinted by permission of Springer Science+Business Media.

[a] Bicarbonate = 0 at pH < 5

[b] PO_4 based on n = 3 years (1971–1973)

[c] DOC based on 1976–1977 (Likens et al. 1983)

[d] DOC charge is assumed at −6 µeq/L per mg/L C

[e] Aluminum is assumed at +3 valence

been found in the region.[7] Mercury concentrations tend to increase with stream discharge, causing output of mercury in streams to be strongly related to streamflow.[8] At Hubbard Brook, the amount of methyl mercury mobilized from the forest floor and flushed in stream water during snowmelt is a large proportion of the annual mercury flux from watershed-ecosystems.[9]

Ozone. Ozone is an atmospheric gas with molecules made of three oxygen atoms (O_3) that is found naturally in the earth's stratosphere, where it absorbs the ultraviolet component of incoming solar radiation that can be harmful to life on earth. Such ozone is sometimes referred to as "good" ozone. Ozone near the earth's surface, in contrast, formed mainly from anthropogenic pollutants such as nitrogen oxides (NO_x), is considered "bad" ozone because of its toxic effects on biota. The chlorine-like odor of ozone at concentrations of 10 ppb (parts per billion) can often be detected by humans, and prolonged exposure to levels above 100 ppb can cause serious respiratory

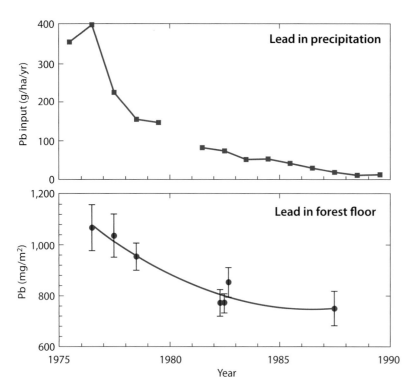

Figure 8.4. Lead input in precipitation and concentration in the forest floor at Hubbard Brook. Lead is measured in the forest floor as the average lead content in 15 × 15 cm block samples; the number of samples taken range from 58 to 119 per year. (Modified from Johnson et al. 1995)

problems for humans and other animals, as well as damage plant tissues and affect ecosystem processes such as forest growth and carbon sequestration.[10]

At Hubbard Brook, daily ozone values throughout the year range between 5 and 10 ppb, although peak values approach 70 ppb in early spring and summer (fig. 8.5). These peak values are sufficiently high to cause damage to plant leaves and other tissues, particularly when the high values last for several hours or more. Furthermore, computer simulations on data from Watershed 6 have suggested that ozone may influence nitrogen dynamics in terms of nitrate loss from the ecosystem.[11] Ozone, however, is only one factor influencing nitrogen flux and dynamics in these ecosystems.[12]

STREAMWATER DISCHARGE AND CONCENTRATION OF CHEMICALS

Nutrients and other chemicals are lost from the watershed-ecosystems in both dissolved and particulate form in stream water.[13] Losses in the dissolved form usually predominate for most

nutrients, but losses of phosphorus and iron in particulate matter are larger than in dissolved form at Hubbard Brook.[14]

Our long-term studies revealed that not only are streamwater concentrations of all nutrients relatively low, but they also vary little with streamflow amount.[15] Although discharge can vary by up to several orders of magnitude during storm events or snowmelt, dissolved substances vary little in concentration and by no more than three- or fourfold for individual solutes during these high-discharge events (fig. 8.6). This relationship may seem contrary to what might be expected, but it reflects the overall stability of the system at Hubbard Brook, including high infiltration of water into soil with little overland flow, even in winter because the soils are usually unfrozen. Also, there is negligible deep seepage, minimal disturbance to the soil surface, and relatively stable stream channels except during very high flows.

In contrast, export of particulate matter is exponentially related to discharge. A single high-intensity storm in November 1966 resulted in the

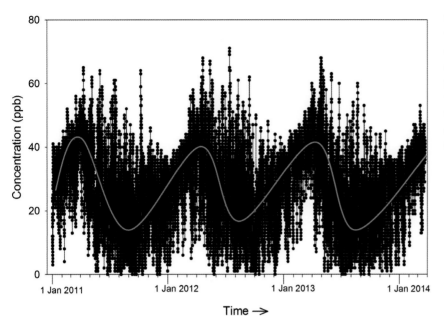

Figure 8.5. Mean hourly ozone levels for January 2011 through March 2014 at Hubbard Brook, calculated from a continuous data stream captured at 1-minute intervals. Mean values range from 5 to 10 ppb (parts per billion), with peaks approaching 70 ppb in early spring and summer. (G. E. Likens, unpublished)

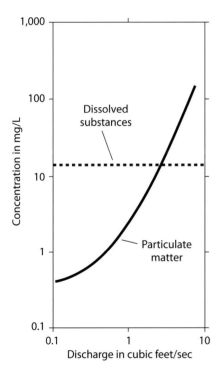

Figure 8.6. Relationship between the concentration of dissolved substances and particulate matter in stream water and in streamflow discharge. Concentrations of dissolved substances are relatively constant over several orders of magnitude change in discharge, whereas particulate matter concentration is exponentially related to discharge. (Modified from Bormann et al. 1969; reprinted by permission of Oxford University Press)

movement of 64% of all particulate matter loss during a two-year period (1965–1967).[16] Nevertheless, the gross export of dissolved substances in stream water at Hubbard Brook during 1964 to 2009 (approximately 119 kg/ha, including dissolved organic carbon (DOC), was about eight times greater than particulate matter losses in stream water during this period.[17]

Discharge rate of water is important to erosion loss of particulate matter, but so is the erodibility of the substrate. Glacial till is relatively resistant to erosion.[18] But we found that it is primarily biotic regulation that minimizes erosion at Hubbard Brook.[19] This effect occurs in three ways: the vegetation takes up liquid water and converts it to water vapor via transpiration (which does not cause erosion); the forest canopy and forest floor intercept precipitation during the growing season, reducing the potential energy of impact of falling raindrops; and because of the high rate of infiltration of water at the soil surface, erosive overland flow is minimized. Also, snow has very low potential energy of impact, resulting in about 30% greater erodibility during the summer.[20] Organic debris dams also play a major role in reducing the export of particulate matter from headwater streams at Hubbard Brook.[21]

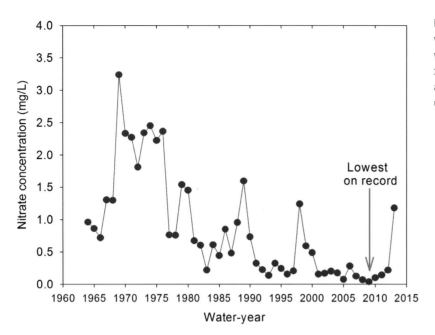

Figure 8.7. Nitrate concentrations (annual volume-weighted averages) in stream water for Watershed 6 from 1964 to 2013. Concentrations in recent years are the lowest on record. (G. E. Likens, unpublished)

The relative constancy of streamwater concentrations of dissolved substances, which is one of the major discoveries of the Study, underpinned our decision to sample streamwater chemistry on a weekly basis. This decision is a compromise between increased variability of chemistry for samples collected at longer time intervals (monthly) and greatly increased cost involved with more samples from shorter time intervals (daily).

The hydrogen ion concentration in precipitation has decreased, making it less acid, by about 80% since 1963. Significant decreases in sulfate, calcium, magnesium, and potassium concentrations in precipitation also have occurred.

Significantly large declines in streamwater concentrations also have been observed since 1963, notably for nitrate, sulfate, calcium, magnesium, potassium, and sodium. The cause for the striking decline in nitrate concentrations in stream water since about 1990, in particular, has been the subject of much debate and research at Hubbard Brook (fig. 8.7).[22] Streamwater values for nitrate, particularly during summer, currently are the lowest on record and often are at or below levels that can be detected. These low

values are especially surprising given that the forest has stopped growing and net uptake of these nutrients in living biomass should be negligible.

The magnitudes of nitrate concentrations in headwater streams at Hubbard Brook vary from year to year, but the annual pattern is remarkably consistent. Concentrations normally are very low during summer, but once leaves fall in autumn, concentrations increase dramatically to peak values in winter, and then decline precipitously with bud break and leaf formation in spring. These large changes in streamwater nitrate concentrations can occur over the course of only a few days.

The declines in calcium and magnesium concentrations are especially interesting, because of their long-term pattern and relevance to policy-management issues. Plotted at larger scales, it is clear that the large declines in concentrations occurred primarily during the early part of the Study: declines in precipitation from 1963 to about 1975, in stream water from about 1969 to 1982 (fig. 8.8). These declines are coincident with national reductions in emissions of particulate matter containing calcium and magnesium and are correlated with reduced deposition of calcium

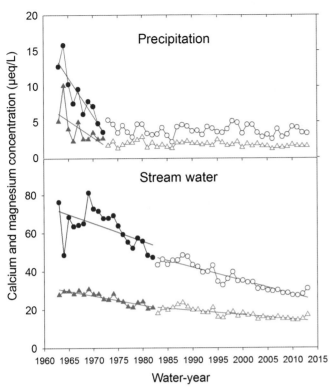

Figure 8.8. Annual volume-weighted concentrations of calcium (blue circles) and magnesium (green triangles) in precipitation and stream water in Watershed 6 from 1963 to 2013. Concentrations decreased more rapidly at the beginning of the record (solid symbols). (Modified from Likens 2013; reprinted with permission of Springer Science+Business Media)

and magnesium in precipitation, and with depletion of calcium and magnesium on exchange sites in soils at Hubbard Brook from acid rain.[23]

The Clean Air Act Amendments of 1970 focused on reducing emissions of particles, which contained calcium and magnesium in addition to pollutants such as sulfur. The declines in calcium and magnesium concentrations in precipitation observed at Hubbard Brook were correlated with reductions in emissions of particles produced by this federal regulation. Equally important relative to the interpretation of these long-term records were federal efforts to monitor the chemistry in precipitation and in streams and lakes for effects of acid rain. The National Atmospheric Deposition Program was initiated in 1978, after most

of the marked decline in these cations had occurred, as recorded in long-term measurements at Hubbard Brook. The pattern of long-term change for these important nutrients and its interpretation would have been very different if the Study were to have started in 1980, after most of the decline had already taken place. Hubbard Brook scientists were active and aggressive in helping to initiate federal monitoring programs for precipitation chemistry throughout the nation, and the long-term records from Hubbard Brook were very important in helping to put the subsequent federal monitoring programs into perspective.

NUTRIENT FLUX AND CYCLING

Starting in 1994, Hubbard Brook scientists published a series of monographs, which synthesized essentially everything we knew about the biogeochemistry of the major elements following decades of research. They prepared and published monographs for potassium, calcium, sulfur, chlorine, and carbon. The compilation and synthesis of information were extremely useful in describing the biogeochemistry of the system and led to new insights about the biogeochemistry of these watershed-ecosystems. These included, for example, the important contribution of dry deposition, particularly of sulfur and nitrogen to the ecosystem, the need to calculate net ecosystem flux values for sulfur to solve a potential "missing" sulfur problem, the need to understand net soil release and its relationship to weathering, and the finding that chloride is not truly conservative (dynamically relatively inert with regard to storage, cycling, and flux) in the Hubbard Brook watershed-ecosystems, because it was found that chloride has active reservoirs in the organic pools of the ecosystem.

Dry deposition. Like wet deposition (through rain, snow, sleet, hail), the transfer of gases and particles directly from the atmosphere, or dry deposition, can be a very significant input of chemicals to a forest ecosystem, but it is difficult to measure this input quantitatively. At Hubbard Brook we have used various methods to estimate the dry deposition of sulfur and

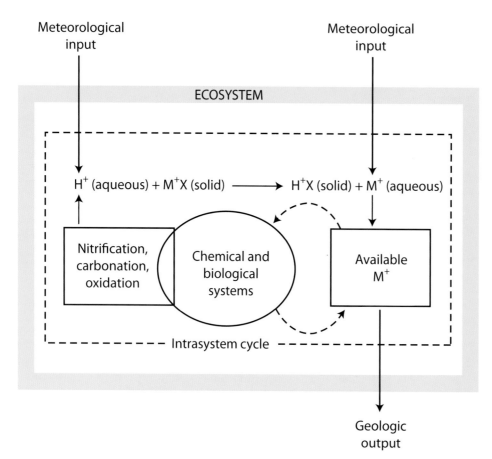

Figure 8.9. A conceptual diagram for weathering at Hubbard Brook. M^+ represents a base cation such as calcium, and X is an ionic exchange substrate (usually a primary silicate mineral, but also such derivatives as soil minerals or soil humates). The diagram illustrates that as H^+ is added to the system from external and internal sources, base cations (M^+) are leached from the system, and primary silicate minerals (M^+X) are transformed into secondary minerals (H^+X). (Modified from Likens 2013; reprinted with permission of Springer Science+Business Media)

other chemicals, but they all have methodological problems.[24] In a promising development, the dry deposition of sulfur was estimated by using four different models based on regional values for sulfur dioxide emissions.[25] Values of dry deposition at Hubbard Brook for 2002 using these models ranged from about 0.5 to 2.5 kg of sulfur per hectare per year, which is a significant component of the atmospheric sulfur input to watershed-ecosystems at Hubbard Brook.

Chemical weathering. Chemical weathering also is difficult to measure quantitatively for a watershed-ecosystem. The chemical composition of precipitation entering watershed-ecosystems at Hubbard Brook is characterized by acid salts, primarily sulfuric and nitric acids, whereas water leaving the ecosystem in stream flow is characterized by neutral salts, primarily calcium and sodium sulfates and to a lesser extent

by nitrates and chlorides. This general change in chemistry is largely the result of chemical weathering reactions within the ecosystem (fig. 8.9). When a hydrogen ion is consumed within the ecosystem, base cations (M^+), such as calcium and magnesium, can be leached from various components of the ecosystem into soil water and stream water, and primary minerals (M^+X) are converted into soluble ions or into secondary minerals or coatings on soil surfaces. The rate at which hydrogen ions are consumed defines the rate of chemical weathering.[26]

Because direct measures of weathering of cations from primary minerals, net long-term storage in the soil organic matter pool, net change in the exchangeable pool, and net change in the secondary mineral pool are all unknown or poorly known at Hubbard Brook, it was recently proposed to use a procedure called net soil release instead.[27] Because weathering and net soil release of nutrients occur within the boundaries of the watershed-ecosystem, these fluxes are referred to as "release" rather than inputs.

Estimates of dry deposition, weathering, and net soil release will be included in the analyses and discussions of the cycling and flux of individual nutrients described below.

Dissolved organic carbon input, lost, and cycled. Dissolved organic carbon can be an important source of energy for microorganisms and aquatic invertebrates and can contribute to biogeochemical dynamics such as the mobilization of mercury. The concentration and flux of dissolved organic carbon (DOC) are highly dynamic as water passes through the forested ecosystem at Hubbard Brook (fig. 8.10). Concentrations of dissolved organic carbon in rain and snow are significant but low, averaging about 1.1 mg/liter. As this water passes through the ecosystem, concentrations and flux greatly increase as carbon is leached from the canopy (throughfall is enriched about tenfold) and particularly the forest floor, then decreases sharply as the carbon is adsorbed in the deeper, mineral horizons of the soil (B and C horizons). Concentrations in stream water (2.5 mg/liter) are approximately double the concentrations in precipitation, but flux is about the same because evapotranspiration has reduced the volume of water by about half as the water passes through the ecosystem.

Calcium input, lost, and cycled. Calcium is an important nutrient for plant and animal growth and structure (for example, cell walls in plants and bones in animals) within ecosystems at Hubbard Brook. Because its abundance is relatively low in these watersheds lying above granitic bedrock, and because it has been markedly depleted by acid rain since the mid-1950s, its importance as a potential limiting nutrient has become even more pronounced.

The complicated flux and cycling of calcium are illustrated for two different 5-year periods, one at the beginning of the long-term record, 1964 to 1969 (fig. 8.11), and one nearer the end, 1987 to 1992 (fig. 8.12).

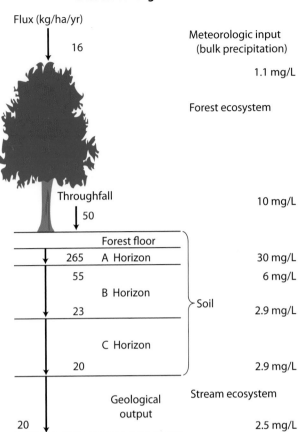

Dissolved Organic Carbon

Figure 8.10. Dissolved organic carbon flux for Watershed 6, showing changes in concentration (mg/liter) on right and flux (kg/ha/yr) on left of dissolved organic carbon (DOC) as it passes through the forest ecosystem at Hubbard Brook. The biotic and soil horizons are active sites of leaching and adsorption of DOC within the ecosystem. (Modified from McDowell 1982)

These diagrams show the details of this complex cycle as well as exchanges with surrounding ecosystems, and a comparison of these data over time. Decades of effort and study by numerous scientists were required to obtain these values.[28]

At first glance, the diagram might appear daunting, but on closer examination it represents a unique snapshot of where calcium is stored in the forest ecosystem and what the major pathways and rates of cycling among these storage pools are. This comprehensive information is of great importance and value in seeking to understand the biogeochemical

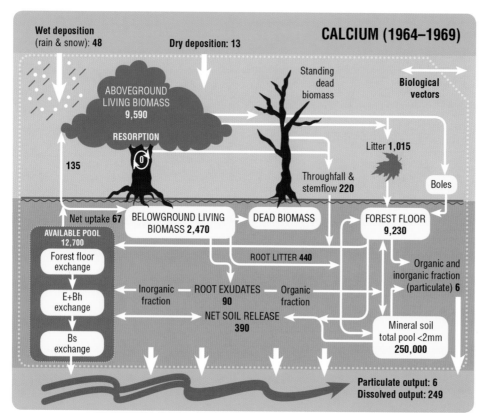

Figure 8.11. Ecosystem pools (boxes) and exchanges (arrows) of calcium for Watershed 6, 1964–1969. This figure provides a unique ecosystem view of the interacting biogeochemical dynamics of calcium in a forest ecosystem, representing a summary of decades of intensive work by many scientists at Hubbard Brook. Average values in mol/ha or mol/ha/yr. Values are modified from Likens et al. 1977. (Modified from Likens 2013; reprinted with permission of Springer Science+Business Media)

and ecological dynamics of an important nutrient like calcium in the forest ecosystem. In this case, the larger amount of acidity in precipitation increased the loss of calcium in stream water, depleting the calcium in the exchange pool of the ecosystem by about 50% during the past 50 years.[29]

We have been able to assemble detailed information of this sort for potassium, sulfur, carbon, chlorine, and calcium for the forest ecosystem at Hubbard Brook, but will use only the example for calcium here. The major insights are as follows:

- Precipitation contributed 1.5 kg calcium/ha each year during 1964–1969 and 0.7 kg calcium/ha/ year during 1987–1992 to the forest ecosystem.
- Atmospheric deposition provides about 50% of the calcium taken up (net) by vegetation annually.

- Storage of calcium in above- and belowground plant tissues was 5.1 kg/ha/yr in 1964–1969 and 1.3 kg/ha/yr in 1987–1992.
- Streamwater loss during 1987–1992 (5.2 kg calcium/ha/yr) is about four times greater than precipitation input, but only 3% of the exchangeable pool of calcium in the soil and about 3.8 times greater than the amount currently stored in the biomass (1.3 kg/ha/yr).
- About seven to eight times more calcium is cycled through the vegetation each year than is lost annually in stream water, showing dramatically how dynamic this element is, yet how effective the ecosystem is at retaining it.
- Some 93% of the calcium is stored in the soil complex, and only 7% is bound in the vegetation.
- Based on the gross uptake of calcium by trees plus loss in stream water, the turnover time of

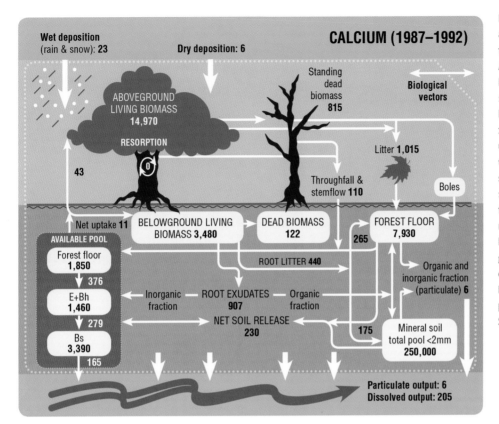

Figure 8.12. Ecosystem pools and exchanges of calcium for Watershed 6, 1987–1992. Above- plus belowground biomass is average of 1987 and 1992 values, with belowground assumed to be 20% of aboveground. Net uptake values are based on the difference in biomass storage between 1982 and 1992. Total forest floor pool from Johnson 1989; total mineral soil pool from Federer et al. 1989. 40.08 grams of calcium/ha = 1 mol calcium/ha. (Modified from Likens 2013; reprinted with permission of Springer Science+Business Media)

exchangeable calcium in the soil is estimated at about 3 years.[30]

- Root exudates release about 20% as much calcium as is lost as root litter.

- The standing crop of calcium was 27.5 g/ha in small mammals, 11.9 g/ha in salamanders, and 6.9 g/ha in birds at Hubbard Brook during the early 1970s.[31] The maximum ecosystem net loss of calcium from birds during 1969–1970 was 3.0 g/ha/yr, compared with a net loss of 9,430 g/ha/yr in dissolved and particulate calcium in stream water during 1969–1970.[32] The bird and salamander populations were relatively large during this period. Increases in large-mammal populations, such as moose with their large amounts of calcium stored in bone and antlers, could greatly change this pattern (see Box 8.2).

- Comparing 1964–1969 and 1987–1992, net

annual biomass storage decreased about 75%, atmospheric deposition by more than 50%, throughfall and stemflow about 50%, and dissolved stream output by approximately 20%.

WHAT ARE THE ANNUAL VALUES FOR NET ECOSYSTEM FLUX?

Many factors can contribute to uncertainty in determining average streamwater concentrations and flux, including precision and accuracy of chemical analyses, measurement of stream flow, and uncertainty in watershed area. Using three methods for calculating flux, annual export values differed by less than 5% for calcium and up to 12% for nitrate at Hubbard Brook during 1997 to 2008.[33]

Flux and cycling of calcium. Combining these values annually and in budgetary format can be very informative, as is shown in a net ecosystem

Box 8.2 The "Bio" in Biogeochemistry

Moose are responsible for a significant amount of nutrient cycling. They forage over large areas (tens of square kilometers) during the summer growing season, and can hardly be called residents of the Hubbard Brook valley during that time. In fact, they spend much of the summer around ponds to the north and south of the valley. But in the winter, after mating season or rut, they gather, or "yard up," in the forest at higher elevations near the spruce-fir and deciduous forest boundary. In the process, they transfer and concentrate valuable nutrients downslope to lower parts of the valley from the upper slopes (sensu Aldo Leopold).[a] Antlers, for example, are grown and shed each year by mature bull moose, and typically are dropped in the winter "yard" along the ridge tops. The bulls begin growing new ones each spring. The antlers themselves are a complex mixture of lipids, proteins, and minerals, most notably containing two of the most necessary and possibly limiting nutrients in the northern hardwood ecosystem: calcium and phosphorus. One exceptionally large pair of antlers collected from the forest (weighing over 20 kg) was estimated to contain the equivalent of 35% of the annual calcium input and 450% of the annual phosphorus input via annual precipitation to the entire 13-ha Watershed 6. One large antler (10 kg) could sequester 45% of the equivalent amount of calcium and more than 20,000% of phosphorus per hectare per year than would be lost in stream water per hectare annually from Watershed 6. The antlers do not decay quickly, but are often chewed on by bears, coyotes, porcupines, squirrels, and other small mammals, slowly releasing and distributing stored nutrients to the ecosystem. Thus, bull moose collect critical elements from a broad, nutrient-rich area and focus them in nutrient-poor zones.

NOTE

a. Leopold 1949.

flux (NEF) diagram for calcium (fig. 8.13). Annual inputs of calcium are from wet and dry deposition, and weathering or net soil release; annual outputs are streamwater loss and net biomass storage (long term). When an element, such as nitrogen, has a prominent gaseous phase, gaseous exchanges must be accounted for as well; calcium does not have a gaseous phase at normal biological temperatures. The difference between inputs and output equals annual net ecosystem flux. Annual net ecosystem fluxes for calcium can be summed to show total ecosystem gain or loss during 1963–2009 and what the major components of these gains and losses are. Calcium was strongly depleted in NEF during 1963–1985 and currently is nearly in balance. This result occurs largely because of markedly diminished net storage in the vegetation and reduced loss in stream water. Total stream loss includes dissolved and particulate calcium.[34] The long-term balance for calcium shows that outputs greatly exceed inputs (Table 8.3).

Flux and cycling of sulfur. Like calcium, sulfur is a nutrient important to plants, animals, and microorganisms in the ecosystems at Hubbard Brook. In excess or in the acid form (H_2SO_4), however, sulfur becomes a major pollutant in the system.

As described in Chapter 10, the flux of atmospheric sulfur has been declining in precipitation collected at Hubbard Brook since about 1970. The biogeochemical dynamics from the long-term study revealed the following.[35]

- Sulfur constituted 71% and 53% of anions in precipitation and 80% and 87% of anions in stream water in 1964–1969 and 1993–1998, respectively.
- Annual atmospheric deposition (wet and dry) of sulfur was about 100 times greater than annual storage of sulfur in vegetation during the early decades of the Study.
- In 1993–1998, net gaseous plus particulate deposition was 18% of total atmospheric deposition or 13% of annual streamwater loss.
- More than 95% of the sulfur is found in the

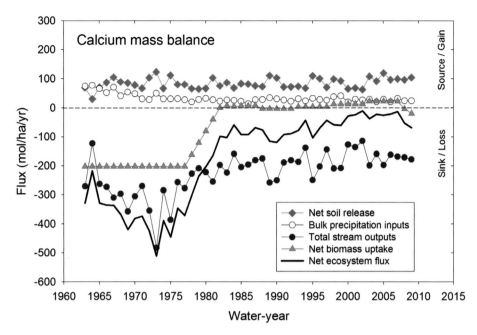

Figure 8.13. Annual mass balances (budgets) for calcium, showing inputs of calcium in precipitation and net soil release and outputs in stream water and net biomass storage. The annual sum of these inputs and outputs gives net ecosystem flux values for Watershed 6 from 1963 to 2009. Net soil release and bulk precipitation are inputs, whereas stream output and net biomass uptake (storage) are losses to the annual budget. (Modified from Likens 2013; reprinted by permission of Springer Science+Business Media)

Table 8.3. Fluxes in the Mass Balance for Calcium, Sulfate, and Dissolved Inorganic Nitrogen for Watershed 6, 1963–2010

Totals are cumulative for the 47-year period.

	Calcium (keq/ha)	Sulfate (keq/ha)	Dissolved inorganic nitrogen (keq/ha)
Precipitation in (P)	3.2	28.6	22.5
Dry deposition (D)	0.0*	5.7	4.7
Atmospheric inputs (P + D)	3.2	34.3	27.3
Net soil release (W)	9.9	6.7	0.4
Total INPUT (P + D + W)	13.1	41.0	27.7
Streamwater out (S)	−21.0	−43.8	−6.5
Net biomass uptake (B)	−3.3	−2.1	−11.4
Total OUTPUT (S + B)	−24.3	−45.9	−17.9
Net hydrologic flux (P − S)	−17.8	−9.5	20.8
Net ecosystem flux (P + D + W − S − B)	−11.2	−4.9	9.8
Ecosystem balance	LOSS large	LOSS modest	GAIN large

* Included in bulk precipitation

Source: Likens 2013, Likens and Buso 2015; reprinted in part by permission of Springer Science+Business Media.

soil complex, and the remaining 5% is in living biomass.

- Based on annual estimates of root exudates, root litter, canopy litterfall, canopy leaching, and net biomass storage during 1993-1998, gross uptake of sulfur is estimated to be about 9.2 kg/ha/yr. Of this amount, only about 1% is accrued in the biomass, and the remainder is returned to the soil.[36]

- Because annual net storage of sulfur in biomass decreased by more than tenfold from 1964-1969 to 1993-1998, streamwater loss of sulfur was more than 13 times greater than biomass storage in 1964-1969 and 145 times greater in 1993-1998.[37]

- Maximum ecosystem net loss of sulfur in birds (left the system and did not return) during 1969-1970 was 0.002 kg/ha/yr as compared to net streamwater loss of 3.5 kg/ha/yr in dissolved and particulate matter.[38]

- Current inputs of sulfur into the available nutrient pool of the ecosystem are dominated by atmospheric deposition and are partitioned as follows: precipitation (50%), net soil release (24%), dry deposition (11%), root exudates (11%), and leached from aboveground vegetation (4%).[39]

- Comparing 1964-1969 and 1993-1998, net annual biomass storage decreased about 92%, atmospheric deposition about 34%, throughfall and stemflow about 33%, and dissolved stream output approximately 10%.[40]

The net hydrologic flux (precipitation input minus streamwater output) for sulfur is consistently imbalanced and the subject of much study at Hubbard Brook.[41] Four hypotheses have been suggested for this budgetary imbalance over time, both positive and negative: (1) dry deposition has been underestimated; (2) sulfate is being desorbed from the soils; (3) release of sulfur from weathering is underestimated; or (4) there is net mineralization of organic sulfur occurring.[42] These components of the sulfur budget are all difficult to measure and have relatively large uncertainties, but net mineralization of organic sulfur seems the most likely source of the imbalance, and especially since 1985 following the decrease in atmospheric deposition. This change from external input to internal release represents a major change in ecosystem dynamics.

By adding the other components of the annual mass balance—that is, dry deposition, weathering/net soil release, and net biomass uptake and producing a net ecosystem flux—the inputs and outputs become roughly balanced (fig. 8.14). This long-term analysis clearly shows the importance of dry deposition and net soil release to the overall budget, and that sulfur is neither gaining nor losing in the ecosystem over time. The very small cumulative net ecosystem flux for the longer time period, 1964-2009, is negative (–5.6 kilo-equivalents (keq)/ha).[43]

Flux and cycling of chlorine. The anion chloride is usually considered a minor, nonessential plant element. It had been thought not to accumulate in forest ecosystems, and thus has frequently been used as a conservative tracer for movement of water and associated chemicals such as sodium chloride.[44] Long-term studies at Hubbard Brook and elsewhere and construction of annual net ecosystem flux balances, however, show that this may not be the case.[45] Chloride deposition from the atmosphere before 1980 was higher than in more recent years, presumably because greater amounts of coal burning released hydrochloric acid into the atmosphere at that time. The average input of chloride in precipitation since 1980 has been about 1.7 kg/ha/yr, and the average streamwater output during 1964-2009 was approximately 3.5 kg/ha/yr, showing a net hydrologic loss from the ecosystem during this period, and strongly suggesting that chloride is not conservative within the ecosystem.[46]

Flux and cycling of nitrogen. Nitrogen is a critical nutrient in ecosystems because plants, animals, and microbial organisms require it for growth and reproduction. When concentrations are low, it can

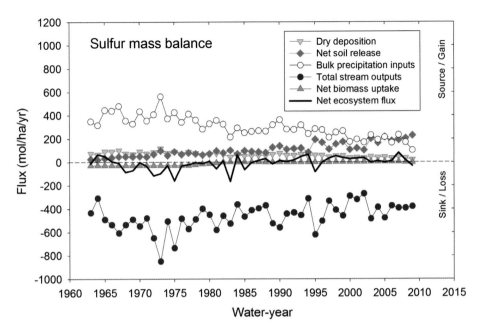

Figure 8.14. Annual mass balances (budgets) for sulfur, showing inputs of sulfate in precipitation, dry deposition, and net soil release and outputs in stream water and net biomass uptake. The annual sum of these inputs and outputs give net ecosystem flux values for Watershed 6 from 1964 to 2009. Annual sulfur inputs and outputs are nearly in balance during the entire period. This result is largely the result of reduced input of sulfate in precipitation combined with reduced loss in stream water. Total stream loss includes dissolved and particulate sulfur. (Modified from Likens 2013; reprinted with permission of Springer Science+Business Media)

become a limiting nutrient for these organisms. When concentrations are very high it can be a pollutant, leading to eutrophication of surface waters.

The net ecosystem flux pattern for dissolved inorganic nitrogen (DIN) is interesting and quite complicated (fig. 8.15). Nitrogen in precipitation and stream water consists of inorganic (nitrate and ammonium = DIN) and organic forms. Stream water also may transport nitrogen incorporated in particulate matter. The ratio of nitrate to ammonium in precipitation is about 2:1 on an equivalent basis, but in stream water the ratio increases to 15:1, indicating the preferential biotic use of ammonium over nitrate within ecosystems at Hubbard Brook. The concentrations of dissolved organic nitrogen in both precipitation and stream water are relatively small.[47]

Gaseous flux of nitrogen is very difficult to measure, particularly at the watershed scale. For example, organic debris dams can be hotspots for gaseous nitrogen loss (a process called denitrification), but it is difficult to scale such point measurements to an entire watershed.[48]

Because nitrogen is an essential plant nutrient, it is stored long term in the biomass of the forest ecosystem. The pool of nitrogen in soil organic matter at Hubbard Brook on south-facing slopes is approximately 4,700 kg/ha. This represents a huge amount of this critical nutrient in the forest ecosystem, and it has accumulated in the soil, primarily from atmospheric sources, during the past 14,000 years. It is approximately 9 times greater than that found in all plant biomass in the forest. The long-term mass balance for nitrogen at Hubbard Brook shows a fundamental shift for the ecosystem, from a net sink/loss from 1964 to 1980 to a net source/gain from 1980 to 2009. This shift is largely driven by the decrease in net biomass accumulation during the latter period, and secondarily by declining precipitation input. Nitrogen oxide emissions contribute to inorganic nitrogen in precipitation; federal regulations starting in about 1995 have resulted in reducing these emissions. Precipitation input of dissolved inorganic nitrogen at Hubbard Brook currently is lower than it was in 1964. Overall, the net ecosystem flux balance for dissolved inorganic nitrogen is positive (9.6 keq/ha) for 1964–2009, meaning that more nitrogen was gained by the ecosystem during this period than was lost.

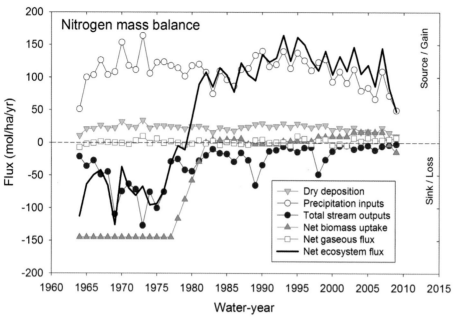

Figure 8.15. Annual mass balances (budgets) for nitrogen. Inputs of nitrate and ammonium in precipitation, dry deposition, and net soil release and outputs in stream water and net biomass uptake are shown for each year during 1964–2009. Estimates of net gaseous flux also are shown. The annual sums of these inputs and outputs give net ecosystem flux values for Watershed 6 from 1964 to 2009. The balance of annual inputs and outputs showed ecosystem losses/storage during 1964–1980 and gains thereafter, largely reflecting changes in biomass storage. Total stream output includes dissolved and particulate nitrogen. (Modified from Likens 2013; reprinted by permission of Springer Science+Business Media)

THE BIOGEOCHEMICAL BOTTOM LINE

The chemistry of both precipitation and stream water at Hubbard Brook has changed significantly with time. For example, hydrogen ion concentrations in precipitation have declined by about 80%, and there are much lower concentrations of sulfate in precipitation and stream water today than in 1963–1964. Nitrate concentrations in stream water today are the lowest on record. In some cases (hydrogen ion and sulfate) the causes for these declines are known (smaller emissions of sulfur dioxide and less formation of sulfuric acid); in others (nitrate in stream water) the causes are subject to ongoing research. Federal legislation has been successful in reducing emissions of sulfur dioxide, and this reduction has resulted in less hydrogen ion and sulfate in precipitation at Hubbard Brook. Long-term monitoring has been crucial in determining and quantifying these

changes. The vital nutrient nitrogen had a net loss from the ecosystem during 1964 to 1980, but now is being strongly retained in the ecosystem by living organisms against its loss in stream water. If this biotic component were destroyed or severely disturbed, nitrogen losses would be greatly increased.

Chloride is not truly conservative at Hubbard Brook and should be used with caution as a tracer of water movement.

Cycles of hydrogen ion, calcium, and sulfur are strongly linked at Hubbard Brook—for example, sulfuric acid leaches calcium from soil (and needles of red spruce). Acid rain has depleted soil exchange pools of calcium by about 50% during the past 50 years, potentially making these watershed-ecosystems more sensitive to continuing inputs of acid rain. Less calcium in stream water because of acid rain also has potential management implications for treatment of water hardness. Some 1,500 mm/yr of stream flow in 1964–1965 would have yielded more than two times as much calcium in stream water than in 2009–2010, or two times more calcium to remove in the treatment of water hardness.[49]

With the exception of chloride concentrations in Hubbard Brook headwaters, which were on the low end, the average long-term concentrations of hydrogen ion, potassium, calcium, sodium, magnesium, aluminum, sulfate, and nitrate in Watershed 6 were well within the distributions of streamwater chemistry found in some hundreds of sites sampled in the White Mountains, attesting to the representativeness of Hubbard Brook's chemistry.[50]

Many new insights and questions have emerged as a result of findings from the long-term biogeochemical studies at Hubbard Brook. For example, the base cation and acid anion concentrations in stream water are rapidly declining, and the declines are expected to level off at very low concentrations in the next decade or so.[51] In spite of these changes, it is likely that the ionic charge distribution in stream water will remain approximately the same because of the slow release of absorbed sulfate from large reserves in the soils that have built up from decades of high loading from the atmosphere. The existence of long-term data allows such questions to be placed into a proper perspective and to enable efficient efforts to study them.

Although much of the biogeochemical information learned from long-term studies at Hubbard Brook has management relevance (for example, air pollution), its primary value is in providing fundamental understanding about how a complicated forest ecosystem works.

9

THE DISCOVERY OF ACID RAIN AT HUBBARD BROOK

It was obvious from the first sample of rainwater collected at the forest on July 24, 1963, that something highly unusual was occurring, but at the time we didn't understand the cause or its significance (fig. 9.1). This first sample had a pH of 3.7, and the annual volume-weighted average pH for the first complete year of sampling of rain and snow (1964–1965) was 4.12. (The annual concentration of hydrogen ions was adjusted, or weighted, for the amount of annual precipitation and converted to a pH value.) These values were between 10 and 100 times more acidic than anyone would have expected for this rather remote forested region. Distilled water in equilibrium with atmospheric concentrations of carbon dioxide produces carbonic acid, which would have a pH of 5.6. A pH value is the negative logarithm of the hydrogen ion concentration in microequivalents (μeq)/liter, and this concentration changes tenfold between each whole number of pH value. For example, the hydrogen ion concentration for pH 5 is 10 μeq/liter, whereas for pH 4 it is 100 μeq/liter. We found that in the 1970s and 1980s, summertime rain was normally more acidic than snow in the winter at the forest (fig. 9.2).

What was the cause of this surprising finding? Was the acidity something unique to the White Mountains of New Hampshire? Could this highly acidic precipitation have an ecological or biogeochemical impact, or both?

Our first paper on this environmental issue was published in 1972, identifying this major problem for the first time in North America.[1] The paper was titled "Acid Rain," and the phrase, as it turned out, caught on with the public and the media. It took a relatively long time (9 years) from the observation of the unusual acidity in that first sample of rain at Hubbard Brook to compile adequate information to be able to publish this first peer-reviewed scientific paper. Initially, we thought that acid rain was produced primarily from emissions of sulfur dioxide (SO_2), mainly as waste products released to the atmosphere from the combustion of coal and oil in electrical power plants. With time, however, we discovered that acid rain was produced from the emissions of

Figure 9.1. Marilyn Fox and Gene Likens collecting a sample of precipitation during the autumn of 1963 in a cleared area in the forest at Hubbard Brook. (Photo from G. E. Likens)

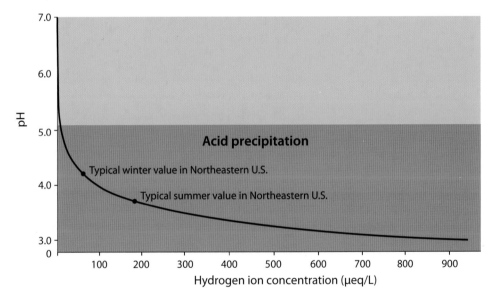

Figure 9.2. Logarithmic relation between pH and hydrogen ion concentration in μeq/liter. Acid precipitation has pH values lower than 5.1 (red area). The colors are analogous to the red (acid)/blue (basic) response of litmus paper. Summer rain tends to be more acidic than winter snow at the forest. "Typical" values refer to conditions in the 1970s and 1980s. (G. E. Likens, unpublished)

both sulfur dioxide (SO_2) and nitrogen oxides (NO_x). Nitrogen oxides also are emitted from combustion of coal and oil, but mostly they are a result of emissions from high-temperature combustion engines, such as automobiles. Both of these gases can be converted in the atmosphere to the strong acids, sulfuric (H_2SO_4) and nitric (HNO_3), and fall to the surface of land and water as atmospheric deposition (fig. 9.3). We also discovered early that other forms of atmospheric deposition in our area had become acidified (see

Box 9.1). Acid rain is now the popular name given to atmospheric deposition of wet (rain, snow, sleet, hail, fog, rime ice, cloud water) and dry (acidifying particles and gases) materials.

When Gene Likens moved to Cornell University in 1969 and collected precipitation in the Finger Lakes region of New York state, his results were very similar to what was being measured in the White Mountains of New Hampshire. This similarity in precipitation chemistry in the Finger Lakes region with the White

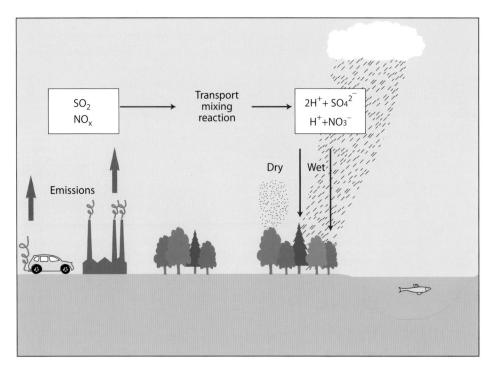

Figure 9.3. Sulfur dioxide and nitrogen oxides are emitted in the combustion of fossil fuels, such as coal, oil, and gasoline, and then are converted to sulfuric acid and nitric acid, which subsequently falls to earth in the form of acid rain. (G. E. Likens, unpublished)

Mountains provided our first clue that acid deposition was a regional problem, not just some local aberration at Hubbard Brook.[2]

FUNDAMENTALS NEEDED TO UNDERSTAND ACID RAIN

To understand more clearly the dimensions of the acid rain problem and to address potential regulatory concerns in response to aggressive deniers of the existence of acid rain, we needed to ask and answer some fundamental scientific questions. One of these was: What was the acidity of rain prior to major human alteration of the chemistry of the atmosphere, for example before the Industrial Revolution? To answer this question, a major effort was launched to establish precipitation collection sites in locations as far away from human activity as could practically be found, thinking that rain in these locations would be most similar to what fell prior to the Industrial Revolution in about 1860. We established 14 sampling stations during 1979 and 1999, including in the southern tip of Chile (Torres del Paine), in the southern tip of South Africa (Cape of Good Hope),

northern Australia (Katherine), Amsterdam Island in the Indian Ocean, and so forth (see locations in fig. 9.10 below). We collected samples of rain at these sites and analyzed them for chemical content, including acidity, for up to 10 years. We discovered that the rain even in these locations contained small amounts of natural organic and inorganic compounds swept into the atmosphere from the surrounding terrain and then deposited with the rain. Our best estimate of what the pH of rain was prior to the Industrial Revolution, based on findings from these remote sites, was a value of about 5.08 (due to mineral acids).[3]

Another question was: Where did the pollutants in our precipitation at Hubbard Brook originate? Because the earliest air pollution problems in the United States had come from particles emitted in the burning of coal, taller smokestacks were constructed in the 1950s to release these particles higher into the atmosphere, thereby allowing greater dispersal and dilution.[4] But the old adage applied: What goes up must come down. The taller stacks did disperse the pollutants, but in the process this converted local pollution problems into regional ones. Still, parties with vested interests, such

Box 9.1. Cloud Water

Because cloud water and fog water can provide ecologically important inputs of water and chemicals to forest ecosystems and, unlike a rain storm, can bathe the surfaces of plants for hours or days on end, we thought it would be important to conduct detailed measurements at Hubbard Brook and elsewhere. To evaluate these atmospheric inputs, cloud water and rainwater samples were collected concurrently at twelve sites throughout North America (in Alaska, Oregon, California, Puerto Rico, Virginia, New York, New Hampshire, and Maine) for several years.[a] We were surprised by some of the findings from this Cloud Water Project.

A cloudwater collector on Mount Washington, New Hampshire. (Photo by K. C. Weathers)

We found that cloud water was *much* more acidic than rainwater in paired collections at the same site, and many times more concentrated in nitrate and sulfate as well as ammonium—essentially all ions that we measured whether nutrient or pollutant associated. Cloud water in the eastern United States was more acidic and had higher concentrations of nitrate and sulfate than cloud water collected in the western United States and Puerto Rico; one cloud water event during August 7–13, 1984, at six non-urban sites (including Hubbard Brook) in the eastern United States had extremely low pH values, ranging between 2.80 and 3.09.[b] As it turned out, immersion in cloud water at Hubbard Brook is not frequent and therefore contributes negligibly to atmospheric inputs on an annual basis. Cloudwater samples were obtained from collectors on the top of Mount Washington in New Hampshire, as well as from the top of Building 2 at the World Trade Center in New York City in 1984 and 1985, which was frequently enveloped in low-hanging clouds.

NOTES

a. Weathers et al. 1988a, b; Weathers et al. 1986.

b. Weathers et al. 1986.

as electrical utilities in the midwestern United States, who were emitting very large quantities of gaseous and particulate sulfur into the atmosphere, denied that they were the source of sulfur pollutants, which were contributing to acid rain in New England. Indeed, we know now that some 70% of sulfur emissions come from electrical utilities, and the biggest of these plants are located in the Midwest. Because wind in the United States tends to flow from west to east, and since sulfur dioxide travels long distances in the atmosphere, these large sources of pollutants to the atmosphere became suspected as the primary cause of acid rain in the eastern United States. So we needed to trace the pollutants from their source to places such as Hubbard Brook. We tried various techniques to follow the emissions of sulfur compounds from these coal- and oil-fired power plants to the depositions of sulfate we were measuring at Hubbard Brook, including the use of isotopic tracers and of airborne sensors in small aircraft and vehicles on the ground to follow plumes of pollutants (Box 9.2).

In the end, we found a very strong, direct

Box 9.2. Tracking Dirty Air

In the early 1980s, it became important to find out whether air pollutants were coming from large sources in the midwestern United States and making their way to New England. We partnered with colleagues from the National Center for Atmospheric Research (NCAR) in Boulder, Colorado, and elsewhere to track pollutants in air masses moving from west to east. Samples collected from a twin-engine Beechcraft airplane and concurrent samples collected on the ground with an instrumented van showed a clear link between SO_2 emissions and both sulfate and acidity of precipitation. Elaborate methods such as these were required to identify the sources of the precursor emissions leading to acid rain in New England. Conclusively demonstrating this direct causal relationship ultimately set the stage for federal legislation to reduce these emissions.

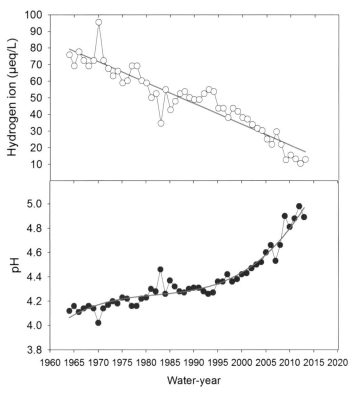

Figure 9.4. Fifty years of acid rain at Hubbard Brook, showing mean annual (volume-weighted) hydrogen ion concentration (µeq/liter) and pH in precipitation at Hubbard Brook from 1963 to 2013. Data measured as water-year, defined as running from June 1 through May 31. (Updated from Likens 2013; reprinted with permission of Springer Science+Business Media)

correlation between the sulfur emitted from the source area for airborne pollutants for Hubbard Brook and the sulfur content in precipitation and stream water measured at Hubbard Brook. This strong correlation provided convincing evidence that sulfur was moving with air masses from the middle of the country to the northeastern United States.

Other questions included: What was the variability of the acidity in different precipitation events? What was the variability of acidity throughout an event? What were the kinds of acids (organic versus inorganic) generating the acidity? All had to be addressed and answered. Decades of research were required to obtain rigorous, credible information so that the critical management question—If we reduce emissions, will the acidity of precipitation lessen?— could be answered with confidence. Indeed, following passage of the Clean Air Act of 1970 and the 1990 amendments to the Clean Air Act by the United States

government to reduce emissions of particles and gases, respectively, sulfur and acidity in precipitation declined concomitantly and significantly.

Although annual acidity of precipitation at Hubbard Brook peaked in 1970, concentrations of hydrogen ion generally have been declining since the earliest measurements in 1963 (fig. 9.4). Annual values are quite variable in the long-term record, and several to many years may show significant upward or downward trends. It required 18 years of continuous measurement from the beginning of the Study before a statistically significant downward trend in acidity at Hubbard Brook was observed in these long-term data. For enthusiasts of monitoring, this is sobering, but it shows the critical value of long-term data. It is important to note that even though the acidity of precipitation has

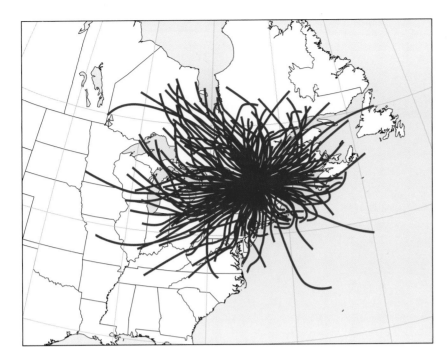

Figure 9.5. Air-mass trajectories for every storm producing rain or snow at Hubbard Brook during the year 2000. (G. E. Likens, unpublished)

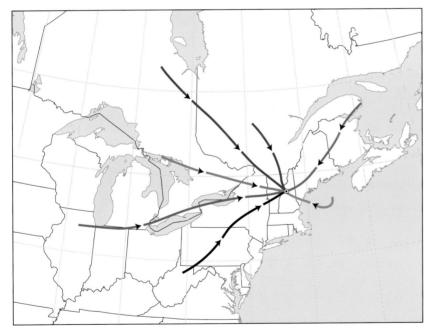

Figure 9.6. Seven clustered 36-hour air-mass back trajectories for the year 2000 at Hubbard Brook, based on the trajectories shown in figure 9.5. (G. E. Likens, unpublished)

declined about 80% since 1965, precipitation is still more acid than it should be if the atmosphere were not polluted. Acid rain has not gone away.

Using meteorologic models, it is possible to trace trajectories for every storm moving through the atmosphere that produces rain or snow at Hubbard Brook.[5] Over the course of a year the back trajectories for these air masses tend to look like a ball of spaghetti (fig. 9.5). With another model, these trajectories can be clustered to give dominant trajectories for these storm tracks (fig. 9.6).[6] Using this approach it was then possible to determine the source area, by state,

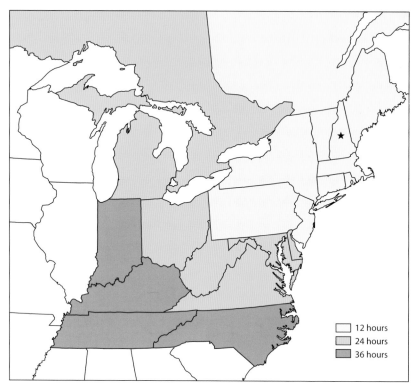

Figure 9.7. States and provinces representing the source areas for emissions contributing to precipitation chemistry falling on the forest at Hubbard Brook (✳), based on 12-hr, 24-hour, and 36-hour air-mass back trajectories. (Modified from Likens et al. 2005; reprinted by permission of The Royal Society of Chemistry)

for precipitation falling on Hubbard Brook and to calculate total emissions from this source area for different periods of time, such as a month or a year (fig. 9.7). This approach allowed calculation of where the pollution comes from, how much was generated in the emission source area for Hubbard Brook, how long it takes for emitted pollutants to get to Hubbard Brook, and what percentage of these emissions was deposited on the watersheds at Hubbard Brook.

A rough calculation suggests that of all the sulfur emitted from the 24-hr source area for Hubbard Brook during 1966–2000 (34 years), less than 0.001% was deposited in rain and snow at Hubbard Brook. Nevertheless, there was a very high direct correlation between sulfur dioxide emissions from the source area and the sulfate concentration in precipitation collected at Hubbard Brook. Even after the sulfur that is deposited in rain and snow passes through the forest ecosystem, there is an equally strong correlation between amount of emissions and concentration of sulfate in stream water in Watershed 6 (fig. 9.8).

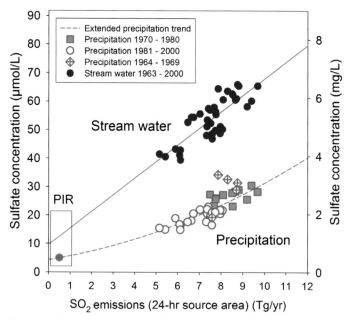

Figure 9.8. Relation between emissions of sulfur dioxide (SO_2) from the 24-hr source area for Hubbard Brook and sulfate concentrations in precipitation and stream water. PIR is pre–Industrial Revolution. (Modified from Likens et al. 2005; reprinted by permission of The Royal Society of Chemistry)

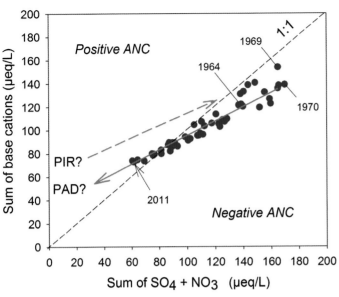

Figure 9.9. Relationship between the sum of base cations (calcium, magnesium, potassium, sodium) and the sum of acid anions (sulfate, nitrate) in stream water of Watershed 6, the reference watershed, during 1964 to 2012. Acidification occurred until about 1969–1970, followed by a significant recovery. ANC is acid-neutralizing capacity; points above the 1:1 line are positive and points below the line are negative ANC; PIR is pre–Industrial Revolution; PAD is post–acid deposition. Each black dot is an annual value. (Modified from Likens and Buso 2012; reprinted by permission of the American Chemical Society, © 2012)

These relationships are particularly remarkable given the distance of Hubbard Brook from large sources (electrical utilities in the Midwest) in the source area, the vagaries of wind patterns, and the small percentage of time when precipitation actually occurs at Hubbard Brook during the year.[7]

From a management and policy point of view this relation means that when emissions are reduced, deposition of sulfur and accompanying acidity will be reduced concomitantly. So federal legislation, such as the Clean Air Act Amendments of 1990, which significantly reduced emissions of sulfur dioxide, and to a lesser extent, NO_x emissions, has been highly successful. Similar successes in reducing acid and sulfur and nitrogen deposition have also occurred in Europe and Canada.

Box 9.3. What Are Base Cations?

Calcium, magnesium, sodium, and potassium are positively charged cations in solution, referred to as base cations. They not only are nutrients for supporting biotic growth, but also, when dissolved, tend to neutralize acids in soil and water. Their loss from the ecosystem results in lower acid neutralizing capacity of the system.

The impact of acid rain on receiving ecosystems can be large, but we were surprised to see how major the effects were on forest soils at Hubbard Brook. As reflected in streamwater chemistry, the watershed-ecosystems were acidifying from acid rain during 1963 to 1970, and then recovering (neutralized by bases in the system) from 1970 to the present (fig. 9.9). Relatively large amounts of base cations, which are also critical nutrients for plants in the ecosystem, were being depleted from exchange sites in the forest soil and lost in stream water during the acidification phase (Box 9.3).

DATA, MAPS, AND CARTOONS

After the discovery of acid rain at Hubbard Brook, long-term scientific research greatly advanced the knowledge base needed for developing policy and implementing federal and state legislation and rules related to air pollution. Atmospheric deposition of various pollutants, such as acidity, sulfur, and nitrogen, is still many times higher than estimated background values, and these pollutants have affected the ecology and biogeochemistry of ecosystems at Hubbard Brook for decades. It is useful to summarize several scientific findings from the Study, which have been important to the science, but also influenced the development of policy and the management of air pollution.

- Changes in sulfate and nitrate concentrations in precipitation and stream (and lake) water

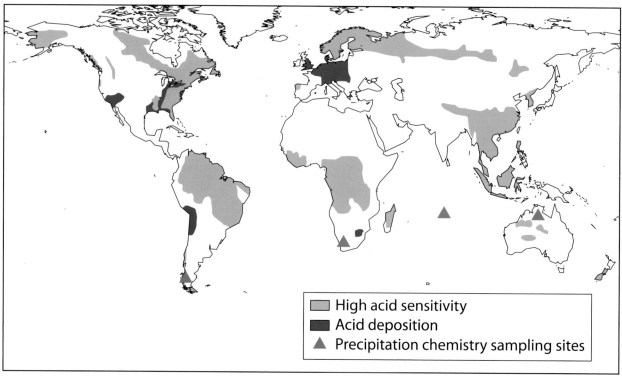

Figure 9.10. A global perspective of the extent of acid deposition and areas most likely to be impacted because of the sensitivity of the landscape to acid inputs. Triangles indicate sampling sites used to examine precipitation chemistry in areas remote from major industrialized areas used in our study. (Modified from Likens et al. 2009; reprinted by permission of Sage Publications)

at Hubbard Brook are strongly and directly correlated with changes in emissions of sulfur and nitrogen oxides (the major precursors to acid rain). As a result of federal legislation that led to a steady decrease in emissions of SO_2 and NO_x, acid deposition also has steadily decreased. When emissions decrease, acidity of precipitation decreases, and vice versa.

- Calcium and other base cations have been markedly depleted in the soils of Hubbard Brook as a result of acid rain inputs to the watershed-ecosystems: As much as one-half of the pool of exchangeable calcium in the soil has been depleted during the past 50 years by acid rain. As a result of losses in soil buffering through depletion of base cations, the forest ecosystem is currently much more sensitive to acid rain impact than previously thought, even though the input of acid rain is declining.

- Nitric acid has increased in relative importance in precipitation at Hubbard Brook, as emissions of SO_2 have declined more than NO_x. Currently, nitric acid is the dominant acid in precipitation at Hubbard Brook.

- Forest biomass accumulation ceased at Hubbard Brook between 1982 and 2002, and was negative during 2002 and 2007 in Watershed 6 due to increased tree mortality. This may have resulted, in part, from the loss of base cations leached by acid rain and the mobilization of toxic aluminum in the soil.

It is highly unusual, if not unique, for scientists to be involved in a major environmental problem all the way from discovery to political action. That was the situation with acid rain at Hubbard Brook—from discovery in 1963 to passage of the Clean Air Act amendments in 1990.[8] Many factors underpinned this progression to political action, not the least of which was the impeccable, long-term data on precipitation chemistry at Hubbard Brook. But other factors, such as cartoons in the media, maps showing a progressively worsening situation, and the title of the first scientific publication on the topic from Hubbard Brook, "Acid Rain," were also important.[9]

Despite being fundamental to understanding this environmental issue, more was required than hard scientific data to convince the public, the media, and the politicians to take action to reduce the emissions causing acid rain. Much activity was required on many fronts, but two examples will illustrate the diversity of approaches that were used to raise awareness and concern.

Cartoons have a long history of being used to sway public opinion because of their often satirical humor and clear and concise message.[10] The topic of acid rain was a perfect example.[11] Although often exaggerated in terms of proposed environmental impact, cartoons seem to have been very effective in raising awareness and concern about the environmental degradation caused by acid rain. A drawing showing the skeleton of a fish dangling from a fishing pole was stark and compelling to many viewers, leaving them with a clearly understandable takeaway message. As is common, politicians were frequent targets for cartoonists.

We, and others, constructed maps to show the increasing intensity and area in eastern North America being subjected to acid rain over time.[12] Maps can be powerful and effective visual tools for conceptualizing a complicated problem and informing the debate for the public and policy makers. This was also the case for acid rain.[13]

Finally, acid rain is not some strange aberration in the chemistry of the atmosphere in the White Mountains of New Hampshire. It occurs in many places throughout the world where there are significantly large emissions of sulfur or nitrogen oxides to the atmosphere (fig. 9.10).[14] Such places occur especially in highly industrialized regions. However, where the receiving system is poorly buffered, the system is sensitive to these inputs of acid in precipitation, and the ecological and biogeochemical impacts may be large.

PART 4

DISCOVERIES FROM LONG-TERM STUDIES AND EXPERIMENTAL MANIPULATIONS

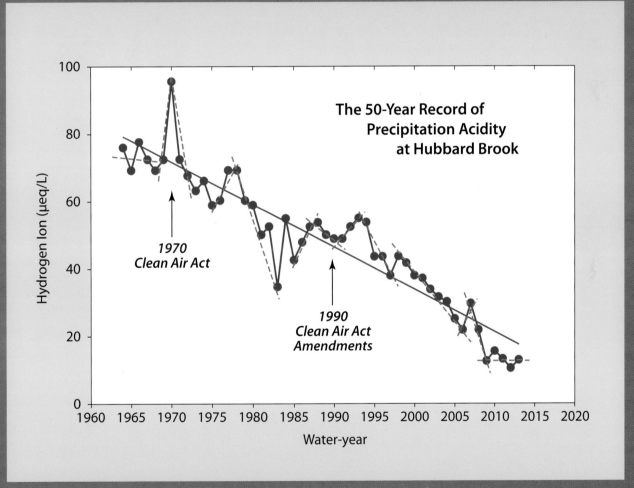

10

THE CONSEQUENCES OF ACID RAIN AND OTHER AIR POLLUTANTS

Most of the early ecological research on the effects of acid rain in eastern North America, including research at Hubbard Brook and in western Europe, focused on sulfuric acid, the predominant acid in precipitation. Sulfuric acid was thought to be the principal cause of lake and stream acidification, which in turn led to adverse effects on fish (fig. 10.1). But subsequent research revealed the environmental effects of air pollution leading to acid rain were much broader. Some of the most serious effects, as now recognized, include acidification of aquatic ecosystems with the elimination of sensitive aquatic organisms; degradation of terrestrial ecosystems, where base cations have been depleted, which has led to a decrease in buffering capacity and mobilization of toxic dissolved aluminum from soils; alteration of fundamental biogeochemical dynamics; corrosion and erosion of buildings, monuments, gravestones, and other structures; and human health effects, particularly respiratory problems.[1]

Acid rain affects ecosystems through not only sulfuric acid, but also nitric acid and other pollutants, such as ozone and dissolved ionic aluminum. Nitrogen and sulfur are also plant nutrients, but nitrogen tends to play a more important ecological role in the ecosystems at Hubbard Brook because it is required by the biota for growth and it has limited availability. Increased nitrogen deposition in precipitation during the first two decades of the Study was therefore particularly important both as a critical nutrient for ecosystem metabolism and as a significant component of the emerging acid rain issue.

Although sulfuric acid was the principal acid in precipitation, Hubbard Brook scientists were among the first to call attention to nitric acid as an important and increasing component of acid rain and its ecological effects.[2] The relative importance of nitric acid increased rapidly from 1964 to about 2002, because sulfur dioxide emissions were being reduced through U.S. Environmental Protection Agency pollution controls, particularly after 1990, and NO_x emissions were increasing (fig. 10.2).[3] The regulatory attention in 1970 focused on reducing

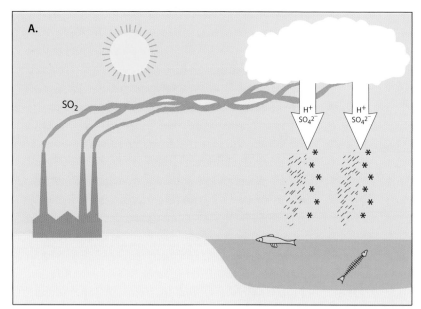

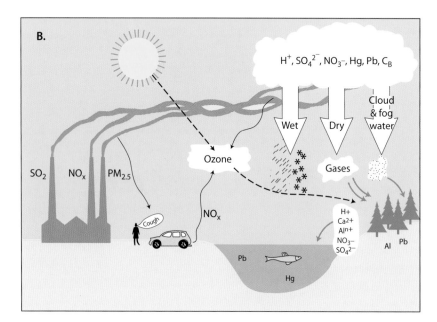

Figure 10.1. The simplistic view of acid rain and ecosystem response that prevailed in the late 1960s to early 1970s (top). Ecological concern focused on sulfuric acid and its effects on aquatic ecosystems, particularly fish. A more recent view of the structural, functional, and temporal complexity of ecosystem response to acid rain, as developed in the 1990s and 2000s, shows how sulfuric and nitric acids, combined with other pollutants, have complicated impacts on aquatic and terrestrial ecosystems (bottom). $PM_{2.5}$ is particulate matter 2.5 μm in diameter, C_B is the sum of base cations (calcium, magnesium, sodium, and potassium). (Modified from Likens 1998; reprinted by permission of Springer Science+Business Media)

emissions of fine particles (soot) to the atmosphere, which also contained sulfur and calcium (the Clean Air Act of 1970). Later, regulatory attention focused directly on reducing acid rain (emissions of sulfur and nitrogen oxides, SO_2 and NO_x, in the Clean Air Act Amendments of 1990). Nevertheless, NO_x emissions were not significantly reduced. More recently larger reductions in NO_x have occurred, largely in response

to national concerns about ozone pollution, in which NO_x emissions play an important role.[4]

THE IMPACT OF ACID RAIN ON SOIL AND FOREST DEVELOPMENT

Soils. After the glacier retreated about 14,000 years ago, soils and then forest vegetation began to develop in the Hubbard Brook valley. It wasn't until

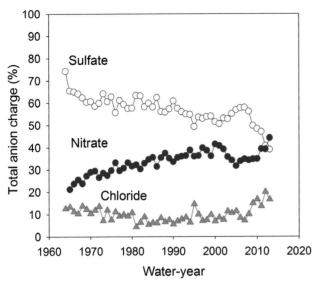

Figure 10.2. Increase of nitrate (contributing to HNO3) relative to sulfate (contributing to H2SO4) and chloride (contributing to HCl) in precipitation at Hubbard Brook since 1960. Sulfate deposition is declining more than nitrate because of federal regulations, which are significantly reducing emissions of SO2. Nitric acid is expected to be the dominant acid in precipitation in future years. (Modified and extended from Likens and Lambert 1998)

about 10,000 years ago that the forest began to form in the valley, and not until about 8,000 to 6,000 years ago that this forest began to resemble the modern-day forest. Concomitantly, soils were developing during this period with increases in humus content and cation exchange capacity, which increased the acid-buffering capacity of these soils. Thousands of years were required to achieve the soil development observed today at Hubbard Brook.

A surprising consequence of acid deposition was that it could affect forest ecosystems. Scientists did not think that forest soils in particular would be vulnerable to acid rain because of their relatively large acid-neutralizing capacity (ability to buffer inputs of mineral acids), built up over thousands of years following the retreat of the glacier. Many in the scientific community were shocked when we reported that acid deposition had leached large amounts of base cations (calcium, magnesium, sodium, and potassium) from the soils at Hubbard Brook relatively

quickly, over mere decades.[5] The depletion of base cations, particularly calcium, from forest soils will be a lasting legacy of decades of acid rain. Furthermore, its impact on biogeochemical flux and cycling and on forest growth is one of the major discoveries of the 50-year research effort at Hubbard Brook.

Forest. The first quantitative measurements of tree biomass were made on Watershed 6 (the unaltered reference watershed) in 1965, and then again in 1977, 1982, and every five years thereafter (fig. 10.3). In 1965, the forest was still recovering from a relatively heavy harvest in 1910–1920. Trees were growing larger each year, and biomass per unit area increased markedly between 1965 and 1982. According to our computer models, we expected tree biomass to continue accumulating for several more decades.[6] Instead, however, it leveled off after 1982, and then significantly declined after 2002 because of increasing tree mortality. What was the cause of this surprising finding?

There are at least four possible explanations for the decline in forest biomass accumulation in Watershed 6 after 1982 (fig. 10.4). First, the decline could have been a natural response to the maturation of the forest, augmented by climate change, or it could mean that our computer models for this maturation process were wrong or didn't adequately account for factors like climate change. Second, it could have been the result of widespread disease in American beech (beech bark disease), which might have led to a decrease in total plant biomass. Third, severe air pollution from ozone in combination with acid rain could have directly and adversely affected tree growth. Ozone can be very damaging to plants, such as sugar maple, and outbreaks of high concentrations of ozone in the Hubbard Brook area are relatively common. Direct effects of acid rain falling on vegetation also can be damaging. The lowest pH we ever measured in rain at Hubbard Brook was 2.85, which produced visible necrotic spots on the leaves of pin cherry (fig. 10.5). Fourth, the forest may have stopped accumulating biomass as a result of some nutrient limitation. Various nutrients can contribute

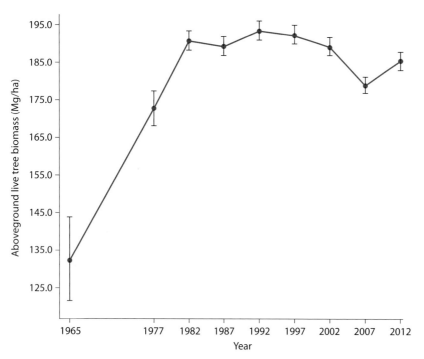

Figure 10.3. Live and dead tree biomass, above- and belowground, in the undisturbed reference watershed (W6) between 1965 and 2012. (Modified from Battles et al. 2014; reprinted by permission of the American Chemical Society, © 2012)

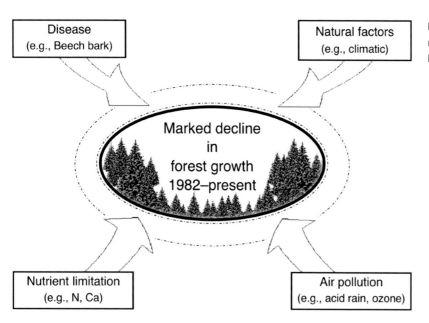

Figure 10.4. Potential causes for the marked decline in forest growth at Hubbard Brook. (G. E. Likens, unpublished)

singly or in combination as limiting factors for plant productivity in forests like those found at Hubbard Brook.[7] In the 1960s–1980s, the expectation was that nitrogen was the limiting nutrient for forest growth at Hubbard Brook, but the large depletion of calcium by acid rain generated the hypothesis that calcium might have become limiting to forest growth. We decided to test the role of calcium in the ecosystem with an experimental watershed manipulation. This was a large and complicated undertaking, funded by the National Science Foundation as a long-term study done in Watershed 1.

Figure 10.5. Pin cherry leaf at Hubbard Brook damaged by acid rain on August 24, 1977. Rain that day had a pH of 2.85. (Photo by G. E. Likens)

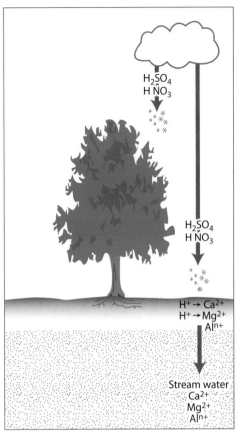

Figure 10.6. Hydrogen ions in acid deposition falling on the canopy of the forest and directly on forest soils are neutralized by exchange reactions, which result in the leaching of calcium (Ca^{2+}), magnesium (Mg^{2+}), and aluminum (Al^{n+}) into stream water, and then lost from the ecosystem. (G. E. Likens, unpublished)

Atmospheric inputs of strong acids leach calcium, magnesium, and other base cations from the forest canopy, forest floor, and possibly even underlying mineral soils, reducing the acid-neutralizing capacity and making the ecosystem more sensitive to further inputs of acidic precipitation (fig. 10.6). The susceptibility of soils to acidification is determined by their cation exchange capacity, percentage of acid saturation, and content of weatherable minerals, such as calcium carbonate.[8] Significant depletion of calcium from labile soil pools and mobilization of toxic aluminum in the rooting zone of the forest soils result from this process of soil acidification.[9] In fact, we calculated from long-term chemical budgets that about 520 kg of calcium per hectare had been depleted from labile soil pools at Hubbard Brook between 1940 and 1995, or approximately 50% of that in exchange pools in the soil.[10]

Acid rain has different effects on coniferous (red spruce) and deciduous (sugar maple) trees (fig. 10.7). Scientists from the University of Vermont discovered that calcium is leached from needles of red spruce by acid rain, making these trees more sensitive to freezing injury during the winter, which can even

cause mortality of the trees.[11] This freezing effect caused by acid precipitation was evident during an exceptionally cold winter at Hubbard Brook in 2003 (fig. 10.8). Sugar maple is affected by acid rain primarily in the rooting zone from depletion of calcium and magnesium (loss of buffering capacity and availability of these nutrients) and mobilization of dissolved ionic aluminum, which can be toxic to fine roots. Aluminum solubility dramatically increases with increasing soil acidification.[12] A number of studies by investigators throughout the northeastern United States and southeastern Canada have shown serious effects on sugar maple and other hardwood species

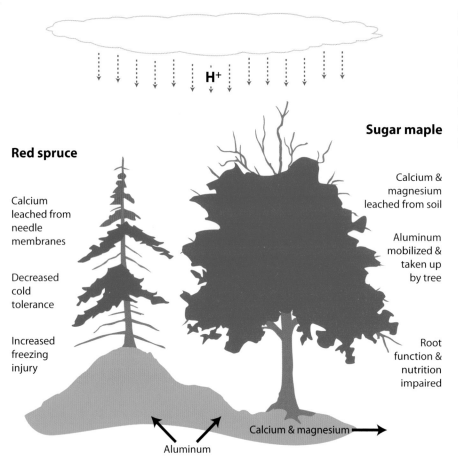

Red spruce

Calcium
leached from
needle
membranes

Decreased
cold
tolerance

Increased
freezing
injury

Sugar maple

Calcium &
magnesium
leached from soil

Aluminum
mobilized &
taken up
by tree

Root
function &
nutrition
impaired

H$^+$

Aluminum

Calcium & magnesium

Figure 10.7. The differential
effects of acid rain on red spruce
and sugar maple in forests of
the northeastern United States
and southeastern Canada.
(Modified from Driscoll et al.
2001; reproduced by permission
of the Hubbard Brook Research
Foundation)

such as American beech by the loss of buffering
capacity and increase in mobilization of aluminum,
which is toxic in the dissolved form.[13]

Very important to the evaluation of acid rain
effects on terrestrial and aquatic ecosystems was the
development of analytical procedures for fractionating
dissolved aluminum found in dilute natural waters
into acid-soluble, non-labile monomeric and labile
monomeric forms. The labile monomeric form
(reactive and toxic) increased exponentially with
increasing acidity, whereas the non-labile form was
relatively insensitive to changes in solution acidity.
These findings were important in evaluating the
biological impact of aluminum and for monitoring the
occurrence of toxic, ionic aluminum in surface waters
such as those at Hubbard Brook.[14]

ACID RAIN, DEPLETION OF SOIL NUTRIENTS, AND SOIL BUFFERING CAPACITY

To test the role of calcium in regulating the structure
and function of a northern hardwood forest
ecosystem, we initiated a watershed-scale experiment
in Watershed 1 in 1999. After much discussion and
debate, the mineral Wollastonite (composed of
calcium and silicate) was chosen as the source of
calcium to add to Watershed 1. Adding calcium
from a silicate mineral would be more similar to the
natural weathering process at Hubbard Brook than
using calcium carbonate. We intended to study the
ecosystem response to this experimental addition
on both terrestrial and aquatic ecosystems, but
particularly on the buffering capacity of the soil.
Wollastonite was mined in the Adirondack Mountains

Figure 10.8. Freezing injury in red spruce needles during the cold winter of 2003 brought about by leaching of "protective" calcium from the needles by acid rain. (Photo by G. J. Hawley)

This watershed manipulation was designed as a long-term experiment (50 years), and results to date have been very interesting and promising from a management perspective. The pH and acid-neutralizing capacity of stream water increased, as did the buffering capacity and the pH of the soil.[15] Soil pH increased, perhaps by as much as a full unit, although it is extremely difficult to measure quantitatively the pH of soil at the watershed scale because it is so variable from place to place within the watershed.[16] Forest soils at Hubbard Brook before deposition of acid rain were already acidic (they had a pH of about 4.5), because naturally occurring organic acids were present, but acid rain over six decades decreased the soil pH to 3.5 or so. Thus, the Wollastonite addition returned the soil pH to approximately what it was before the advent of acid rain.

Positive responses of sugar maple and red spruce, the two target species for this experiment, were dramatic (Table 10.1). Moreover, tree biomass and leaf area of sugar maple increased significantly in the treated watershed following the addition of calcium silicate (fig. 10.10).[17] This surprising result suggests calcium may have become limiting to vegetation growth because of large acid rain–caused losses of this nutrient from the soil.

Significantly less winter injury was observed in red spruce growing in the treated watershed (fig. 10.11). Also, more evapotranspiration and less stream flow occurred for three years after treatment.[18] The Wollastonite addition also contributed to an increase in leaf and wood production and a decline in fine-root biomass—that is, the trees refocused primary production from belowground to aboveground, especially at higher elevations, following this experimental manipulation.[19] Nevertheless, major questions remain to be answered from this watershed-scale manipulation: How long will the alleviation of acid rain effects last? Will sugar maple fully recover? In sharp contrast to other adjacent watershed-ecosystems, there has been an unexplained large increase in nitrate output in stream water from

of New York, ground to a fine powder, and then pelletized with the aid of a small amount of a binder, sodium-lignon-sulfonate (so that it could be handled), and shipped to Hubbard Brook. The pellets were then distributed evenly on Watershed 1 in October 1999 by helicopter (fig. 10.9). Based on our measurements, approximately two times as much calcium bound in Wollastonite was added to Watershed 1 as the amount calculated to have been leached from this watershed in the previous 50 years by acid rain.

Figure 10.9. Adding 1,189 kg/ha of calcium to Watershed 1 in 1999. Large bags filled with pelletized Wollastonite (CaSiO3) were lifted by helicopter and then released from the air; the white spots at lower left are Wollastonite pellets on the forest floor. (Photos from the Hubbard Brook Research Foundation)

Table 10.1. Response of Sugar Maple to Calcium Silicate Addition
Calcium silicate (Wollastonite) added in Watershed 1, compared with an adjacent untreated reference area.

Sugar maple trees
   Increased concentration of foliar calcium
   Crown condition healthier (2005)
   Higher seed production (2000 and 2002)
   Fungal (mycorrhizal) colonization of fine roots (a sign of root health) higher (56% of root length) than in the reference site (3.5% of root length)
Sugar maple germinants (sprouts from seeds)
   Larger by about 50% (2003) and foliar chlorophyll concentrations significantly higher (0.27 g/m^2 vs. 0.23 g/m^2 of leaf area)
Sugar maple seedlings
   Density increased significantly after 2002 relative to the reference, untreated Watershed 6
   Survivorship of 2003 cohort higher
   Foliar and fine-root calcium concentrations were higher by about twofold; manganese concentration was lower by about twofold
   Mycorrhizal colonization of fine roots was greater (22.4% of root length) than in the reference site (4.4% of root length)

Source: Juice et al. 2006

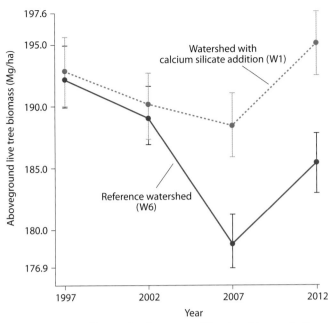

Figure 10.10. Changes in aboveground live-tree biomass in the watershed treated with Wollastonite (Watershed 1), compared with the untreated watershed (Watershed 6), 1996–2012. (Modified from Battles et al. 2014; reprinted by permission of the American Chemical Society, © 2014)

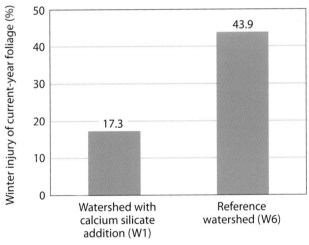

Figure 10.11. Foliage of red spruce in Watershed 1 amended with Wollastonite had significantly less winter damage than in the nearby reference watershed during the cold winter of 2003. (Adapted from Hawley et al. 2006; reprinted by permission of Canadian Science Publishing, © 2006)

Watershed 1 since 2011. The cause for this dramatic change in biogeochemistry is currently under investigation.

OBSERVED CHANGES IN PRECIPITATION AND STREAMWATER CHEMISTRY

Another major consequence of long-term acidification and recovery of the forest ecosystem at Hubbard Brook has been the highly significant and rapid decline in concentrations of solutes (dissolved materials) in both precipitation and stream water since about 1970.[20] Our long-term records clearly show that both precipitation and stream water are on robust trajectories to become extremely dilute in only a few years in the future. For example, values for electrical conductivity (a measure of all ionically charged species in solution) are expected to approach that of demineralized water by 2018 for bulk precipitation and by 2030 for stream water (fig. 10.12). Because such dilute values are unrealistic for natural waters, theoretically expected baselines of 3–5 microsiemens (μS)/cm have been calculated, where concentrations are projected to stabilize in the near future.

The tentative explanations for these surprising trends are that concentrations of solutes in bulk precipitation have been declining since about 1970 because of U.S. air pollution regulations, which have reduced emissions of acid-forming gases and fine particulate matter containing base cations, such as calcium. These reductions have lowered concentrations of hydrogen ion, sulfate, and base cations in precipitation, while at the same time the acidity in precipitation has leached away large quantities of base cations, particularly calcium, from the soil. As a result, over time, concentrations of base cations, hydrogen ion, sulfate, and nitrate in stream water have declined as sources in the soil have been depleted and as emission sources have been reduced. Although these results represent a management success story because acid rain regulations have been so effective, demineralized water in precipitation and streams would represent a clear environmental hazard. Such dilute waters would present problems,

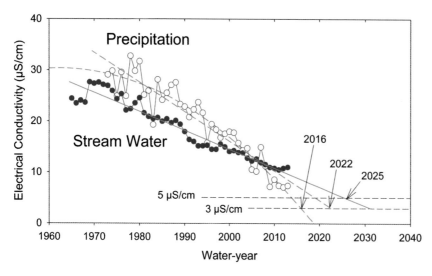

Figure 10.12. Changes in volume-weighted annual electrical conductivity (EC) concentration, a measure of the total ionic solute concentration, showing the striking decline in both precipitation and streamwater concentrations of Watershed 6 since the 1970s. Theoretical baselines (minimum values expected) of 3 and 5 μS/cm, which are extremely low for natural waters, are projected to occur for precipitation and stream water, respectively, within only a few years. Electrical conductivity is measured in microsiemens (μS)/cm. (Adapted from Likens and Buso 2012; reprinted by permission of the American Chemical Society, © 2012)

including osmotic regulation and nutrient availability, for aquatic organism survival and for overall ecosystem function.[21] It will be most interesting and important to learn how this change plays out in the near future. These trends in precipitation and streamwater chemistry provide another important example of why long-term monitoring is so important.

ACID PRECIPITATION AND TERRESTRIAL ANIMALS

Research in the Netherlands and other parts of northern Europe revealed that birds living in areas receiving high levels of acid precipitation produce eggs with thinner shells and eggs that are more likely to break before they reach hatching, resulting in reproductive failure. This effect has been attributed to acid rain leaching calcium from the soil, making it difficult for birds to find sufficient invertebrates with high calcium levels, primarily snails, for the calcium needed to build eggshells.[22] Given the overall impact of acid precipitation at Hubbard Brook since the mid-1950s and the importance of reproductive success to bird populations, this seemed like an important hypothesis to explore as part of the Study. To date, however, the results are negative. Among the thousands of bird nests examined over several decades of intensive avian study at Hubbard Brook,

no evidence of eggshell thinning or breakage during the incubation process has been found. Furthermore, eggshell thickness and reproductive success of the most intensively studied species, the black-throated blue warbler, living in calcium-rich forests in Vermont were found to be no different than those living in the relatively calcium-poor forests at Hubbard Brook.[23] Despite the lack of evidence for an acid rain effect on birds at Hubbard Brook, several recent studies from elsewhere in eastern North America have found evidence that calcium depletion caused by acid rain may be having an influence on bird populations.[24] The latter has been experimentally tested by a large-scale addition of pulverized limestone (calcium) to forests in Pennsylvania, which positively affected bird productivity (clutch size and rate of renesting) but not the physical characteristics of eggshells.[25] So the possibility remains that calcium may become an important limiting factor for birds in the future in North America, at least in some localities. Further studies of this topic are certainly warranted.

ACID PRECIPITATION AND AQUATIC ECOSYSTEMS

Acid rain has been shown to have many effects on sensitive aquatic organisms (fig. 10.13). As the acidity increases, invertebrates such as snails, amphibians,

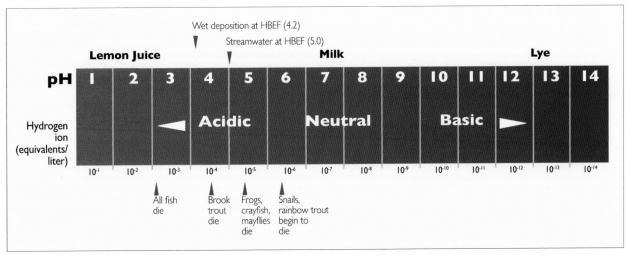

Figure 10.13. The relative acidity levels that are lethal for various aquatic organisms. The location of common products on the pH scale is shown along the top edge. Very high pH levels, such as 9 or higher, would also result in mortality for most organisms. (From Driscoll et al. 2001; reprinted by permission of the Hubbard Brook Research Foundation)

and fish begin to die because of the lethal direct and indirect effects of the acid environment. At Hubbard Brook, sensitive mayfly species and all fish (native brook trout, sculpins, and black-nosed dace) no longer occur in the episodically acutely acidified upper reaches of headwater streams. These losses are the result of the combined effects of acidification and physical barriers, such as organic debris dams and waterfalls.

Rapid snowmelt usually releases a pulse of acidity into these headwater streams because of the effects of melting on differential chemical release from snow. The first 20% or so of meltwater entering stream channels in the spring may be significantly more acidic than the remaining meltwater or when the snow melts more slowly. This sudden drop in pH to values of 4.0 or below, because of preferential leaching of acids from the snowpack, and the associated increase in toxic dissolved aluminum concentrations, frequently is severe enough to be deadly to fish and aquatic invertebrates.

Overall, acid rain and associated effects such as mobilization of toxic aluminum have had serious consequences on the ecosystems of the Hubbard Brook Experimental Forest. These changes to the structure and function of both terrestrial and aquatic ecosystems will continue to play out in coming decades in terms of impacts on ecosystem services and resilience of these ecosystems to other disturbances. The legacy effect of the major long-term disturbance of acid rain continues to be the subject of intensive study and scientific debate, even though the acidity of precipitation has declined markedly since 1970.

11

THE EFFECTS OF FOREST HARVESTING AND OTHER DISTURBANCES

Whole-Watershed Manipulations

Experimentation is a powerful scientific tool. It is used successfully in all fields of science and at multiple scales, such as the laboratory vessel, a field plot, or an organism. It is used less frequently, but no less powerfully, at the scale of an entire ecosystem—that is, a lake, a stream, or a watershed. Watershed-scale experimental manipulations have been the hallmark of the Hubbard Brook Ecosystem Study since 1965, and the results have been featured prominently in leading textbooks on ecology and biology. So, if we consider our forest as the laboratory, then its watersheds and streams are our "vessels" for experimental manipulation.

It is very difficult, if not impossible, to replicate natural ecosystems and therefore to conduct controlled experiments at the ecosystem scale. We use paired watersheds or upstream/downstream reaches of streams in experimental studies. However, because they are natural systems with all of their inherent variability and are therefore not exact replicates, we refer to them as treated and reference units, instead of treated and control units.[1] It is critical to have such reference units for comparison during an experimental manipulation, particularly when several years are involved during which conditions might change. This need is evident when environmental conditions change, even abruptly, such as when floods or droughts occur during an experimental manipulation. The effects of these conditions and factors need to be separated from the experimental conditions and factors being tested.

This experimental approach for studying ecological and biogeochemical dynamics at the watershed scale, pioneered at Hubbard Brook, has been used to examine and elucidate a number of ecosystem processes, including ones with major management implications. In this chapter, we consider the impacts of forest harvesting (and deforestation) on nutrient dynamics and water quantity and quality.

Figure 11.1. View looking upslope from the V-notch gauging weir in the deforested Watershed 2 during summer 1967, after all of the trees had been cut down and left in place. The small building at left is the gauging station for the weir. (Photo by G. E. Likens)

WATERSHED MANIPULATION EXPERIMENTS AT HUBBARD BROOK

The first experimental watershed manipulation focused on effects of deforestation, with no timber products removed, on water quantity and quality. In general, we wanted to know what were the effects of such a major forest disturbance on the amount and quality of water exiting a watershed, and how such a major disturbance would affect nutrient dynamics. Several watershed manipulations have now been completed at Hubbard Brook and in the surrounding White Mountains to test hypotheses relative to these questions, particularly regarding forest harvest by clear-cutting.[2]

For this experiment, the forest on Watershed 2 was completely deforested (trees cut but not removed from the site) during the winter of 1965–1966 and then treated with herbicides during the summers of 1966, 1967, and 1968 to inhibit regrowth of vegetation (fig. 11.1). The design of this first experimental manipulation was to maintain deforested conditions for three years, not to replicate some forest harvest scheme. The trees were cut with chainsaws, and no wood products were removed, which means all vegetation was left on the ground inside the watershed. Care was taken to avoid disturbance to the stream channel, including not felling trees into it. After the herbicide applications, the vegetation was allowed to recover, beginning in 1969.[3]

In 1970, a second watershed experiment was initiated in Watershed 4. In this manipulation, trees were commercially harvested, again with chainsaws,

Figure 11.2. View of cut watersheds on the south-facing slope of the Hubbard Brook valley in summer 1984. From left to right: Watershed 5 (whole-tree harvested watershed), where the logging road and skid trails appear as white lines; Watershed 4 (strip-cut harvest with a buffer strip of uncut trees along the drainage stream; and Watershed 2 (deforested watershed). The small clearings in the forest outside the experimental watersheds contain meteorological stations. (Photo from Northern Research Station, U.S. Forest Service)

in a series of strips along elevational contours, from the gauging weir to the top of the watershed. Over a five-year period starting in 1970, strips 25 m wide were harvested, leaving 50-m strips uncut. In 1972, adjacent 25-m strips were harvested, and in 1974 the final 25-m-wide strips were harvested. At this point, all trees on the watershed had been cut and the tree boles

removed. A narrow buffer strip of trees was left uncut along the stream channel, partly to provide shade to the stream but mostly to test whether this buffer would reduce the loss of nutrients moving in drainage water to the stream (fig. 11.2). After this progressive strip, clear-cutting approach on Watershed 4 was completed in 1974, the forest on the watershed

Figure 11.3. The whole-tree harvest experiment on Watershed 5 used this feller-buncher for cutting trees and stacking them into piles. Once cut, the boles, branches, twigs, and foliage (when present) were all fed into a chipper, and then the chipped wood products were trucked to a local mill, giving meaning to the term whole-tree harvest. (Photo from Northern Research Station, U.S. Forest Service)

was allowed to recover and regrow without further disturbance.

Watershed 5 was commercially harvested as a whole-tree logging operation in the winter of 1983–1984, under contract with a logging firm. Large machines called feller-bunchers cut the trees off at the base of the bole with large pincers, and placed these entire trees into piles on the watershed (fig. 11.3). A forwarder was used to drag the trees to a chipper at the base of the watershed, where the entire tree was chipped. The chips were then hauled to a paper mill in large tractor-trailers. After this whole-tree harvest, the forest was then allowed to regrow.

Watershed 2 was planned as an experiment relative to deforestation, whereas Watersheds 4 and 5 were experiments built around variations in commercial harvesting approaches to test management questions. For example, could the environmental impact of clear-cutting be reduced while still harvesting timber (Watershed 4)? What were the ecological and biogeochemical impacts of removing all aboveground forest material (Watershed 5), an increasingly common practice at that time in New England?

THE WATERSHED 2 DEFORESTATION EXPERIMENT

The trees on Watershed 2 were cut with chainsaws during the winter of 1965–1966, with the timber and debris left where they fell (fig. 11.4). The following summer, the herbicide bromacil was applied aerially to prevent regrowth of root sprouts of the dominant tree species and growth of seedlings, most of which were from buried seeds of pin cherry. During each of the following two summers (1967 and 1968), the herbicide 2,4,5-trichlorophenoxyacetic acid was added individually to scattered regrowths of stump sprouts to prevent regrowth of vegetation.[4]

Water response. We had expected an increase in streamwater yield from this experiment, because transpiration of water vapor would be dramatically reduced by the experimental deforestation. As it turned out, 1964–1965 was exceptionally dry, and 1967–1968 was very wet, which strongly justified the need for a paired reference watershed. Annual stream flow during 1967–1968 and 1968–1969 increased by 39% and 28%, respectively, over what we would have expected if the watershed were not deforested. The greatest increases occurred during June to September when

Figure 11.4. Felled trees on the ground in Watershed 2, spring 1969, after herbicides had been applied during the summers of 1966, 1967, and 1968, looking upslope from just above the gauging weir. Richard Holmes is standing at right. (Photo by C. J. Krebs)

stream flow was 414% (1966–1967) and 380% (1967–1968) greater than expected (fig. 11.5).[5] Runoff also started earlier after snowmelt during each of the three years of the deforestation.

Nutrient response. What was completely unexpected, however, was the large chemical increase, particularly of nitrate (NO_3^-), in stream water after deforestation (fig. 11.6). Nitrate concentrations increased sharply in June 1966, reaching a peak of more than 60 mg/liter in the fall of that year, declined somewhat over the winter, and then rose to values greater than 80 mg/liter during late summer of 1967. This result came as a great surprise to us and to the scientific community, and it also had important implications for public health and water management. The U.S. Public Health Service had posted a national standard stating that drinking water with more than 40 mg of nitrate per liter had undesirable health effects, particularly for pregnant women and infants.

Concentrations of hydrogen ion, aluminum, calcium, magnesium, potassium, and sodium in stream water also increased dramatically in response to the deforestation. The pH of stream water in deforested Watershed 2 dropped almost one full unit, from about 5.3 to about 4.3, or a tenfold increase in hydrogen ion concentrations (fig. 11.7). At these low pH values, aluminum in the soil dissolved and increased in stream water in concordance with the increasing acidity. The timing of the response of the base cation increase was calcium > magnesium > potassium > sodium (fig. 11.8).

Because nitrate became the dominant anion mobilized in response to the deforestation, it became the primary anion to combine with base cations in stream water exiting the system. Correlations between calcium and magnesium concentrations and nitrate concentrations in stream water were very high (0.90 and 0.86, respectively) following deforestation (fig. 11.9).

The initial response from the forestry community to the deforestation experiment was that the use of herbicides to suppress forest regrowth might have been inappropriate relative to normal forestry practices. Since Watershed 2 was done as

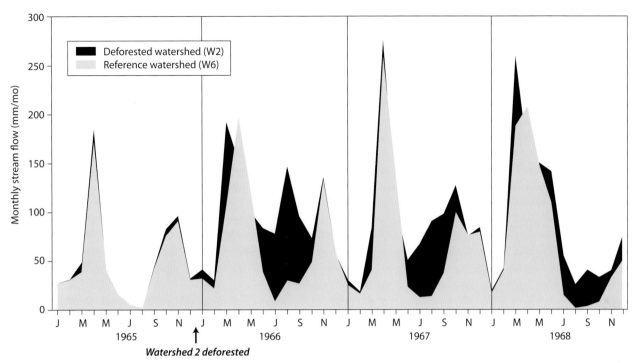

Figure 11.5. From 1966 through 1968, the stream flow in the deforested watershed started earlier after snowmelt and was greater in volume than in the reference watershed. (Modified from Likens et al. 1970)

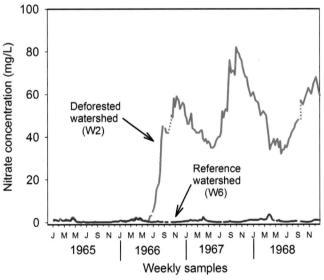

Figure 11.6. Nitrate concentrations in stream water exiting the deforested watershed increased dramatically in the winter of 1965–1966 after experimental treatment compared with the reference watershed. Concentrations of nitrate didn't increase until the growing season after disturbance, and then increased to even higher values in the second summer after deforestation. (G. E. Likens, unpublished)

an experiment and not as a commercial harvest, questions were raised about what would happen in a commercial forest harvest. In response to this concern, eight commercially clear-cut areas in the White Mountains, including the ungauged Watershed 101 at Hubbard Brook, were studied in the early 1970s to evaluate the effects of commercial harvest on water yield, nutrient loss, and site productivity.[6]

Watershed 101 was clear-cut in the autumn of 1970, and skid trails were constructed for the removal of all merchantable timber.[7] Increased nutrient export in stream water from the eight study areas in the White Mountains region ranged from 2% to about 50% of the value found from Watershed 2, but the pattern of loss was the same in all of the clear-cut watersheds: accelerated nutrient loss in stream water occurred for at least three years and peaked in the second year after cutting.[8] The harvesting of timber biomass (the merchantable logs) removed additional nutrients from these clear-cut areas. In the Gale River clear-cut in the White Mountains, in the first two years after

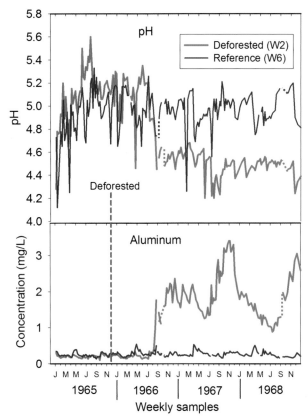

Figure 11.7. The pH and dissolved aluminum concentration in stream water leaving the deforested and reference watersheds before and after experimental cutting. (Modified from Likens et al. 1970)

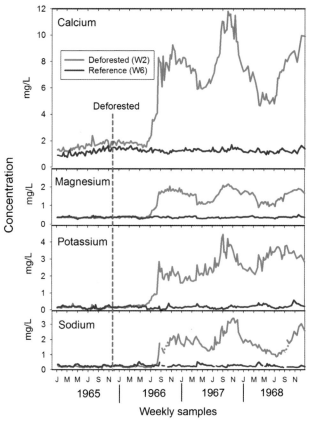

Figure 11.8. Concentrations of calcium, magnesium, potassium, and sodium ions in stream water of the deforested watershed before and after experimental cutting, compared with those in stream water of the reference watershed. (Modified from Likens et al. 1970)

deforestation, two times more calcium and about the same amount of nitrogen were lost in combined stream water and logging materials than were lost from Watershed 2 in stream water during the first two years following deforestation.[9]

Conversely, calcium is retained as a result of decomposition. Long-term research at Hubbard Brook has shown that calcium content increases in decaying wood, such as tree boles, over time. Thus, accumulations of decaying wood may help prevent loss of critical nutrients from the forest ecosystem.[10] These results have management implications for whole-tree harvest, where all aboveground forest materials are removed in a logging operation, and in salvage logging following major disturbance from wind storms.

The magnitude of chemical response in the Watershed 5 and especially the Watershed 4 experimental harvests was much less than in Watershed 2, but again the patterns were the same. As a result, it is possible to interpret the response of the water cycle and biogeochemistry in all of these experiments to the major disturbance of the forest ecosystem, whether by deforestation or by clear-cutting.

We found that major disturbance to the forest effectively places a block on the nitrogen cycle by eliminating the uptake of ammonium (NH_4^+) and nitrate by green plants (fig. 11.10). Instead, nitrogen incorporated in the dead organic matter of the ecosystem was decomposed (mineralized) to

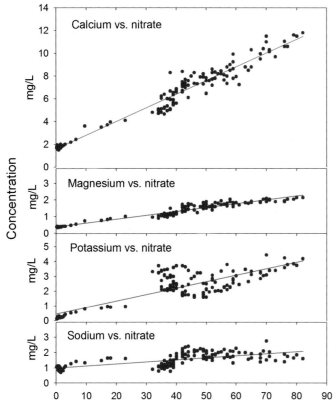

Figure 11.9. Concentrations of calcium, magnesium, potassium, and sodium ions were strongly correlated with nitrate concentrations in stream water on Watershed 2 following deforestation. (Modified from Likens et al. 1970)

ammonium, nitrified to nitrate, and lost in stream water. Because clay and especially organic matter in the soil tend to be negatively charged, positively charged ammonium is attracted and held in the soil, whereas nitrate becomes mobile for transport in water. Hydrogen ions are produced in the nitrification process and are also lost in stream water or exchanged for cations, such as calcium, magnesium, sodium, and potassium in the soil, which are then lost in stream water in combination with nitrate.

Major forest disturbances such as clear-cutting or deforestation set in motion a series of ecosystem reactions leading to increased stream flow, modest increase in erosion, loss of particulate matter, such as leaves and branches, and greatly accelerated loss of dissolved nutrients. Nitrogen is mobilized by

mineralization, or microbial conversion of organic matter containing nitrogen to the inorganic form (ammonium), and by nitrification, where ammonium is converted to nitrite and nitrate (fig. 11.11). This pattern was common to all of the experimentally manipulated watersheds and to commercially clear-cut watersheds studied in the White Mountains.

Using stable isotopes, it was possible to track some of these complex biogeochemical interactions of nitrogen that occurred following clear-cutting. Researchers found that greater amounts of isotopic nitrogen in leaves coincided with elevated nitrate in stream water, indicating that the increased nitrification that occurs after clear-cutting may also cause an enrichment of plant-available ammonium in the soil.[11]

These results raise the question whether natural disturbances such as soil freezing (which damages fine roots) and ice storms (which damages aboveground branches and twigs) would produce similar results. Long-term data available from Hubbard Brook help answer this question. In response to major disturbances such as soil frost and ice storms for watersheds that also had been experimentally manipulated (Watersheds 1, 2, 4, and 5), nitrate concentrations tended to increase sharply, and these increases were all greater than in Watershed 6, the reference watershed (fig. 11.12).

The magnitude of these responses is variable and complex. For example, there was only minimal response of nitrate concentration in stream water in Watershed 5 to the ice storm or soil frost. Why these deforested or clear-cut watershed ecosystems respond so differently to subsequent disturbances is the subject of intense current research and discussion among the scientists at Hubbard Brook. More research is needed to sort out the complex interaction and legacy effects of various disturbances and various intensities of disturbances on the biogeochemistry of these watershed-ecosystems.[12]

The overall net retention of nitrogen by the unmanipulated watershed-ecosystems at Hubbard Brook is large and consistent, even after major natural disturbances, which tend to mobilize nitrate

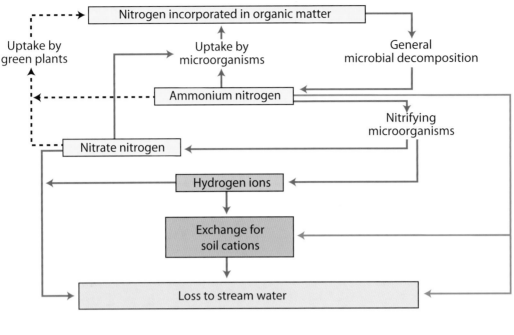

Figure 11.10. Partial effects of a major disturbance, such as deforestation or clear-cutting, on the cycling and flux of nitrogen in the forest ecosystem at Hubbard Brook. (G. E. Likens, unpublished)

——— Accelerated by treatment
- - - - - Blocked by treatment
——— Relatively unaffected by treatment

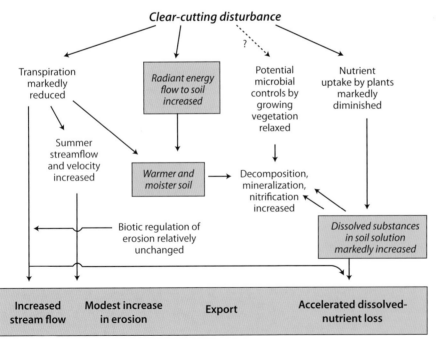

Figure 11.11. Effects of major forest disturbance, such as clear-cutting, on the water and nutrient dynamics of a northern hardwood forest ecosystem. (Modified from Bormann and Likens 1979; reprinted by permission of Springer Science+Business Media)

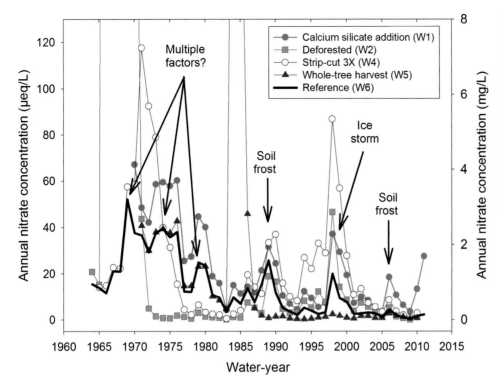

Figure 11.12. Nitrate response in stream water in relation to different kinds of disturbances such as deforestation, clear-cutting, soil frost, ice storm, and caterpillar defoliation in several experimentally manipulated watersheds and in the reference watershed. (G. E. Likens, unpublished)

as explained above. Currently, with loss of nitrate in stream water being so small, the net retention of nitrogen in the unmanipulated Watershed 6 ecosystem is more than 90%, primarily because of aggressive uptake and storage of nitrogen by plants and microorganisms. When this biotic function is weakened or destroyed, such as by clear-cutting, retention of vital nutrients, such as nitrogen, is greatly diminished.

There is no lack of ideas (hypotheses) to test about the function of entire watershed-ecosystems at Hubbard Brook. For example, do individual tree species differ in their responses to disturbance? Or what would be the response of excluding all large herbivores from a watershed-ecosystem? Nevertheless, although powerful as a scientific tool, entire-watershed manipulations are expensive and require the long-term dedication (at least 10 years) of large areas suited for such studies.[13] Also, the number of gauged, paired watersheds available in the Hubbard Brook valley is now limited, making it difficult to do additional whole-watershed experiments.

∎

The results of these watershed-scale experiments revealed previously unknown effects of major disturbance on nutrient cycling and flux, which in turn raised questions about the environmental consequences of harvesting northern hardwoods by clear-cutting. Research at Hubbard Brook suggested that long-term environmental effects could be minimized by reducing the size of areas cut to maximize the biotic potential for recovery; not cutting the same area more often than once every 75 years, to allow the system to recover nutrient capital before cutting again; protecting stream channels from disturbance; and minimizing physical logging disturbance by avoiding very steep slopes and wet conditions.[14] Currently, forest harvest by clear-cutting has fallen out of favor in northern hardwood forests because of environmental and aesthetic concerns. When clear-cutting is implemented, the U.S. Forest Service recommends cutting no more often than once every 100–120 years.

HOW DOES THE FOREST ECOSYSTEM RECOVER AFTER HARVESTING AND OTHER DISTURBANCES?

Major catastrophic disturbances, such as a powerful hurricane or an extensive forest harvest, severely disrupt the forest ecosystem through removal or damage to trees and other vegetation (fig. 12.1). As illustrated by the whole-watershed experiments at Hubbard Brook, deforestation and clear-cutting result in reduced transpiration, increased stream flow, increased loss of dissolved nutrients, and more erosion.[1] How do ecosystems recover from such events and what mechanisms are involved? Answers to such questions are important to inform managers about ecologically appropriate forest harvesting rotations and other management practices.

ECOSYSTEM RECOVERY FROM DISTURBANCES

The recovery of forest ecosystems after disturbance has been studied extensively at Hubbard Brook.[2] This process involves rapid regrowth of vegetation, successional turnover in plant species composition, and changes in ecosystem properties such as decomposition, net primary productivity, and biogeochemistry.[3] The recovery process can be divided into four phases, which reflect the life history characteristics of the organisms involved (mostly plants) and the prevailing environmental conditions (Table 12.1, fig. 12.2).

In the reorganization phase immediately following the disturbance, fast-growing and opportunistic plants, such as raspberry, ferns, and asters, along with tree seedlings, become established and ecosystem (plant) biomass is low. Although forest biomass increases steadily over the long term, there is a net loss in total biomass (both living and dead organic matter) in the first 15 years (fig. 12.3). This loss is due to an imbalance between decomposition of organic matter in the forest floor and deadwood components and the growth (net accumulation) of new living biomass. Early in this successional process, decomposition is high because of higher temperatures reaching these unshaded areas, but eventually the nutrients released by this process become available to rapidly growing plants, especially pin cherries. The ratio of dead to living biomass changes throughout ecosystem

Figure 12.1. Robert S. Pierce surveying the effects of the whole-tree harvest on Watershed 5. The large machinery left skid tracks on the forest floor, even though most of the work took place in the winter. (Photo by G. E. Likens)

Table 12.1. Developmental Sequence of the Northern Hardwood Forest Following Catastrophic Disturbance, Such as Major Hurricanes or Clear-Cut Harvesting

Time since major disturbance	Successional events and dominant species
2–3 years	**Reorganization phase.** Raspberry and young pin cherry from the seed bank in the soil are the dominant species, along with a mix of grasses, ferns, and herbs.
6–80 years	**Aggradation phase.** Characterized early by dense pin cherry, which, after 25–35 years of regrowth, senesce and die. Thereafter, mid-successional species such as red maple, quaking aspen, and paper birch are released and grow. Between 35 and 80 years, yellow birch predominates, with a mix of sugar maple and American beech; as shade-tolerant species, these grow well under the dense forest canopy.
80–100 years	**Transition phase.** Slower-growing and more shade-tolerant species such as sugar maple and American beech begin to replace the mid-successional species. Yellow birch persists as a major dominant tree in the forest canopy.
100+ years	**Shifting-mosaic steady-state phase.** A mix of yellow birch, American beech, and sugar maple is maintained by disturbances, which create gaps allowing mid-successional species to become established and persist. Red spruce and balsam fir replace sugar maple and beech at higher elevations.

Sources: Bormann and Likens 1979; Van Doorn et al. 2011

149

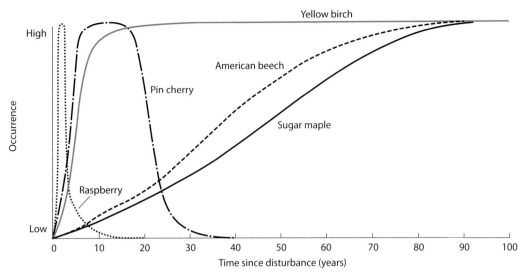

Figure 12.2. Turnover of major plant species in the successional sequences after a major disturbance in the northern hardwood forest. (Modified from Marks 1974)

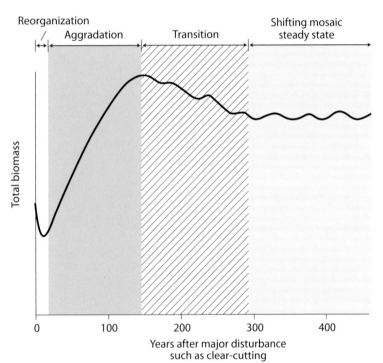

Figure 12.3. Model indicating hypothetical changes in biomass (living plus dead organic matter) of components of the northern hardwood forest ecosystem following major disturbance, such as caused by hurricane or a clear-cut harvest. (Modified from Bormann and Likens 1979; reprinted by permission from Springer Science+Business Media)

development, with dead biomass predominating in the first third of this period.[4]

Changes in plant species composition continue to occur, with the slower growing, shade tolerant tree species gradually replacing the initial fast-growing plants and becoming dominant in the forest over relatively long periods or until some major new disturbance occurs. Small disturbances such as those caused by wind or ice storms continue to occur periodically, creating gaps of various sizes where younger trees are growing. This late stage consisting of shifting patches of old and young trees represents the steady-state mosaic. Ecosystem biomass in this final stage is hypothesized to fluctuate around a long-term mean, but as we have seen, various environmental perturbations such as atmospheric pollutants, acid rain, and disease may affect forest biomass and the level where it plateaus or fluctuates over the long term.

At Hubbard Brook, the patterns of recovery of the aboveground living biomass on the experimentally deforested watersheds have been remarkably similar and conformed closely to expectations from the model in fig. 12.3. Initial vegetation growth was rapid, and biomass accumulated at a rate of about 2–3% per year, with the more mature forest declining in rate of biomass accumulation as it became older (fig. 12.4). Thus the vegetation on the experimental watersheds

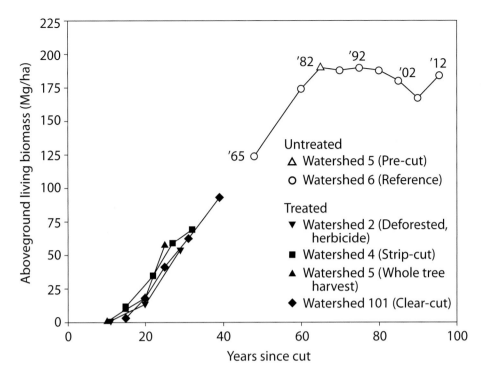

Figure 12.4. Recovery of aboveground live biomass (trees greater than 10-cm diameter at breast height) on five experimentally deforested or harvested watersheds at Hubbard Brook. Solid symbols indicate forest growth following experimental treatment; open symbols are measurements from untreated (older) forests. (Modified from Campbell et al. 2013)

recovered at similar rates irrespective of the type of disturbance or experimental treatment.

THE ROLE OF PIN CHERRY IN FOREST RECOVERY

A look around the forest on Watershed 2 in summer 1965 (before its experimental deforestation) showed essentially no pin cherries, but in the following summer, after it had been deforested, large numbers of pin cherry seedlings appeared and began to grow rapidly. Because the experiment was designed to suppress regrowth for three years, these seedlings were then killed by an aerial application of bromacil, a relatively new herbicide at the time. In each of the following two summers, new crops of pin cherry seedlings also appeared, and were killed this time with a different herbicide, 2,4,5-trichlorophenoxyacetic acid. And then, in 1969, yet more pin cherry seedlings appeared and began to grow on the deforested watershed. Where did all of these seedlings come from? What role do they play in the recovery of these northern hardwood forests?

In this case, the pin cherry seedlings grew from seeds that had fallen to the ground or been dispersed by animals during a previous flush of pin cherry growth. For Watershed 2, this must have been after the major cutting of the forest in the early 1900s, or perhaps for some local areas, after the 1938 hurricane. Pin cherry seeds buried in the forest litter remain viable for long periods of time, often 50 years or more.[5] They then germinate quickly after a disturbance, such as logging or the creation of large gaps due to strong winds, and within one to two years they become the tallest and most abundant tree species in these disturbed areas (figs. 12.5 and 12.6). By 35 years after a major disturbance, most of the pin cherries trees have died, eventually disappearing from the developing forest. The species, however, remains in the ecosystem as seeds buried in the soil until the next major disturbance. The environmental factors causing pin cherry seeds to germinate under these conditions are not well understood, but presumably changes in soil temperature, light or nutrient levels associated with the formation of a large gap in the canopy, or a major disturbance such as a clear-cutting are responsible.[6]

The rapid growth of pin cherry after a large

Figure 12.5. Pin cherry growth on Watershed 2 following deforestation that took place in 1965–1966. Upper photo shows a view down across the watershed in 1972. Lower photo is a view upslope from the weir in 1984; a view of the watershed following the cutting is shown in figure 11.1. (Photos by W. A. Reiners, G. E. Likens)

disturbance results in the uptake of nutrients such as potassium, calcium, and especially nitrogen, which otherwise would be lost from the system in stream water. Fast-growing plants, such as pin cherry and raspberry, take up and transpire liquid water, thereby reducing erosion and nutrient loss; provide dense shade and cooling to the forest floor, reducing decomposition rates; and incorporate large quantities of nutrients in the production of new biomass.[7] Thus, the presence of such plants that can respond quickly to a disturbance serves an important role in forest recovery by regulating ecosystem nutrient loss.[8]

The important role of pin cherry is made possible in part by its production of large numbers of fruits that are dispersed widely by birds or cached by small mammals, distributing and spreading them throughout the surrounding forested landscape. In some parts of the undisturbed forest at Hubbard Brook in the early 1970s, seed density was estimated at up to 494,000 per hectare, or about 5,000 per square meter![9] Although consumed by small rodents, many of these seeds persist in the soil for long periods of time after production of fruits at a site ceases. Although about 40% of the seeds become inviable within just a few years of production, many can still sprout 50 or more years later, and thereafter their viability declines. By 120 to 130 years, only a few remain viable.[10] It is remarkable that not all pin cherry seeds germinate at the same time, allowing for survival of some until a future disturbance occurs. The mechanism underlying this phenomenon is not known.

Now, approximately 50 years after the deforestation and herbicide experiment, only the rotting remnants of a few pin cherry boles remain in Watershed 2, but the buried seeds dropped from those trees during their 35-year presence on the watershed await their time to germinate after the next disturbance.

The size and intensity of the disturbance, such as a single treefall or a large wind storm, which make gaps in the canopy, and land-use history can determine the initial vegetation response to disturbance. For example, raspberries, blackberries, and ferns are likely to dominate in relatively small openings or on

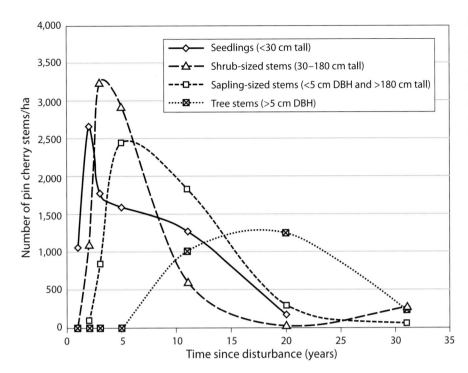

Figure 12.6. Growth stages of pin cherry on Watershed 2 in years following clear-cutting and then three years of herbicide application. (Modified from Reiners et al. 2012; reprinted by permission of Springer Science+Business Media)

previous agricultural land, whereas pin cherries may dominate in larger openings in forested areas.

APPLICATION OF THESE RESULTS FOR FOREST PRACTICES

How frequently can a forest be harvested without seriously degrading the site? This is a major question for managers and property owners concerned about the long-term sustainability of a forest ecosystem or woodlot. The results of our harvesting and deforestation experiments address this difficult question and can provide guidelines for achieving intelligent harvests of a northern hardwood forest.

Critical for the development of such guidelines is to consider the nutrient economy of the forest site, particularly for nutrients with limited availability, such as calcium, phosphorus, and nitrogen. As discussed earlier in this book, nutrients are continually being added to the watershed-ecosystem from meteorologic vectors such as rain and snowfall, and by weathering release. And they are lost by geologic and meteorologic vectors—for example through streamwater drainage or gaseous loss. In addition,

when a forest is harvested, large quantities of nutrients are removed from the site in logging products.[11] All of these fluxes must be accounted for in a budgeting analysis to determine whether the system is gaining or losing or neutral relative to the available nutrient capital for the ecosystem.

Nutrients can be lost or gained after a clear-cut harvest (fig. 12.7). Nutrient loss is high immediately after harvest because of the removal of wood products, followed by an increased loss of dissolved and particulate matter in stream water and in volatile compounds to the atmosphere. With the rapid reestablishment of vegetation, which takes up nutrients and stores them in biomass, the forest begins to recover and indeed exhibits high rates of primary productivity in these early growth stages.[12] Overall, nutrient recovery appears to take 60 to 80 years. This rate suggests that these northern hardwood forests can be sustainably harvested at 80–100 year intervals when proper forest management practices are followed. Harvesting at more frequent intervals is likely to deplete nutrient reserves and severely curtail the ability of the forest to recover.[13]

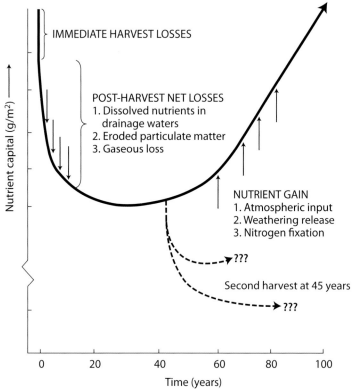

Figure 12.7. A conceptual model of ecosystem recovery from harvest disturbance based on research at Hubbard Brook. Nutrient recovery through natural processes may take at least 60 to 80 years. More frequent harvesting is projected to lead to a major loss of nutrients and slower recovery. (G. E. Likens, unpublished)

These findings lead to the following suggestions for improving forestry practices involving clear-cutting of blocks of northern hardwoods forest.

- Cutting should be limited to sites with strong recuperative capacity—those with modest slopes and more fertile soils.

- Blocks of forest to be cut should be small enough to minimize excessive amounts of stream flow, dissolved nutrients, and eroded material loss and to allow reseeding of early successional species, such as aspen and ash, from adjacent uncut forest areas. The actual size will depend on tree species present, slope, and other site characteristics.

- A logging operation should minimize damage to the forest floor and stream channels by logging in winter when soils are snow covered, strategically locating logging roads and skid trails, and protecting and minimizing impact on associated aquatic ecosystems, such as streams or lakes.

- Noncommercial ecologically valuable species, such as pin cherry and raspberry, should be protected, because they are important in the recovery of the forest, and they are a source of food and shelter for wildlife.

- Finally, operations should allow sufficient time between cuttings for the system to regain nutrient capital through natural processes, allowing for sustainable use of forest resources.[14]

Partly as a result of these findings at Hubbard Brook, the cutting rotation time now recommended for the White Mountain National Forest is 100 to 120 years. But clear-cutting blocks of forest is now less common in the northern hardwood forests of New England than it was 30 to 40 years ago. Newer forest management practices, such as ecological forestry, are now being used more frequently, retaining more ecosystem functions and providing more services from forest ecosystems.[15]

13

HOW STREAM ECOSYSTEMS ARE INTEGRATED WITH THEIR WATERSHEDS

Many scientists and engineers have often thought of streams and rivers simply as conduits (such as inert Teflon pipes) for the transport of water and various compounds. They are, however, ecosystems in their own right, with all of the interacting components and functions of any other ecosystem. As open systems, stream ecosystems receive energy from solar radiation and from organic matter input from surrounding terrestrial ecosystems. Researchers at Hubbard Brook were among the first to construct a comprehensive annual energy budget for a headwater stream (Bear Brook), and it was found to be much more interesting and complicated than a Teflon pipe (fig. 13.1). One interesting and unexpected discovery was that the living components of this stream ecosystem were depauperate of algae, an important autotroph in many aquatic ecosystems. Despite extensive searching and disbelief by fellow scientists, it was determined that algae were indeed exceedingly rare, if they existed at all, in headwater stream ecosystems at Hubbard Brook.[1]

Then, in the mid-1980s, algae were found in the gut contents of stream invertebrates (researchers had looked for algae in gut contents earlier). By the mid-1990s, mats of algae were commonly found on rocks and boulders in the headwater streams at Hubbard Brook (fig. 13.2), and currently blooms of filamentous algae are relatively common during the short period between snowmelt and canopy leaf-out, the so-called vernal window. The potential drivers for this dramatic long-term change in algal presence and abundance are likely the changing light conditions as the forest canopy has become less dense (allowing more sunlight to reach the stream channel) and the changing chemistry of stream water (less acid). But no cause and effect tests have yet been conducted, and there may be other possible explanations for this dramatic long-term change. Solving this puzzle—why stream algae have appeared and then increased during this long period—is the subject of much current discussion and research at Hubbard Brook. This change is particularly important, because it represents a state change, from a heterotrophic ecosystem (where

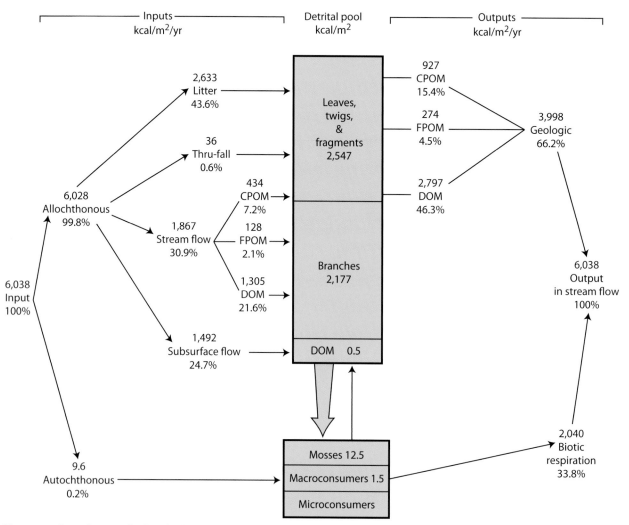

Figure 13.1. Annual energy budget for Bear Brook, a tributary of Hubbard Brook, in the late 1960s, showing major inputs and outputs of energy in kcal/m²/yr and mean organic matter storage in the stream channel. The stream ecosystem has a complicated energy budget and is dominated by inputs of organic matter from the adjacent terrestrial ecosystem. CPOM is coarse particulate organic matter (> 1 mm in size), FPOM is fine particulate organic matter (< 1 mm), and DOM is dissolved organic matter. Particulate matter is mostly fragments of leaves, bark, branches, etc. (Modified from Fisher and Likens 1972; reprinted by permission of Oxford University Press)

energy is obtained from ingestion of organic matter) to at least a partly autotrophic ecosystem (where energy is captured directly from solar radiation or oxidation of inorganic chemical compounds by organisms). The causes of such state changes are of great interest in ecology and biogeochemistry.[2]

The annual energy budget for Bear Brook compiled in the late 1960s revealed that organic matter produced outside the stream ecosystem (referred to as allochthonous) contributed more than 99% of the energy for this stream ecosystem at Hubbard Brook. Primary production by mosses, the only autotroph present in the early years, provided less than 1% of the energy input to Bear Brook, making the stream ecosystem highly heterotrophic. Most (90%) of the forest litter falling into the stream channel (1,200 kg/yr) is decomposed in place primarily by fungi and bacteria.[3]

Figure 13.2. Filamentous algal growth on rocks in a headwater stream at Hubbard Brook, May 7, 2013. Algae were absent from these streams when the Study began, but have appeared and increased during the past 30 years. (Photo by D. C. Buso)

WHAT HAPPENS WHEN AN ENTIRE STREAM IS EXPERIMENTALLY MANIPULATED?

With this understanding of the energetics of a stream ecosystem as a framework, a relatively large number of whole-stream experimental manipulations have been conducted at Hubbard Brook to investigate various components of stream ecosystem structure and function, and to understand better the integration of a stream ecosystem and its terrestrial drainage basin (Table 13.1).

We now briefly describe four case studies (acidification, alkalinization, organic debris dam removal and replacement, and movement and dispersal of stoneflies and salamanders) to illustrate the use of experimental manipulations of entire stream ecosystems and their components, and to provide some of the results from these studies at Hubbard Brook.

Experimental stream acidification. Following the discovery of acid rain and the realization that headwater streams were much more acidic than many stream organisms could tolerate, we wondered how the full assembly of stream organisms and stream food webs could function under these stressful conditions. Although the natural experiment (acid rain) was under way, we decided to experimentally manipulate entire stream segments in different ways to unravel the effects of acidification on stream ecosystem structure and function (Box 13.1). For example, acidifying (to pH 4) a headwater stream at Hubbard Brook from April to September 1977, revealed the following important effects of acidification on ecosystem structure and function:

(1) The stream water remained acid only from the point-source addition for several hundred meters downstream, because chemical reactions occurring within the stream ecosystem neutralized the acidity rather quickly. The acidifying effects thus extended far shorter distances downstream than expected, and the resilience of the stream ecosystem was greater than expected.

(2) The chemistry of stream water was changed

Table 13.1. Experimental Manipulations of Entire Stream Ecosystems at Hubbard Brook

Topic	Purpose of experiment	Principal references
Organic debris dams (removal and addition)	Test the functional role of organic debris dams in sediment transport and biogeochemical dynamics	Bilby 1979; Bilby and Likens 1980; Hedin et al. 1988; Warren et al. 2007, 2013
Add phosphorus	Test the functional role (nutrient limitation) of phosphorus in stream ecosystems	Meyer 1979; Meyer and Likens 1979; Meyer et al. 1981
Add nitrogen	Test the functional role (nutrient limitation) of nitrogen in stream ecosystems	Sloane 1979; Bernhardt 2001; Bernhardt et al. 2002; Richey et al. 1985; Meyer et al. 1981
Add organic carbon	Test functional role of dissolved organic carbon in the energetics and dynamics of stream ecosystems	McDowell 1982; Bernhardt and McDowell 2008; McDowell and Likens 1988
Add nutrients and particulate matter	Test effects of nutrient additions and particulate organic matter on microbial dynamics	Stelzer et al. 2003
Light and shade	Test effects of light (solar radiation) and shade on primary productivity in stream ecosystems	Thornton 1974; Hendrey et al. 1979
Acidification and alkalinization	Test the effects of acidification and alkalinization in stream ecosystems	Hall and Likens 1980, 1981; R. J. Hall et al. 1980, 1982, 1985; R. O. Hall et al. 2001; Likens et al. 2004a
Add stable isotopes (^{15}N and ^{13}C)	Test nutrient dynamics and invertebrate dispersal in stream ecosystems using stable isotopes to clarify nutrient pathways and to trace invertebrate movement.	Tank et al. 1998; McCutchan and Likens 2002; Macneale et al. 2005

Box 13.1. Experimentally Altering the Chemistry of a Headwater Stream

To understand better the consequences of acidification for stream ecosystems, an experiment was performed that involved acidifying a portion of a forest stream. For this, researchers continuously dripped dilute (0.05–1 N) concentrations of sulfuric acid into a small waterfall on Norris Brook from a manually operated valve on a large container from April 18 to September 20, 1977. Constant care and monitoring were required during this period to maintain a consistent pH of 4.0 (range 3.9 to 4.5) in the stream water below this addition point, as flows fluctuated with rainfall and drying events. The pH of the upstream reference segment above the drip point ranged from 5.9 to 6.4 during the study.

In another experimental manipulation, researchers added the stable isotope of nitrogen ^{15}N, using a peristaltic pump to drip a labeled solution of $^{15}NH_4Cl$ into the stream for 4 to 8 weeks at a point where mixing throughout the water column would be thorough. The rate of addition was adjusted to correspond to the discharge when that was desired or necessary. Similar procedures were followed in all of the nutrient addition experiments done in stream ecosystems at Hubbard Brook.

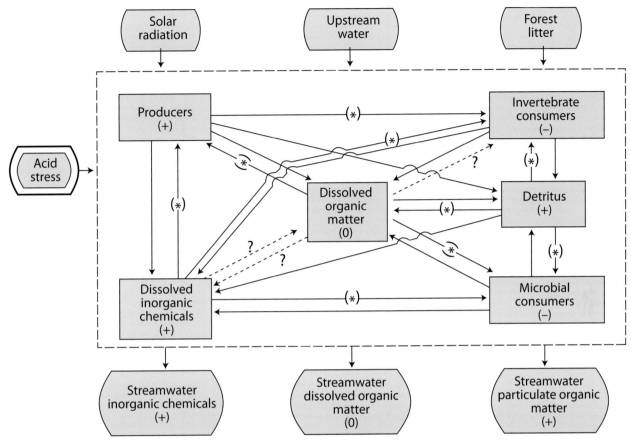

Figure 13.3. Effects of experimental acidification (acid stress) on the living and nonliving components of a stream ecosystem. Dashed line indicates ecosystem boundary. Inputs to the ecosystem are at the top, outputs at the bottom. Boxes represent biomass of each component; arrows represent the flux of inorganic and organic nutrients among the boxes. Asterisks indicate fluxes most significantly affected by addition of hydrogen ion. Plus signs represent an increase and minus signs a decrease in response; zero indicates no change. (Modified from Hall et al. 1980)

by the experimental acidification, including an average 181% increase in toxic dissolved aluminum concentrations, and an increase in calcium and several heavy metals, such as manganese and copper.

(3) Total phosphorus, which normally occurs in very low concentrations in stream water at Hubbard Brook, was mobilized from sediments during the experimental acidification.

(4) Changes in important ecosystem functions, including stresses on food-web dynamics, were observed. Stream invertebrates were depleted and export of dissolved inorganic solutes and particulate organic matter was increased as a result of the acidification (fig. 13.3).

(5) Drift of mayfly nymphs and black fly and midge larvae increased in the acidified segment. In general, macroinvertebrates drifting in the acidified section increased 3.9-fold on the first day and 13-fold on Day 3 relative to the reference area of the stream (fig. 13.4).[4]

Experimental stream alkalinization. As acid rain was leaching calcium from the soil at Hubbard Brook, we wondered what the role of stream ecosystems was in storing and exporting that calcium. To test the

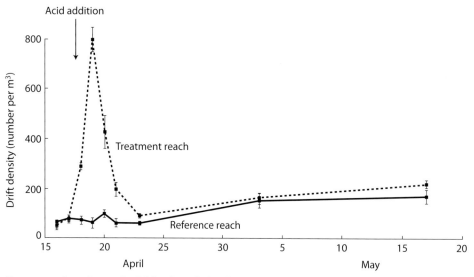

Figure 13.4. Experimental acidification of a headwater stream resulted in a dramatic increase in stream invertebrates moving downstream. Stream invertebrates often drift downstream when they are stressed, losing their position in the stream channel to the force of the current. The graph shows the total number of drifting invertebrates collected in nets per 24-h period from April 16 through May 17, 1977, for the experimentally acidified (dashed line) and upstream reference reaches (solid line) of Norris Brook. Closed circles represent the mean for two drift nets; vertical bars represent the range in the number of organisms collected. (Modified from Hall et al. 1980)

response of increased calcium and alkalinization in acidified headwater streams at Hubbard Brook, a solution of calcium chloride and sodium bicarbonate was dropped continuously into a headwater stream segment below the gauging weir on Watershed 3 for two months during 1997. A second stream segment below the confluence of Watershed 4 and Watershed 1 received only calcium chloride for the same two-month period. The target for the calcium increase in each stream segment was 120 μeq/liter higher than upstream reference areas. In the stream segment receiving both the calcium chloride and sodium bicarbonate, the bicarbonate was added as a buffer to raise the pH and alkalinity of the stream water. This approach was successful, as the pH in the buffered stream segment ranged from 5.6 to about 7.0 during the experiment. Some of the major results of this experimental manipulation included:

- Net uptake of calcium occurred in the stream ecosystem and was positively related to the pH of the stream water. The higher the pH, the greater the uptake of calcium. Over the two-month addition, about five times more calcium was taken up in the buffered segment than in the unbuffered stream.

- Leaves that were experimentally added to the stream decayed within weeks, but over that time, calcium concentrations in leaves in the buffered segment retained approximately twice the amount of calcium as leaves in the reference reach above the experimentally treated segment.

- After the experimental addition was stopped, calcium was slowly desorbed from the sediments of the buffered stream segment. Most of the added calcium was lost downstream in drainage water within a couple of months.

- These streams can be a significant landscape sink for calcium over short time periods.[5]

In a subsequent experimental manipulation, calcium silicate was added directly to a headwater

Figure 13.5. The field crew adding calcium silicate to a gridded section of stream below the gauging weir of Watershed 1 in June 1999. (Photo from G. E. Likens)

stream below the weir on Watershed 1, before Wollastonite was added to the entire watershed. Pelletized Wollastonite (calcium/silicate) was added evenly by hand to a 50-m segment of stream below the gauging weir in Watershed 1 during June 1999.[6] The Wollastonite was added quantitatively to the stream channel and adjacent bank using a grid (fig. 13.5). This collaborative research project had 11 senior scientists, postdoctoral associates, graduate students, and support staff. The amount of calcium added was about 50% greater than the amount calculated to have been depleted from watershed-ecosystems at Hubbard Brook during the previous 50 years by acid rain.

This one-time application of a calcium-rich mineral to a stream segment, which constituted only 5% of the entire stream above the experimental application, dampened extreme chemical concentrations in stream water and produced a relatively long-lasting, 100-day increase in streamwater pH and acid neutralizing capacity. Such results indicate that calcium/silicate additions may be an important management tool for restoring the acid buffering capacity (increase in alkalinity) of acidified surface waters.[7]

ORGANIC DEBRIS DAM REMOVAL AND REPLACEMENT

Organic debris dams are an obvious structural feature of stream ecosystems at Hubbard Brook, but we were curious about how these piles of logs and other organic debris, wedged behind boulders and damming the flow of water, might affect stream function. These dams form when large organic matter, such as the bole of a tree or a large branch falls across the stream and impedes flow (fig. 13.6). Other materials, such as

Figure 13.6. Organic debris dam (foreground), consisting of leaves being held in position by a log in a headwater stream. (Photo by D. C. Buso)

leaves, collect upstream of this framework and form an almost watertight structure with a pool upstream where sand, silt, and other fine sediments can settle and accumulate. As streams get larger in the drainage network, it takes ever larger trees to form and stabilize an organic debris dam against the eroding force of the current.[8]

Organic debris dams tend to occur at high frequency and at regular intervals of stream distance at Hubbard Brook: 20 to 40 dams per 100 m in first-order streams, 10 to 15 in second-order streams, and 1 to 6 in third-order streams (ordering of streams is described in Chapter 3).[9]

To examine the physical and functional relationships between stream channel characteristics and the

formation of debris dams, 24 organic debris dams were experimentally removed from a 175-m stretch of stream above the weir on Watershed 5 in 1977.[10] The results of dam removal were dramatic relative to stream ecosystem function, including the following ecosystem losses measured at the gauging weir.

- The output of dissolved organic carbon (less than 0.50 μm in size) increased by 18%.
- The export of fine particulate organic carbon (0.50 μm–1 mm in size) increased by 632%.
- Coarse particulate organic matter (greater than 1 mm in size) increased by 138%.

When large organic debris was experimentally added to streams where dams had previously been removed, organic debris dams formed in approximately the same locations where dams had occurred formerly.[11] Organic debris dams form when the primary structural member becomes wedged on some physical feature, such as protruding large rocks or boulders in the streambed. This supporting structure allows organic debris dams to persist during high flows. So, in hindsight, it is not too surprising that the dams reformed where they had been previously.

Based on the results of various studies of organic debris dams, it is possible to predict how dams would be lost and then reform in seriously disturbed watersheds, such as by clear-cutting of forest, and at different stages of forest recovery following disturbance. The size of trees, their strength in resisting strong currents during high flows, and when trees might start to fall into the stream channel is largely determined by the stage of succession of the forest (fig. 13.7).[12] Using these concepts and the understanding obtained from our many studies of organic debris dams in different situations and locations, it has been possible to predict the pattern of loss and recovery of these dams in the Hubbard Brook area after disturbance, for first-, second-, and third-order streams.

First-order streams are small and have much less waterpower to erode dams than larger streams, so smaller-sized organic materials can form permanent dams. Increasing stream order (size) requires larger and stronger trees to form and then fall to maintain permanent dams, but these trees are not abundant until later stages of succession. Critical times in this progression of redevelopment can be summarized as follows (numbers refer to those in fig. 13.7).

(1) Clear-cut logging of the drainage network, particularly along the stream channel, greatly reduces the maintenance and delays the reformation of organic debris dams.

(2) Slash from the logging enters the stream channel and causes an increase in organic debris dam formation, but this is only

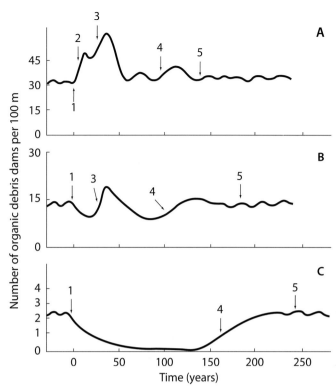

Figure 13.7. Expected changes in the occurrence and frequency of organic debris dams over time in first-order (A), second-order (B), and third-order (C) streams at Hubbard Brook (note differences in values on vertical axis). (Modified from Likens and Bilby 1982)

temporary because of the small size of the organic materials available to form the dams. The absence of overstory trees because of the clear-cutting greatly limits the availability of trees large enough to form dams.

(3) Falling pin cherry trees enter the stream channel when they begin to die and fall over some 40–50 years after clear-cutting, and form debris dams, but the small size of these trees and their inadequate strength limit the time they can persist as dams.[13]

(4) Mature late-successional tree species (sugar maple, American beech, and yellow birch) begin to fall across the stream channel, forming debris dams, and also forming bridges, but these eventually drop into the channel to form debris dams.

(5) Return to precut, steady-state condition. Based on this model, it would take about 100 years for organic debris dams to reform in a drainage network of this type following clear-cutting.[14]

The occurrence, function, and persistence of organic debris dams illustrate one of the strong interrelationships between terrestrial and aquatic ecosystems.

DISPERSAL OF A STREAM INSECT LABELED WITH A STABLE ISOTOPE OF NITROGEN

Streams form networks of aquatic habitats, but stream ecologists have wondered for nearly a hundred years how stream organisms maintain their position or migrate upstream against the pervasive downhill current in most streams. Likewise, to what degree are populations of stream organisms connected in these networks, and at what scale does dispersal occur? The distance, direction, and frequency with which animals move have consequences for the persistence and resilience of their populations, as well as for their ecological roles within stream networks. These questions have been studied in salamanders and in stream invertebrates at Hubbard Brook. We illustrate the complexity of this problem and the surprising dispersal patterns that can occur in stream invertebrates, using studies of a small stonefly (*Leuctra ferruginea*) in streams of the forest (fig. 13.8).

Researchers had known that larvae of this stonefly live for about ten months in a stream and feed on organic detritus, but little was known about the adults. Through painstaking observation, we discovered that both the females and males feed on terrestrial vegetation and that, surprisingly, this feeding in the adult stage is critical for egg maturation. Individuals are small (the forewing of a female stonefly is about 7 mm long), and they can be seen flying near streams, but how far and where they traveled was a mystery. What was the chance, for instance, that a female depositing eggs just below the water surface in one stream had grown up in an adjacent stream?

To have a reasonable chance of measuring where

Figure 13.8. Adult stoneflies (*Leuctra ferruginea*) at Hubbard Brook. (Photo by K. H. Macneale)

and how far adult stoneflies flew after emerging from their natal streams, it was necessary to mark in some way hundreds of thousands of individual stonefly larvae. The tool for doing this marking was a nonradioactive isotope of nitrogen (^{15}N), added to the stream, which would enrich the food resources of the larvae with this isotope and in turn provide us with an identifiable marker. These additions were timed so that the food resources would be enriched during at least the last month of larval growth (mid-June through late August). To validate whether the marking worked and to establish the location of the stream the labeled adults came from, adult stoneflies emerging from the stream in late summer and early fall were captured with a number of emergence traps resting on the stream sediments. Individual stoneflies captured while flying within the labeled area of the stream and those upstream from above the labeled area (reference area) were then analyzed for the presence of ^{15}N. Adult stoneflies could then be uniquely identified as being from the ^{15}N labeled reach or not, and their movements measured.[15]

Adults were collected along isotopically labeled streams, on transects through the forest, and along

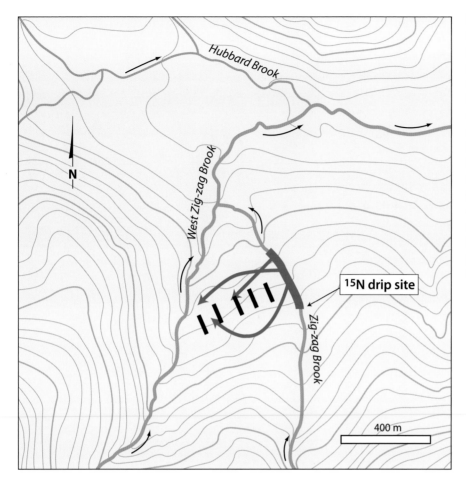

Figure 13.9. Results of an experiment to determine dispersal patterns of a stream-dwelling stonefly. A stable isotope of nitrogen (^{15}N) was added to Zig-Zag Brook, which labeled the detrital material in a section of the stream on which these insects feed (solid red area). The short black lines show the location of sticky traps used to collect flying adult stoneflies. Red arrows indicate the direction of movement between stream channels. (Modified from Macneale et al. 2005; reprinted by permission of John Wiley and Sons, © 2005)

an unlabeled tributary that flowed into the labeled stream (fig. 13.9) and analyzed to determine whether they contained ^{15}N. The results were surprising and ecologically important.[16]

- Of 966 individual adult stoneflies collected while flying, 20% were labeled with ^{15}N.
- Stoneflies had flown a mean distance of 211 m upstream. The greatest net distance a male and female had flown upstream was 730 m and 663 m, respectively.
- Mature labeled females were captured along an unlabeled tributary some 500 m from any labeled stream.
- The results demonstrate that the forest was not a barrier to dispersal between and among streams in different watersheds, and thus dispersal of

adults could link stonefly populations across a landscape within one generation.

MIGRATION AND DISPERSAL OF STREAM SALAMANDERS

Direct observation, particularly on dark, rainy nights, was used to discover movement patterns of stream salamanders along headwater streams and in adjacent terrestrial habitats. Despite the reputation for being sedentary, stream salamanders are capable of significant movement, but little is known about how these movements relate to large-scale patterns of dispersal and gene flow in networks of interconnected streams within a landscape. We conducted mark and recapture studies at Hubbard Brook to understand how the movement decisions of salamanders, along with the topography of headwater watersheds, influence dispersal along streams and resulting patterns of genetic divergence.

The northern spring salamander has an upstream bias in movement—most individuals of this species do not move far (tens of meters), but those that do move tend to head upstream (fig. 13.10). In contrast, the northern two-lined salamander has a downstream bias—those individuals that move tend to move downstream, with the flowing water. Using molecular techniques, we found that genetic differences between paired populations separated by one kilometer along different headwater streams increased with steeper watershed slope in the spring salamander, but decreased with slope in the smaller, two-lined salamander. These patterns were surprising, but consistent with predictions based on the observed movement behavior of the two species—an upstream bias in the spring salamander and a downstream bias in the two-lined salamander. This research provided new evidence about the importance of animal movement behavior to understand large-scale patterns of dispersal and genetic divergence in complex landscapes.[17]

∎

Research on stream ecosystems at Hubbard Brook has been foundational to stream ecosystem ecology and biogeochemistry, contributing significantly to the historic success of the overall Hubbard Brook Ecosystem Study. Although difficult to do in a natural setting and requiring much effort, experimental manipulations have helped us to understand better the structure and function of stream ecosystems and their integration with the terrestrial systems in which they are functionally and physically embedded. Such manipulations are particularly useful in identifying

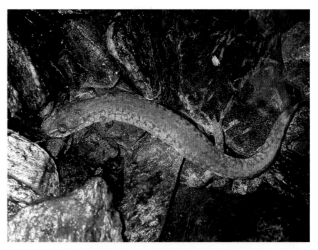

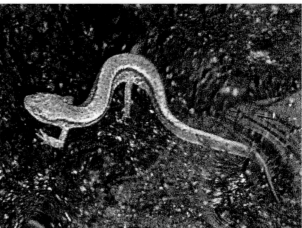

Figure 13.10. Two common stream-dwelling salamanders at Hubbard Brook, the northern spring salamander (top), and the two-lined salamander. (Photos by M. P. Ayres and D. C. Buso)

and testing the mechanisms that control function in these ecosystems. Future research is needed on the changing dynamics of attached algae, energetics, and chemistry in these streams, particularly as atmospheric and terrestrial inputs to these ecosystems are changing.

14

WHAT CAUSES POPULATION CHANGE IN FOREST BIRDS?

Birds are conspicuous, fascinating, and charismatic components of the forest ecosystem, and investigations began early in the Study to determine their roles in ecosystem processes.[1] Initially, it was important to know, as precisely as possible, how many birds of all species occurred on a given area of forest. To do this, an intensive monitoring procedure was started in 1969, and has continued during the breeding period (May through August) every year since. The result is one of the longest records of forest bird population trends in North America.

One of the surprising findings from this long-term record was an overall decline in the number of birds living within the forest, including dramatic declines for some species. Over the same period, however, the abundance of other species either remained stable or even increased. So why have some species declined and not others? Why the divergent patterns among species? And, more generally, what environmental factors and processes determine the numbers of organisms, birds in this case, that live in this forest ecosystem, and how do they operate to produce the patterns of abundance that we see in the forest?

The decline in bird numbers at Hubbard Brook first became evident in the late 1970s to mid-1980s, coinciding with a growing concern that many bird populations, especially migratory songbirds, seemed to be declining across eastern North America.[2] Since then, the possible causes and forces underlying changes in bird abundance have been hotly debated in both the scientific literature and the popular press (fig. 14.1).

Because the vast majority of birds breeding at Hubbard Brook are migratory and spend up to three-quarters of the year away from this area, many in tropical and subtropical regions, learning what was happening to these migratory species both in summer at Hubbard Brook and in the nonbreeding period became essential (fig. 14.2). Research from Hubbard Brook therefore expanded to include studies of migratory songbirds in their tropical winter grounds and, indirectly, during their migrations, something

Figure 14.1. Newspaper and magazine headlines from the 1980s and 1990s reporting apparent declines in the abundance of North American migratory songbirds. Tropical deforestation affecting these migratory species in winter was prominent among the suggested causes. (R. T. Holmes, unpublished)

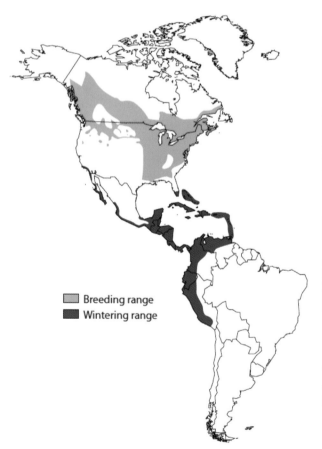

Breeding range
Wintering range

Figure 14.2. The American redstart, with a map showing its breeding and wintering ranges. Like many other species of migratory songbirds, the redstart breeds in North America and winters in tropical regions of the Caribbean, Central America, and northern South America. (Map modified from Sherry and Holmes 1997; photo by F. Jacobsen)

Box 14.1. How Are Birds Counted?

Multiple methods have been used to obtain quantitative estimates of the numbers of birds present in the forest each season. These included timed censuses, mapping of territories, finding nests (upper right), and capturing, color-banding (lower right), and then following individual birds.[a] The field map here shows territorial boundaries of black-throated blue warblers and their nest locations on one sector of the study area (grid scale = 50 m).

- Timed transect census
- Territory mapping
- Nest finding
- Locations of individually-marked birds

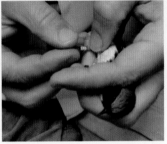

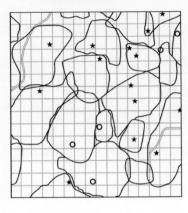

Key
★ Nest fledged
○ Nest failed

NOTE

a. Holmes and Sherry 2001.

Sources: Gridded map from R. T. Holmes, unpublished; photos by N. L. Rodenhouse

that had never been attempted previously on such a large geographic scale.

HOW HAVE BIRD POPULATIONS CHANGED SINCE 1969?

The longest continuous record of bird numbers at Hubbard Brook is from a 10-ha study plot adjacent to the lower part of Watershed 6. Using standardized census methods, we determined the number of breeding adult birds of all species actually feeding, nesting, and carrying out their lives within this circumscribed area of forest during the summer period (Box 14.1).

Since 1969, the total number of breeding adult birds on the study plot has varied from a high of 214 individuals in 1972 to a low of 71 in 2002 (fig. 14.3). Numbers increased in the first three years of the study, but then declined gradually through the late 1980s, rising briefly in the early 1990s before dropping again. Since then, the number of birds breeding on the plot has remained relatively stable at about 80–110 per 10 ha through about 2004, followed by a slightly increasing trend.

Looking at trends of species grouped by migratory status, it is clear that the Neotropical migrants (including flycatchers, warblers) have declined most. Species that migrate short distances in the winter (such as hermit thrushes and dark-eyed juncos) and those that are permanently resident (woodpeckers, nuthatches, chickadees) occur at relatively low abundances, and their populations have remained relatively constant over the 47 years of this study.[3]

Despite the general downward trend in the total number of birds on the study plot, trends of individual species have differed widely. Several species, such as least flycatchers and wood thrushes, declined sharply

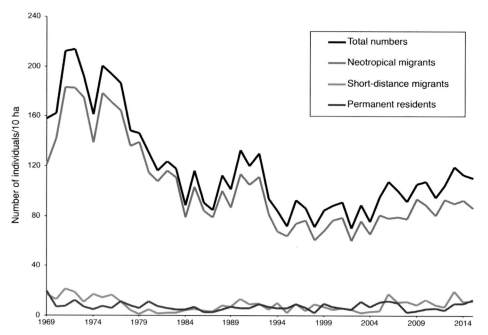

Figure 14.3. Total number of birds breeding on the bird study area during 47 consecutive years, 1969 to 2015. The black line shows the total number of individuals (males plus females) for all species combined. Neotropical migrants (red) show marked decline, while the numbers of short-distance migrants (green) and permanent resident (blue) species have remained relatively stable. (Modified and updated from Holmes 2007)

in the 1980s and 1990s, and in fact, these two species have disappeared from the study area. American redstarts reached a peak in abundance in the late 1970s, followed by a gradual decline through about 2010. No redstarts have bred on the study area since 2012. Declines in these three species, all Neotropical migrants, account for most of the drop in total bird numbers over the 47 years of study. In contrast, other species, such as the black-throated green warblers and yellow-rumped warblers have gradually increased in abundance, whereas the trends for the majority of species, including the black-throated blue warbler (fig. 14.4) have fluctuated some from year to year, but have remained relatively stable (fig. 14.5, Table 14.1).

These findings raised a number of questions: Why have some species decreased while others have increased? What causes the year-to-year differences in abundance for any given species? Why are the numbers of some species relatively stable in the long term? In an attempt to answer these questions, we developed a conceptual model, which was used to formulate hypotheses for testing and to guide the research on the ecology of these migratory species in both breeding and wintering areas (fig. 14.6).

HOW BREEDING SEASON EVENTS AFFECT BIRD NUMBERS

Research at Hubbard Brook has identified multiple factors that influence bird numbers in the forest, but the two most important are (1) the interaction of species-specific preferences for breeding habitat with the changing structure of the forest, and (2) the temporal and spatial variation in habitat quality (food availability, weather, predation risk) that determines how many young the birds can successfully fledge each year.[4] How do these influence the abundance of birds in the forest?

Habitat change and species-specific habitat preferences. Birds, like many organisms, have evolved preferences for specific habitats. Each species has a particular habitat or set of habitats that it prefers and where it survives and reproduces best. One finds meadowlarks in grasslands, for example, song sparrows in shrubby areas along forest edges, and red-eyed vireos in tall, more mature deciduous forests. In central New England, the plant communities develop and change predictably over time, from open land following recovery from agriculture or forest harvesting, to stages with shrubs and scattered

Figure 14.4. The black-throated blue warbler, a species closely associated with the northern hardwood forest, has been the subject of intensive study at Hubbard Brook for several decades. (Photo by F. Jacobsen)

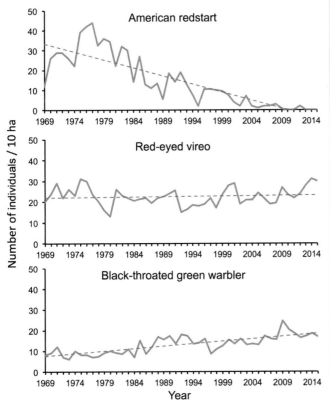

Figure 14.5. Population trends for three bird species at Hubbard Brook, 1969–2014. Some species (such as the American redstart) have declined in numbers, others have remained relatively stable (red-eyed vireo), and a few have increased in abundance (black-throated green warbler). (Modified and updated from Holmes and Sherry 2001; photos by F. Jacobsen, top and middle, and P. D. Hunt)

Table 14.1. Summary of Trends in Abundances of the 22 Most Common Bird Species Breeding on the Bird Study Area, 1969–2013

Declining	More or less stable	Increasing
Least flycatcher*	Yellow-bellied Sapsucker*	Blue-headed vireo
Wood thrush*	Downy woodpecker	Black-throated green warbler*
Veery*	Hairy woodpecker	Yellow-rumped warbler
Philadelphia vireo*	Winter wren	Ovenbird*
American redstart*	Hermit thrush	
	Swainson's thrush*	
	Red-eyed vireo*	
	Black-throated blue warbler*	
	Blackburnian warbler*	
	Scarlet tanager*	
	Rose-breasted grosbeak*	

* Neotropical migrants (species that winter at tropical latitudes)
Sources: Holmes and Sherry 2001; Holmes unpublished data

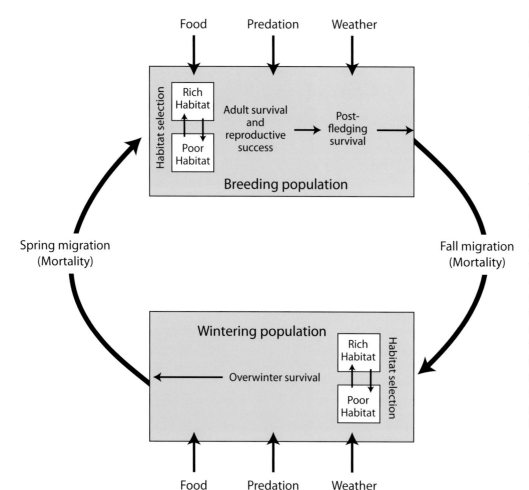

Figure 14.6. Conceptual model used to guide research on the factors and processes that affect migratory bird populations in breeding and wintering areas. The overall hypothesis was that, in summer, habitat quality (determined by food availability, predation, and the effects of weather) largely affects reproductive success, whereas in winter, habitat quality primarily influences survival. (Modified from Sherry and Holmes 1995; reprinted by permission of Oxford University Press)

tree saplings, and finally to mature, well-developed forests. As this successional change in vegetation and associated ecosystem components occurs over decades, the array of bird species living in these habitats also changes.

Although it is not always obvious to the casual observer, this successional change also continues to occur within what might appear to be already well developed forests. When the bird research began in 1969, the forest had been growing for 50–60 years after a major harvest in the early 1900s. The forest contained a mix of trees of different ages: some were older trees that had not been harvested during the logging period, trees that were too young to have been cut at that time, and some that had started to grow after the 1938 hurricane. In other words, it was a well-developed, mixed-age, second-growth forest, with a more or less continuous canopy of large trees and a distinct understory and shrub layer. By the late 1980s, this structure began to change as the large older trees died and fell to the ground, creating gaps in what had been a closed canopy layer. More sunlight reaching the lower strata of the forest resulted in increased growth of the understory shrubs and saplings. This process of gap formation, a natural feature of older forests, was accelerated at Hubbard Brook by the arrival of beech bark disease in the 1980s. Not only did the large beech trees die and fall over, but the large fallen trees also sent up numerous sprouts (suckers) from their roots. These young beech saplings formed a dense shrub layer, which lasted for 15 to 20 years.

These changes in canopy structure and in the distribution of foliage in different strata changed the suitability of this forest for the least flycatcher, American redstart, and wood thrush. Their numbers each spring gradually dwindled, as presumably returning birds moved on to settle elsewhere. The result was a marked decline in their numbers on the study plot. Other species, such as the black-throated green warbler, responded differently to these changes; their numbers have gradually increased over the 47 years of study. The changes in the abundance of these and other species occurred not only on the study plot but also throughout most of the Hubbard Brook valley and for some, regionally in north-central New England, as habitats along the gradient of forest succession changed throughout this entire region.[5]

These observations illustrate important links between these bird species and the physical and biological features of their environment. Changes in habitat structure as a forest matures affect food (insect) abundance and availability for birds, nest site availability, risk of predation, and other conditions that determine the suitability of that habitat for each bird species. Also very important is that each species responds differently to this changing environment depending on its morphology, physiology, behavior, and evolutionary history.[6]

These bird species-specific preferences for forest conditions, combined with a forest ecosystem changing through time, help to explain the major drop in total bird numbers at Hubbard Brook over the past 47 years. But what about the majority of the species living in the forest, which seem able to thrive over a wider range of conditions? The numbers for many of them have remained relatively constant over time. So other questions arise: Why do the populations of these species remain more or less stable? What are the mechanisms that limit or regulate their population sizes?

Breeding habitat quality and bird reproductive success. The size of a population at a given place and time results from the sum of the new birds produced through reproduction or immigrating into the area, minus the individuals lost due to death or emigration. Evaluating these processes requires information about demographic parameters such as reproductive success (fecundity, or the number of young successfully fledged per female per season), survival, and the environmental factors that affect them (Box 14.2).

Results from demographic studies of bird populations at Hubbard Brook indicate that the number of young fledged is a particularly important factor, one that may override mortality, indicating that what happens on the breeding grounds is of great importance in affecting the numbers of these

Box 14.2. How Are Bird Reproductive Success and Survival Quantified?

Determining how well birds reproduce each season (fecundity), and how many survive from one year to the next requires study of known individuals throughout each season and from year to year. Such information has been collected at Hubbard Brook for two songbird species, the American redstart and the black-throated blue warbler. The black-throated blue warbler has been the focus of intensive study for the longest time (now nearly 30 years), not only because it is relatively common in the forest, but also because it builds nests in shrubs low to the ground, allowing researchers easy access to nests. These warblers also feed mostly in the lower strata of the forest where foraging behavior can be readily observed and food availability measured.

Each year, all male and female warblers living within designated plots in the forest are caught and fitted with unique combinations of colored plastic leg bands, which allows observers to identify them by sight. Males are captured in mist nets, where they are attracted by the species' song played from loudspeakers, while females are captured in nets placed near their nests. When in the hand, birds are measured, weighed, and banded, and sometimes their blood is sampled for genetic, physiological, and isotope studies. They are then released, the whole process after capture taking no more than 10 minutes.

By observing these banded individuals during their daily activities, researchers obtain information about the size and characteristics of the territories defended by males, the locations and number of nests built by females, the number of eggs laid, the number of young fledged, the length of stay in the breeding area (over-summer survival), the return rates of individuals from one year to the next (annual survival), and the extent to which individuals returned to their territories of previous years (site faithfulness). Obviously this task of capturing, marking, and observing a large number

Female black-throated blue warbler being removed from a mistnet for banding and measuring. (Photo by S. A. Kaiser)

Female black-throated blue warbler provisioning nestlings with leaf hoppers. Females build nests in understory shrubs, usually hobblebush, at heights averaging about 0.5 m above the ground. This female has an orange band on its leg, which in combination with two to three other bands (not visible here) allow for identification of this individual. (Photo by N. L. Rodenhouse)

of birds (for black-throated blue warblers, this means about 30–40 pairs per year in the early years of the Study, 80–100 pairs or more yearly during the past decade) is a time-consuming and sometimes arduous task, involving the principal investigators, graduate students, and a large number of hard-working student field assistants.

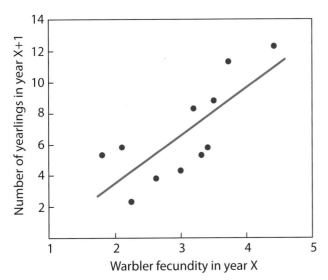

Figure 14.7. Fecundity (the number of young fledged per breeding season) of black-throated blue warblers in one year (year X) is strongly correlated with the number of yearling breeders in the next breeding season (year X+1). Thus, the reproductive success of this species in one year strongly influences the recruitment (numbers) of birds in the population in the subsequent season(s). (Modified from Sillett et al. 2000; reprinted by permission from AAAS)

long-distance migrant species. This conclusion is based on data largely from our studies of the black-throated blue warbler, showing that the number of new (one-year-old) individuals in the breeding population is strongly correlated with the reproductive success (number of young fledged per female) in the preceding summer (fig. 14.7). This finding shows that successful reproduction can override whatever might happen to these birds during the nine months between breeding seasons. Even though there may be high mortality during migration and over the winter, the number of young fledged in one year is still evident in the next. This relationship, which has also been documented for the American redstart at Hubbard Brook and for several other long-distance migrant songbirds, is surprising, especially since these birds face numerous hazards to their survival during long migrations and an extended winter period in tropical regions, which might be expected to diffuse or even negate such a relationship.[7]

If reproduction is so important, the question then arises: What determines the number of young produced in any given year? Analyses of the reproductive success of the black-throated blue warbler have shown that 93% of the variation among years in the number of young fledged per season can be attributed to annual differences in three variables: food abundance related in part to weather, predation rates on eggs and young, and the effects of local population density.[8] Unlike forests in many other parts of North America, the largely unfragmented forests at Hubbard Brook are not suitable habitat for the brown-headed cowbird, which often parasitizes nesting songbirds. Nest parasitism by cowbirds is therefore not a factor affecting birds at Hubbard Brook. Similarly, in spite of the high incidence of acid rain at Hubbard Brook, there is no evidence of eggshell thinning or breakage in the thousands of eggs that have been examined in bird nests in the forest, which would be expected if calcium were a critically limiting nutrient for birds.[9] So how do food, nest predation, and population density affect bird reproductive success?

Food limitation. One of the explanations for why birds migrate is that they move south in autumn to avoid the harsh weather and poor living conditions of the northern winter, and they return north in the spring because they can exploit an abundant (some would say superabundant) supply of food, as well as use the longer days in which to gather food and raise more young. Research at Hubbard Brook, however, has found that food, most of which consists of insect adults and larvae, is not superabundant for birds in this forest, at least in most years (fig. 14.8). Indeed, the numbers and availability of insects vary markedly among years (as we have seen for annual differences in caterpillar abundance), and can often be in short supply, negatively affecting bird reproductive success.

That food supply might be limiting for birds has been documented through both observations and experiments with black-throated blue warblers. In years with relatively more abundant food, mainly larval Lepidoptera, these warblers fledged a greater

Figure 14.8. Male black-throated blue warbler feeding an adult moth to its nestlings. Forest moths and especially their larval caterpillars are primary foods for songbirds during the summer breeding period, but are often in limited supply, affecting bird reproductive success. (Photo by N. L. Rodenhouse)

number of young than when caterpillars were scarce. This effect was confirmed by an experiment in which caterpillar numbers were reduced by spraying a biological insecticide, *Bacillus thuringiensis,* or Bt, from an airplane over an area of approximately 10 ha of northern hardwood forest outside the Hubbard Brook valley. These naturally occurring bacteria reduce the numbers of leaf-chewing insects, specifically caterpillars. The result of this experiment was that warblers living on this caterpillar-reduced plot fledged fewer young than those breeding on the unsprayed area. Also, on the unsprayed plot, after successfully fledging young from their first nests, more warblers built second nests and laid second clutches of eggs,

thus increasing the number of young they produce each season.[10]

Another test of food as a limiting factor involved providing black-throated blue warblers with extra food. For this experiment, live insect larvae were placed daily on small platforms near nests. Most warbler individuals readily found and took the food provided. The fecundity of birds given supplemental food was significantly greater than that of unfed females, mostly because the experimental birds laid and successfully fledged a second clutch after fledging the first. The birds thus had fledged more total young when food was more abundant.[11] Food availability was also found to influence the breeding effort through

Box 14.3. Bird Mating Behavior, Paternity Patterns, and Food Availability

Research at Hubbard Brook has shown that forest birds shift their reproductive behavior in response to changing food resource conditions. In black-throated blue warblers, for example, a male is typically paired with one female, which nests in the male's territory, and the male and the female both feed offspring in the nest for a period of time once they have fledged.[a] Males, however, often seek additional matings with females on neighboring territories, with the result that the young in a single nest often have been sired by more than one male.[b] Indeed, based on genetic analyses, about 43% of nestlings in this population of warblers at Hubbard Brook are sired by males outside their own territories, and more than half of the nests contain at least one nestling sired by such "extraterritorial" males.[c]

These patterns of extrapair paternity (EPP) vary with habitat quality, specifically food availability. The frequency of EPP is lower when food is abundant and higher when food is scarce.[d] When food is plentiful, either naturally or when experimentally supplemented, males spend more time defending their own territories and guarding their own social mates, which results in lower levels of EPP and higher reproductive success. The latter is also boosted by the fact that females on these territories with high food abundance more often fledge a second brood during the season. In contrast, when food is scarce, males are less vigilant in territorial defense and mate guarding, and they may have to wander farther to find food. In doing so, they encounter and opportunistically mate with more females, whose mates are either not present at the time or are less vigilant in their guarding activities. As a result, EPP rates are higher when food is limited. Thus, bird reproductive behavior and effort are altered by changing resource conditions, a finding that may be important to understanding how such species will respond to environmental change in the future.[e]

NOTES

a. Bigamous associations of two females within the territory of one male occur infrequently (0–15%). See Holmes et al. 1992, 2005; Chuang-Dobbs et al. 2001.

b. Chuang et al. 1999; Webster et al. 2001.

c. Kaiser et al. 2015.

d. Kaiser et al. 2015.

e. Kaiser et al. 2014, 2015.

its effects on bird mating systems and reproductive behavior (Box 14.3).

But what causes annual differences in caterpillar abundances? One possibility is weather. Caterpillar numbers and biomass at Hubbard Brook are related not just to local weather but also to climate cycles acting at a global scale, specifically the El Niño/La Niña Southern Oscillation (ENSO) weather system.[12] ENSO is caused by water temperature changes in the central Pacific Ocean, far distant from Hubbard Brook. Between 1986 and 1999, caterpillars were more abundant and black-throated blue warblers fledged more young in La Niña years than in El Niño seasons. At Hubbard Brook in La Niña years, the springs and summers are relatively warm and wet, whereas in strong El Niño years they are cooler and drier, which apparently decreases caterpillar growth, abundance, and/or availability. Not only did warblers fledge more young in La Niña years, but also the body weights of their nestlings near the time of fledging were higher than in El Niño years. This result indicates that the young birds were better fed and were in better

Box 14.4. What Are the Important Predators on Bird Eggs and Young?

Field observations, coupled with photographic studies of predators at natural nests and at artificial (wicker basket) nests baited with quail eggs, have identified many species that prey on bird eggs and nestlings. These include several species of rodents (red squirrels, eastern chipmunks, flying squirrels, white-footed mice), fishers, blue jays, and even black bears.[a] Video taping at nests during the incubation and nestling periods has confirmed that eastern chipmunks and red squirrels are the two most frequent predators of nests in the lower to middle strata of the forest. Video recording showed American marten (1 record) and sharp-shinned hawks (several records) taking eggs or nestlings from warbler nests.

Eastern chipmunk preying upon black-throated blue warbler nestlings. (Photo from R. T. Holmes)

Finally, in another study to determine the impact of predation on nest success, baffles were placed around tree boles below active nests of American redstarts, preventing predators from climbing up the nest tree. In this experiment, 74% of the nests with baffles were successful, compared with 50% for those without baffles. This result indicates that predators approaching nests from the ground, such as chipmunks and squirrels, are the most important predators at nests of redstarts, black-throated blue warblers, and other species that nest in the lower strata of the forest.[b]

NOTES

a. Reitsma et al. 1990; Sloan et al. 1998; R. T. Holmes unpublished.

b. Sherry et al. 2015.

condition, allowing them to survive better once they had fledged from the nest. These findings suggest that local weather, driven at least in part by global weather systems, influences caterpillar abundance, which in turn affects warbler fecundity, fledgling survival, and ultimately the number of birds returning in the following year.

Effect of nest predators. Depredation of eggs or nestlings is another important and widely recognized factor that affects the number of young successfully fledged by forest birds. At Hubbard Brook, for

example, 17% to 42% of black-throated blue warbler nests are lost each year to predators. Although the predator community at Hubbard Brook is diverse, the red squirrel and the eastern chipmunk are the two most important species of predators at nests, especially for birds like the black-throated blue warbler, American redstart, ovenbirds, and thrushes that build their nests close to or on the ground (Box 14.4).

Effect of population density. The third major factor to influence the number of young fledged each year by forest songbirds is density of the breeding population, defined as the number of adult birds or territories per unit of area. Field data and experimental studies of black-throated blue warblers show that females fledged fewer young that were lighter in weight and were less likely to survive in years when the density of nesting pairs was higher.[13] Although the mechanism operating here is not fully understood, one possibility is that adult birds at high densities are distracted by the presence and territorial activities of close-by neighbors, which means they spend less time feeding or attending and protecting their offspring, the result being fewer young successfully fledged.

Such a crowding effect represents a type of negative density dependence, one that lowers reproductive success when population numbers are high and promotes higher success when numbers are low.[14] This process results in adjusting or regulating the numbers in the population at a level presumably related to availability of resources such as food or space. This and other forms of density dependence are important, if not essential, mechanisms in the regulation of population size for the warbler populations at Hubbard Brook.[15]

HOW EVENTS IN THE NON-BREEDING SEASON AFFECT BIRD NUMBERS

Thus far, the focus of this chapter has been on birds during their breeding season at Hubbard Brook and the factors affecting the number of young they produce. The breeding season, however, lasts for only 2 to 3 months of the year. What happens during the other seasons? Do events during that time affect bird numbers as we've measured them at Hubbard Brook?

The importance of winter conditions is most obvious for species that remain at Hubbard Brook throughout the year, such as several species of woodpeckers, nuthatches, and black-capped chickadees. High mortality during the winter probably accounts for their consistently low numbers. These species often endure harsh conditions during the cold and snowy winters in New Hampshire, and their numbers decline after unusually prolonged or severe winter storms, such as those that occurred in 1969 and 2003.

But what about the species that migrate to distant wintering sites, such as the black-throated blue warbler and American redstart? Much could happen to these birds during the other 9 to 10 months that could affect their general body condition and health, and ultimately their survival. As mentioned above, it has often been assumed that migratory birds move south to avoid the harsh conditions of temperate and higher latitudes, and therefore might benefit from those seemingly benign environments during the winter. But migration itself can be hazardous, and even though the climate at lower latitudes may appear benign, not all habitats and conditions there are suitable or equally favorable for overwintering birds. Starting in the late 1980s, it was suggested that deforestation and degradation of bird habitats in the tropics might be responsible for observed declines in the numbers of breeding Neotropical migrants in North America. This hypothesis stimulated researchers, including us at Hubbard Brook, to examine what happens to migratory songbirds during the nonbreeding period. Where do these migrants go in winter? What habitats do they occupy? Are resources plentiful in these warmer climates or might they be limiting? How severe is the risk of mortality during migration and the overwintering period? And, ultimately, do these faraway events affect the populations inhabiting the northern hardwood ecosystem at Hubbard Brook?

In 1986, to address these questions and to understand the possible importance of the nonbreeding period to the numbers of migrants breeding at Hubbard Brook, we initiated studies of the two migratory species that we had been studying at Hubbard Brook, the black-throated blue warbler and American redstart, on their winter grounds in Jamaica. This Caribbean island was chosen as the site for our studies largely because both species had been reported as widely distributed and relatively abundant there in winter. Very little, however, was known then (and even now) about exactly where birds from a particular summer breeding locality, like Hubbard Brook, spend the winter. Even though many

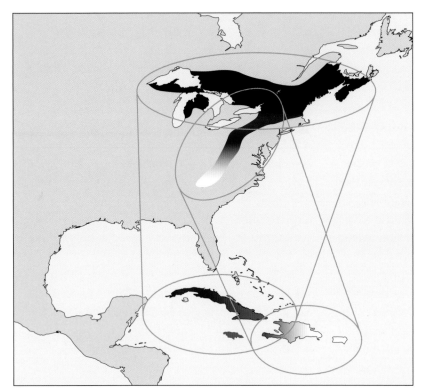

Figure 14.9. Black-throated blue warblers from the northern part of the breeding range tend to winter primarily in Cuba and Jamaica; those from the more southern areas migrate farther east in winter. This was determined with the use of naturally occurring stable isotopes in feathers that vary geographcially across North America. (From Holmes 2007, based on data from Rubenstein et al. 2002)

tens of thousands of these two warbler species have been banded in North America over the past several decades, only a dozen or so banded individuals of each species have been recovered in any wintering locality. And, most of those recovered had been banded during migration, so that locations of their breeding sites were unknown.[16]

As it turned out, the choice of Jamaica was a good one, as was the decision to study what happens to these birds in the nonbreeding season. Using stable isotopes in feathers as markers, individuals of both species breeding in northeastern North America were found to winter mostly in the Caribbean.[17] The method used is based on the fact that naturally occurring (nonradioactive) stable isotopes of carbon and hydrogen vary geographically across North America. These isotopes are in the foods eaten by birds, and are incorporated into body tissues, including feathers that grow during the annual molt, which occurs at or near the breeding site. Analyses of isotopes from warbler feathers sampled in winter grounds help to identify

breeding regions of those individuals. We found that black-throated blue warblers from the northern part of their range, including Hubbard Brook, wintered mostly in the western islands of the Greater Antilles, including Cuba and Jamaica, whereas those from more southerly areas along the southern Appalachians wintered farther east, in Puerto Rico and Hispaniola (fig. 14.9). Thus, even though there have still been no recoveries on the winter grounds of any birds banded at Hubbard Brook (nor of birds banded in Jamaica recovered in their northern breeding areas), it seemed reasonable to compare the biology of the summer populations at Hubbard Brook with those wintering in Jamaica.

Patterns of winter habitat use. In Jamaica, both black-throated blue warblers and redstarts were found to occupy a variety of habitats very different from those in their breeding areas. These wintering sites ranged from second-growth scrub to mangroves to wet rainforest, as well as agricultural sites such as coffee and citrus plantations. Some migrants occurred in gardens and shrubs around houses, even in cities like Kingston and Montego Bay. We found that certain habitat types were preferred by each species, as indicated by their higher densities there and by the fact that they returned to those habitats in subsequent years at a high rate.[18] Redstarts were most abundant in coastal mangroves at low elevations, while black-throated blues were most common in wet limestone forests in the more mountainous regions of the island (fig. 14.10). Individuals of both species were spread out within their respective habitats, and were often aggressive toward other conspecifics (members of the same species). Indeed, the males and females of these species occupied separate areas or territories in these winter habitats.[19] In redstarts, males were clearly dominant over females and often aggressively excluded females from the more preferred (wetter) habitats. The result was that some habitats (mangroves) contained mostly males, while the females mostly occupied drier, scrubby sites.[20]

As at Hubbard Brook, the warblers in Jamaica were captured in mist nets and given unique combinations of colored leg bands. Subsequent observations of these identifiable individuals led to the discovery that these birds remained very localized on their territories from their time of settlement there in October until they departed in late March or April, and that they often returned year after year to the same local areas. Similar strong site fidelity had been known for songbirds in their breeding territories, but to find it in the winter for these two species was unexpected. An extreme example of this faithfulness to a site in Jamaica was an exceptionally long lived American redstart that returned from North America to the same small patch of habitat in Jamaica for nine consecutive winters (which meant nine round trips!), a remarkable feat, especially considering that these individuals weigh only 8 to 9 grams (less than a third of an ounce) and have to travel about 2,500 km (1,700 miles) each way.

Environmental factors affecting migratory birds in the nonbreeding season. The research in Jamaica has elucidated the impact on migratory songbird

Figure 14.10. A black mangrove swamp near Black River, Jamaica, a high-quality habitat for American redstarts in winter (top), and a wet limestone tropical forest at Copse Mountain, near Betheltown, Jamaica, a habitat preferred by black-throated blue warblers. (Photos by N. Cooper, A. Peele)

population dynamics of events occurring during the nonbreeding part of the annual cycle. Four significant discoveries came from these studies.

(1) Migratory birds are affected by habitat quality in winter, especially food availability. Several lines of evidence have led to the conclusion that quality of the habitats occupied in the winter is important for these migrant species and that food is the major limiting resource. First, in both species studied, males and females were found to defend individual territories aggressively in winter.[21] Such use of time and energy to defend an area is usually indicative of a limiting resource (why otherwise spend time and energy in aggression when there is no reward?). Second, bird densities differed across a range of habitats in Jamaica, with birds being more abundant in those habitats in which food (primarily insects) was more abundant.[22] Third, redstarts in winter spent more time and effort in searching for and capturing food than they did in summer at Hubbard Brook, even though they were living in the presumably more benign environment of the warm tropics.[23] Thus, birds seem to have to work harder in winter to fulfill their daily needs. These migrants are not "on vacation" in these winter grounds!

Altogether, these findings of territorial spacing, intraspecific aggression, density differences, foraging patterns, and physiological condition of migrants in their winter grounds indicate that birds compete for higher quality sites and that food is limiting during this season. The loss of higher quality habitats in the winter grounds, for example through deforestation, habitat degradation, or climate change, will have major consequences for migrant species by influencing their physiological condition, survival during northward migration, and as described below, even future reproduction.

(2) Habitat quality in winter can influence reproductive success in the subsequent summer. Another important discovery from this research was that events in the nonbreeding period can have important effects on these bird populations during their breeding season at places like Hubbard Brook.[24]

In other words, there is a carryover effect from one season to another, so that what happens in summer is not independent of what happened months earlier in the Caribbean wintering grounds. An initial test of this hypothesis was made possible by the discovery that redstarts feeding in different habitats in winter had different levels of a naturally occurring isotope of carbon in their blood. By analyzing the ratios of these isotopes in blood samples obtained from redstarts as they arrived at Hubbard Brook, it was found that those first to arrive had wintered in the preferred wetter types of habitats (such as mangroves), while those arriving later had occupied poorer quality winter habitats (such as scrub lands). Winter habitat seemed to have affected the timing of migration and arrival times of birds at their northern breeding grounds.

The importance of this finding was determined in subsequent research, which revealed that individual redstarts coming from poorer winter habitat and arriving later on the breeding grounds not only initiated breeding later but fledged fewer young than birds arriving earlier in the spring.[25] The effects of habitat quality in the winter thus carry over from one part of a species' annual cycle to another, illustrating how events in one season influence those in another.

(3) Mortality is highest during migration. Prior to the research at Hubbard Brook, little information was available on the survival of long-distance migratory birds during different parts of their annual cycle. The data set assembled by Hubbard Brook researchers has allowed such estimates to be made for the first time.[26] This data set was made possible because of the site faithfulness of these birds to both breeding and winter sites, which allowed direct measurements of the summer and over-the-winter periods. Individuals banded in May at Hubbard Brook that were observed and therefore alive in early August, for example, provided a measure of survival over the summer period. Similarly, the proportion of banded individuals seen on winter study sites in October and resighted there in March provided an estimate of overwinter survival. Annual survivorship could be calculated as the number of birds banded either

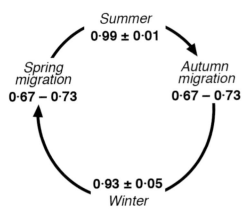

Figure 14.11. Survival probabilities of black-throated blue warblers differ during the phases of the bird's annual cycle. Values are average survival during each of the four major parts of the bird's annual cycle: "Summer" is the 3-month breeding season at Hubbard Brook, May to August; "Winter" is the 6-month stationary period in Jamaica, October to March; and the two migration intervals represent chances of survival during the 6 weeks of migration in spring and in autumn. Most mortality occurs during the migratory periods. (From Sillett and Holmes 2002; reprinted by permission of John Wiley and Sons, © 2002)

in May of Year 1 that returned in May of Year 2 or, alternatively, from October to October in Jamaica. By combining data on survival from the stationary (over-summer, over-winter) and migratory periods, it was possible to extrapolate and estimate survival during the migratory periods (fig. 14.11). An important caveat here is that these were samples of the population from one breeding site and one winter site, and that the individuals being observed at Hubbard Brook were not the same individuals being observed in Jamaica. In the absence of better data, however, they can be used to represent "a breeding population" and "a wintering population" of a long-distance migratory species, such as the black-throated blue warbler.

These data and survivorship analyses showed that individuals tend to survive well during each of the stationary periods during the year, but not during the migratory periods. Another way to illustrate these results is the following: starting with a cohort of 100 adult birds present in the breeding area at the beginning of the nesting period in the month

of May, 99 will be alive at the end of the summer; of these 99, an estimated 30 will die during southward migration in the autumn, resulting in 69 arriving on the wintering area; over the winter, 5 more of this cohort will die, leaving 64 to depart from the winter grounds on northward migration in the spring; and finally, of these 64, 45 individuals will arrive back on the breeding grounds in early May. An average annual survivorship of 40–60% is considered typical for these long-distance migrant songbirds.[27]

Although the estimated mortality during migration may not seem very impressive, it is, in effect, at least 15 times greater than mortality during the stationary periods and accounts for most of the deaths over the course of a year. In the hypothetical example above, 49 of the 55 individuals dying during a year's time do so during migration; this represents 85–90% of all deaths in the population. Hazards during migration, such as adverse weather (storms, hurricanes), encounters with large buildings and wind turbines, as well as the availability and quality of habitats used for shelter and food-gathering at migratory stopover sites, all have important impacts on the survival and thus the abundances of these migratory species (fig. 14.12). As far as we know, these estimates of mortality and survival for black-throated blue warblers during the different phases of their annual cycle are the only ones available to date for any long-distance migratory songbird.

Even though most bird deaths occur during migration, it is important to note that these deaths may in fact be preconditioned by events affecting the health and condition of these individuals before they migrate. Evidence from studies of American redstarts indicates that ecological conditions in their Caribbean wintering grounds, mainly annual variation in rainfall, which determines insect abundance, strongly affect the ability of birds to maintain their body weight and physiological condition over the winter period, which in turn influences their survival during the subsequent migration.[28] Even though these findings have greatly expanded our understanding of the connectivity between seasons for migratory species, more

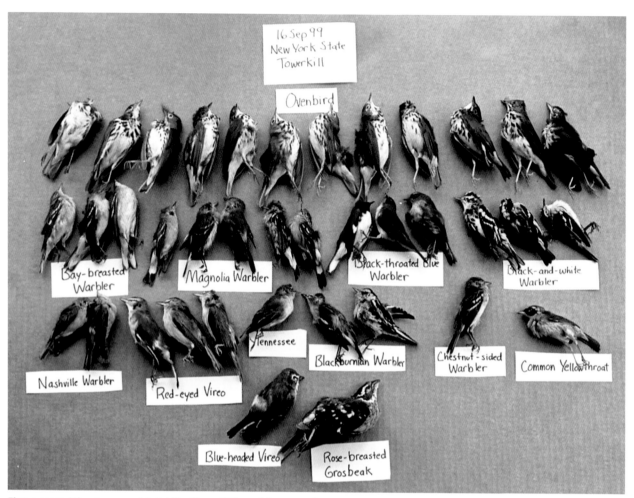

Figure 14.12. Migratory songbirds killed by colliding with a tower in upstate New York during autumn migration. (Photo by W. R. Evans)

investigations are needed to determine the relative importance of population regulatory mechanisms in these and other migratory species and where they operate during the annual cycle.

(4) Migrant birds from different breeding areas mingle in the winter grounds. Finally, another important result from the stable isotope analyses of the black-throated blue warbler feathers has been the finding that individuals from one breeding population or local area apparently do not usually winter together, but spread out to many different localities in the nonbreeding period. This conclusion is based on the discovery that individuals within single study sites in Jamaica did not come from a single breeding region,

but derived from a wide range of breeding longitudes, in this case ranging from Nova Scotia to Michigan.[29] The converse is probably also true—that individuals from one local wintering site in Jamaica spread out to many different sites in the breeding range. This mixing of populations from different regions has important implications for conservation and management purposes. For instance, any loss of winter habitat would have a broad and very diffuse effect on the number of birds in any one northern breeding site like Hubbard Brook. Likewise, the loss of a particular patch of northern forest would have little effect on the numbers of birds in any one local area in Jamaica or other wintering area. Such mixing therefore makes identifying the impact of habitat loss or degradation in either season difficult, and complicates conservation planning.

These discoveries of winter breeding area effects on migratory populations and the delayed effects on population abundance and reproductive output in the following breeding season illustrate the importance of events during the non-breeding season to the overall survival and success of migratory bird populations.

∎

The long-term avian research at Hubbard Brook has revealed that bird abundance in these northern hardwood ecosystems is determined by multiple factors, operating at different times and places. Changes in forest structure and probably the corresponding availability of certain types of food resources have led to the gradual decline and eventual disappearance of three formerly abundant Neotropical migrant species that have preferences for early to mid-successional forests for breeding. Other changes in local abundances result from a complex interaction of biotic and abiotic factors operating in both breeding and nonbreeding seasons. Food limitation in both breeding and wintering areas, nest predation during breeding, and for migratory species, high mortality during migration are among the most significant limiting processes. Finally, these results powerfully demonstrate that ecosystems such as those being studied at Hubbard Brook are not closed, but can be significantly influenced by outside events, sometimes far away, just as nutrient flux and vegetation in the watershed-ecosystems are affected by air pollutants generated at great distance from Hubbard Brook.

15

SCALING UP
Ecosystem Patterns and
Processes Across the Valley

The small watershed-ecosystem approach used at Hubbard Brook has proven highly successful for measuring and assessing the flux of materials within and across ecosystem boundaries, and has led to new insights into the structure and functioning of forest and aquatic ecosystems. These topographically defined watersheds, 12 to 68 ha in size, are sufficiently large for measuring the input, output, and cycling of materials and nutrients, yet small enough to allow for experimentation, including whole-watershed manipulations. This scale of investigation seemed highly suitable for our purposes and sufficient to inform management questions.[1] Moreover, investigations at even smaller scales, for example on small plots within or adjacent to the watersheds, have been made to determine and evaluate the mechanisms and processes occurring within the watershed-ecosystems.

But how representative are these plots or even the small watersheds of larger units of forest? Can findings from the small watersheds be extended to larger areas, such as the entire forest or even the surrounding region?[2] To address such questions, researchers at Hubbard Brook expanded the scope of their studies to consider ecosystem components and processes at larger spatial scales (Box 15.1). These forest- and valley-wide studies deal with topics such as spatial patterns of tree species composition and forest structure, primary production, stream nutrient dynamics, and the distributions and abundances of animals. Although many of these studies are still in progress, we have selected several examples from terrestrial and stream ecosystems to illustrate the value of considering ecosystem processes at a variety of spatial scales.

TREE DISTRIBUTIONS AND ASSOCIATIONS ACROSS THE HUBBARD BROOK FOREST

Although the forest at Hubbard Brook is generally representative of northern hardwoods, it is not uniform over the entire valley. The tree species composition and physiognomy (physical structure such as height and branching pattern) vary markedly with elevation, slope, aspect (compass direction), soil depth, drainage, and disturbance history. On Watershed 6, the reference watershed, the forest consists of about equal frequencies of sugar maple, American beech, and yellow birch, followed by red spruce and paper birch.[3] Vegetation sampling across the entire forest, however, has shown that yellow birch is the most abundant and frequently encountered tree species. On forest-wide surveys, its mean basal area (cumulative cross-section area of a specified tree species at 1.4 m aboveground per hectare) averaged 9.0 m²/ha, followed by sugar maple (4.9), red spruce (3.5), American beech (3.3), red maple (2.1), paper birch (2.1), balsam fir (1.5), and all other tree species combined (2.8).[4] This species composition at the scale of the entire forest differs from that on Watershed 6 and other nearby small watersheds, as well as from the pre-settlement forest, in which red spruce and American beech were the dominant species. Expanding measurements to the forest-wide scale clearly captures the greater heterogeneity and increased diversity in the vegetation that exists across the Hubbard Brook landscape.

From the relative frequency and sizes (basal areas) of trees sampled systematically across the forest, it is possible to analyze statistically the relationships of each tree species with factors such as topography and soils, and to identify distinct forest types or species associations. The resulting patterns show a mosaic of tree species associations across the forest (fig. 15.1). The most widespread of these is the northern hardwood association (57% by area), characterized by an approximately equal mixture of sugar maple, American beech, and yellow birch as the dominant trees in the canopy. This association occurs mostly in the central and outer portions of the valley, including the south-facing watersheds. The other forest types are variants of the northern hardwood type, containing different but distinct proportions of many of the same tree species. The high hardwood forest type (13% by area) is dominated by yellow birch but with relatively frequent red spruce, sugar maple, and American beech, and occurs at slightly higher elevations. The fir-birch forest type (12% by area), characterized by more balsam fir and paper birch, is found along the upper slopes of the valley on both the south and north ridges. Just below these high-elevation sites on steep slopes with shallow soils are patches with a larger component of red spruce, and these are referred to as the spruce-fir-birch forest type (8% by area). At lower elevations and mainly along the main channel of Hubbard Brook, eastern hemlock becomes the

Hubbard Brook Experimental Forest

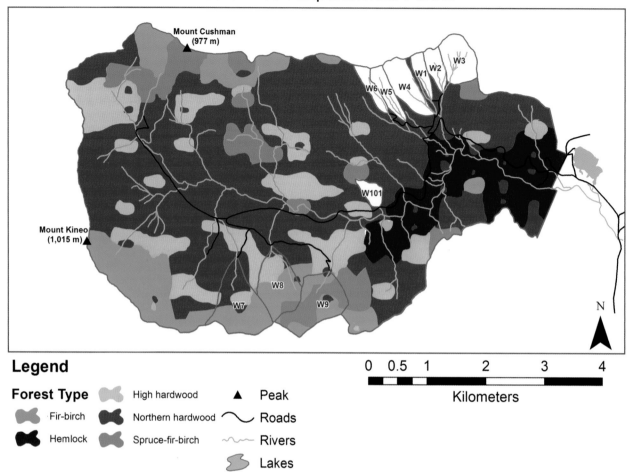

Legend

Forest Type

- High hardwood
- Fir-birch
- Northern hardwood
- Hemlock
- Spruce-fir-birch

▲ Peak
⌒ Roads
〜 Rivers
Lakes

0 0.5 1 2 3 4
Kilometers

Figure 15.1. Forest tree associations in the Hubbard Brook valley, determined from forest-wide sampling in 2005–2006. (J. J. Battles, unpublished)

most characteristic tree species. This hemlock forest association (10% by area) also includes some sugar maple and yellow birch, and mixes with red oak and some white pine around Mirror Lake at these lower elevations. Overall, 80% of the forest is dominated by northern hardwood tree species in one combination or another, and 20% by spruce-fir.[5]

Data collected across the forest-wide survey plots provide an important baseline for evaluating future changes in the vegetation that might be attributable to climate change or other major disturbance such as a pathogen outbreak. They also help to interpret differences in the distributions and numbers of

animals, as well as stream chemistry, both of which are strongly influenced by vegetation.

SPATIAL PATTERNS OF PLANT BIOMASS AND NET PRIMARY PRODUCTIVITY

Using the vegetation inventory data from the forest-wide surveys and from Watershed 6, one can determine how forest biomass and primary productivity vary across the forest (fig. 15.2). Both plant biomass and productivity decline with increasing elevation, because of shallower soils and more extreme climatic conditions on the upper slopes of the valley. Plant biomass is greater on the south-facing

A. Aboveground plant biomass

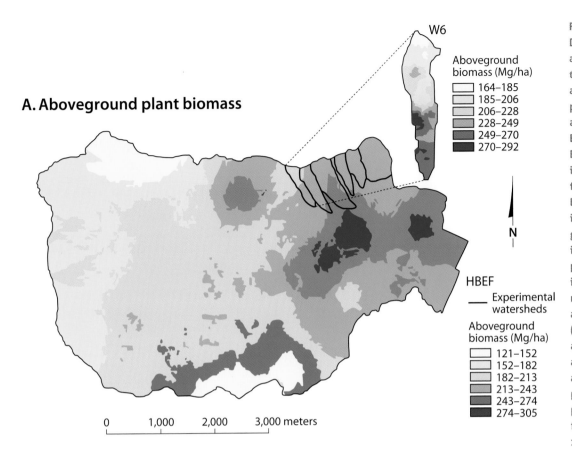

W6

Aboveground
biomass (Mg/ha)
- 164–185
- 185–206
- 206–228
- 228–249
- 249–270
- 270–292

N

HBEF
—— Experimental
watersheds
Aboveground
biomass (Mg/ha)
- 121–152
- 152–182
- 182–213
- 213–243
- 243–274
- 274–305

0   1,000   2,000   3,000 meters

Figure 15.2.
Distribution of
aboveground
tree biomass and
aboveground net
primary productivity
across the Hubbard
Brook forest (1997).
Biomass is also shown
in the upper diagram
for Watershed 6,
based on a complete
inventory of trees
greater than 10 cm
in diameter. Net
primary productivity
is defined as the
net rate of biomass
accumulation
(plant growth) on
an annual basis,
and is measured
as grams of carbon
per square meter
per year. (Modified
from Fahey et al.
2005; reproduced by
kind permission of
Springer Science+
Business Media)

B. Net primary productivity

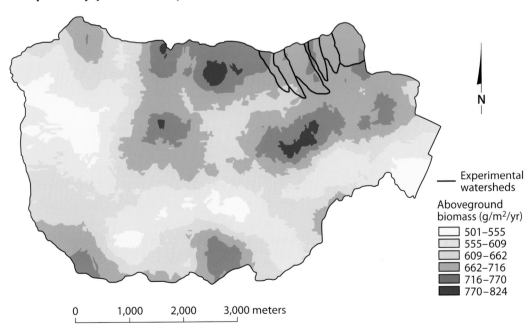

N

—— Experimental
watersheds
Aboveground
biomass (g/m²/yr)
- 501–555
- 555–609
- 609–662
- 662–716
- 716–770
- 770–824

0   1,000   2,000   3,000 meters

slopes, where stands dominated by sugar maple occur on relatively rich soils. In general, mixed hemlock and hardwood stands support higher biomass and productivity than pure hardwoods, whereas forests at the highest elevations with the more extreme environment and the poorest soils are dominated by balsam fir and support the lowest biomass.[6]

These patterns of plant biomass and productivity across the Hubbard Brook landscape reflect variation in environmental factors, such as climate, soil characteristics, nutrient availability, and tree species composition.

FOREST ANALYSIS FROM REMOTE SENSING

Most of the large, spatially extensive data sets reported in this chapter are the result of investigators spending from weeks to years in the field making hands-on measurements, and then processing and analyzing the data collected, which may take weeks or months. Another way of assessing the composition and function of ecosystems over large areas is to use remote sensing technology, such as airborne imaging spectroscopy to estimate concentrations of nitrogen in tree foliage across the landscape, and LIDAR, which uses pulses of laser beams for measuring characteristics of forest structure.

Foliar chemistry at the landscape scale: airborne imaging spectroscopy. Nitrogen is an important nutrient for plant nutrition and growth. High concentrations of nitrogen in leaves generally indicate high rates of uptake and storage, as well as photosynthesis and primary productivity. To document the distribution pattern of leaf nitrogen across the Hubbard Brook valley, remote sensing images were obtained by an aircraft imaging spectrometer, or "hyperspectral" sensor, that measures reflected sunlight in 360 continuous channels over visible and infrared wavelengths. By examining relationships between field measurements of foliar nitrogen concentration and the ways in which light is absorbed and reflected by the forest canopy, spatially explicit estimates of nitrogen concentrations in foliage can be estimated (fig. 15.3).[7] The resulting

patterns reflect a combination of the distribution of individual tree species and local factors that influence site fertility.[8] For example, the highest concentrations of foliar nitrogen at this large scale occurred in the experimentally manipulated (clear-cut and deforested) watersheds on south-facing slopes, which are in an earlier successional stage and thus represent plants that are adapted for high rates of growth. The lowest foliar nitrogen concentrations were mostly on the ridge tops at the higher elevations, partly because conifers, which are more frequent at high elevations, have inherently lower foliar nitrogen than leaves of deciduous trees.

When these patterns are compared with ground-based measures of net primary productivity (which we saw above in fig. 15.2), the patterns are generally similar, although differing in many details. Differences reflect a variety of factors, including the growth strategies of individual species and site-level properties such as soil moisture, disturbance history, and the availability of other nutrients that can also influence growth. Further study is needed to understand fully how these and other important variables interact to produce the observed patterns of foliar nitrogen and plant growth.

Spatial variation in forest structure: measurements using LIDAR. Variation in forest structure, particularly the vertical distribution of foliage, can influence ecosystem properties, such as the rates of photosynthesis and the patterns of biodiversity. Measurements of vertical structure in forests have traditionally been made from the ground, using methods involving cameras or other sensors that focus upward on overhead leaf layers or that record the amount of canopy cover (for example, the percentage of sky visible through the canopy). These techniques are often laborious and limited in their spatial applicability. A remote sensing technique involving laser technology has been developed that allows for the measurement of the vertical canopy structure (canopy height, canopy cover, canopy complexity, foliage height diversity) over large spatial scales (Box 15.2).

%N
1.0
3.0

1   0   1   2   3 km

Figure 15.3. Spatial variability in foliar nitrogen concentrations (% nitrogen by leaf mass) across the forest, estimated using high spectral resolution remote sensing from SpecTIR LLC's ProSpecTIR-VS instrument. Data were collected in 2012 from a fixed-wing aircraft, and were calibrated using field-measured foliar nitrogen concentrations. Areas of high foliar nitrogen in the upper right (red) correspond to experimentally manipulated watersheds where early successional plant species with higher foliar N concentrations still persist. (Map prepared by L. C. Lepine and S. V. Ollinger, based on Ollinger et al. 2008 and additional data collected in 2012)

LIDAR methodology applied at Hubbard Brook has quantified foliage structure across the entire forest, showing how canopy structure varies across the forest.[9] Both canopy cover and vertical stratification, two similar measures of canopy structure, decline with elevation, but they also change with slope and aspect (fig. 15.4). Tree foliage tends to be concentrated at lower and middle strata (6–15 m) of the vertical profile at low and middle elevations, while the overstory vegetation (high canopy) is more open. Overall, about 55% (1,762 ha) of the forest could be characterized as having two foliage profile layers, 34% had essentially one layer, and less than 1% of the forest was tall enough to exhibit three distinct layers.[10]

The ability of LIDAR to measure and quantify forest structure at the landscape scale has many potential uses in ecological research and for forest management. At Hubbard Brook, LIDAR measurements have been used to measure fine-scale topography, allowing for detailed studies of soils and hydrology.[11] It also has been used to examine how the heterogeneity of forest structure affects bird distributions and to predict bird habitat use at the forest-wide scale over multiple years.[12]

DISTRIBUTION OF INSECTS, BIRDS, AND MAMMALS ACROSS THE HUBBARD BROOK FOREST

Understanding the environmental factors that determine the distributions and abundances of

Box 15.2. Using LIDAR to Measure Forest Structure

Light detecting and ranging (LIDAR) is a remote sensing technology that measures distance by illuminating a target with many laser beams and detecting the time it takes for beams to reach the earth's surface (or canopy top) and to be reflected back to the sensor. It was first used for mapping the surface of the moon, but the technology has proven useful for measuring land topography, elevations, or the structure of forest canopies, as has been done for Hubbard Brook. Measurements are made from an aircraft or satellite containing the appropriate LIDAR sensor.

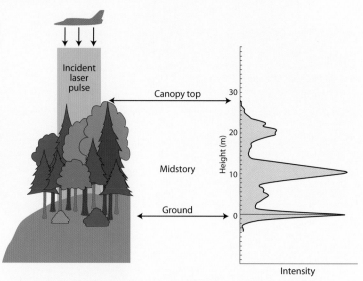

Schematic diagram illustrates how LIDAR measures foliage density at different levels within the forest, as well as total canopy (tree) height. (Modified from diagram by R. O. Dubayah)

animals has long been a challenge to ecologists. Most studies of spatial patterning of forest animal populations have considered distributions on relatively small areas, from a few square meters to several hectares for invertebrates or from tens to perhaps hundreds of hectares for vertebrates. Patterns at broader scales have been difficult to obtain. At Hubbard Brook, with the network of sampling points across the forest, it has been possible to sample distributions at a much expanded scale. Such data allow for assessment of year-to-year differences in distribution and abundance, as well as of correlated environmental parameters (for example, the composition of the vegetation). We cite here several examples of these patterns to illustrate this approach.

Caterpillars. Long-term sampling on the bird study plot west of Watershed 6 shows caterpillars (Lepidoptera larvae) to fluctuate strongly from year to year within this small 10-ha area, but little was

known about the spatial extent of these patterns (see Chapter 5). Were caterpillar numbers uniform across the landscape within a year, or did they vary from one part of the forest to another? Sampling across the forest has shown that caterpillar abundance is indeed highly variable both spatially and temporally (fig. 15.5). It is interesting that the peak year of caterpillar abundance on the forest-wide sampling was 1999, whereas numbers on the local plot scale in that year or the next were not unusually high. These findings illustrate the differences that arise when sampling at different scales, and raise questions about the best method and scale for sampling, as well as what factors determine these differences in local caterpillar abundances.

Birds. Birds have been studied intensively on a 10-ha study plot west of Watershed 6 since 1969. But again, how representative are data from such a relatively small area? How do distributions of

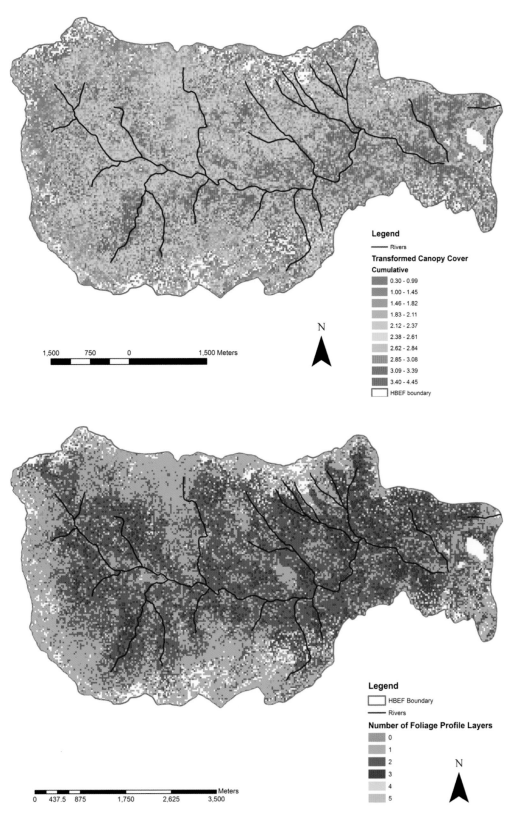

Figure 15.4. Relative density of vegetation across the Hubbard Brook forest, measured by LIDAR (top). Denser foliage cover (red) occurs in the lower parts of the valley and along larger stream courses where trees are taller. The canopy is less dense at higher elevation (green). (A. S. Whitehurst, unpublished.)

Vertical distribution of foliage at Hubbard Brook (bottom). The majority of the forest (55% of the pixels, covering an area of 1,762 ha) was best described as having two foliage layers (dark green), located mostly in the interior of the forest at low and middle elevations. Forest with tall stature—three or more layers (purple, yellow)—occurred in less than 1% of the area, mostly in the lower portions of the valley along the larger stream courses. (Whitehurst et al. 2013)

Legend

— Rivers

Transformed Canopy Cover
Cumulative

0.30 - 0.99
1.00 - 1.45
1.46 - 1.82
1.83 - 2.11
2.12 - 2.37
2.38 - 2.61
2.62 - 2.84
2.85 - 3.08
3.09 - 3.39
3.40 - 4.45

☐ HBEF boundary

N

1,500   750   0   1,500 Meters

Legend

☐ HBEF Boundary

— Rivers

Number of Foliage Profile Layers

0
1
2
3
4
5

N

0   437.5   875   1,750   2,625   3,500   Meters

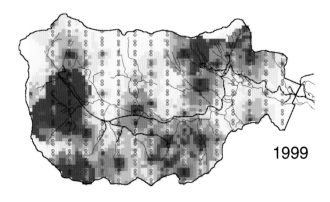

1999

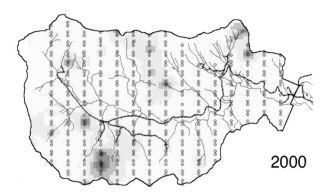

2000

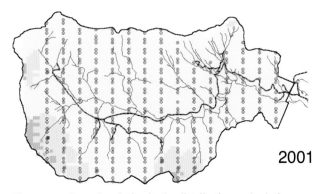

2001

Figure 15.5. Annual variation in the distribution and relative abundance of Lepidoptera larvae on understory foliage, 1999 to 2001. Darker colors represent higher caterpillar abundances. Dots are the sampling points systematically arrayed across the forest. (P. J. Doran and R. T. Holmes, unpublished)

individual bird species relate to differences in forest vegetation as the scale of sampling increases? To address these questions, the presence and abundance of all species of birds have been recorded during the breeding season at each of 371 sampling points across the forest since 1999.[13] Whereas only 22 species breed regularly on the 10-ha plot, these forest-wide censuses show a total of 66 species breeding across the larger 3,160-ha Hubbard Brook forest. Furthermore, each species has a unique distribution across this landscape. Some are relatively evenly distributed (yellow-rumped warbler, black-throated blue warbler), while others are clustered or more localized in their distributions (fig. 15.6). These species-specific dispersion patterns reflect evolved preferences of each species for specific habitat variables and especially vegetation type (deciduous vs. coniferous) and structure.

Rodents. Eastern chipmunks and red squirrels are two of the most conspicuous mammals present in the forest, partly because of their relatively high abundances but also because they are active during the day and are easily seen and heard. They are also the most important predators on bird eggs and nestlings. To determine their spatial distribution and relative abundances and hence their potential predation pressure on bird populations across the forest, their presence and numbers were quantified on the forest-wide sampling points.[14]

Both species occur widely throughout the forest, but especially on the south-facing slopes (fig. 15.7). In 1999 and 2001, years following high mast (seed) production by American beech and sugar maple, both species were abundant throughout the forest, but differed in relative distributions across the landscape. Chipmunks occurred at higher densities at middle and higher elevations on the south-facing slope, whereas red squirrels were more common in the more mixed deciduous/coniferous forests in the west-central region of the forest. Both populations had extremely low, and almost unmeasurable, densities in 2000, a year following nearly zero seed production by forest

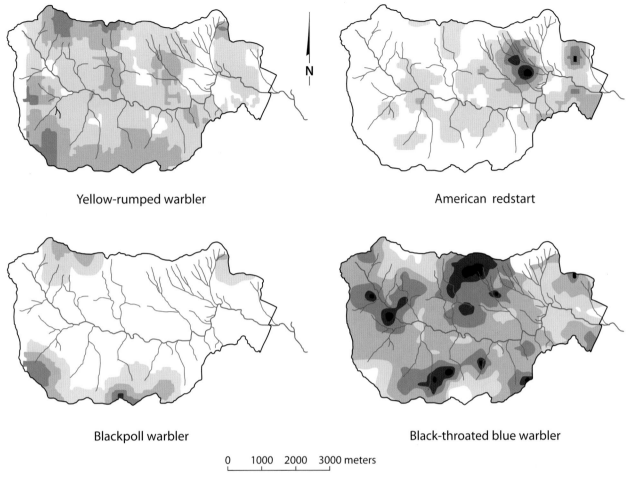

N

0   1000 2000 3000 meters

Fig. 15.6. Bird species differ in their distributions and abundances across the Hubbard Brook valley. Yellow-rumped and black-throated blue warblers are widely distributed across the forest landscape, with the former being relatively evenly dispersed and the latter locally dense (darker blotches) indicating areas of high-quality habitat for that species. Blackpoll warblers occur only in the more coniferous areas at upper elevations, and American redstarts in deciduous areas in lower central parts of the valley. (Modified from Holmes 2011; reprinted by permission from Elsevier)

trees, which resulted in high mortality and thus low overwinter survival of these rodents.

MAPPING DISTURBANCE EVENTS ACROSS THE VALLEY: THE SOIL FREEZING EVENT OF 2006

As discussed in Chapter 4, disturbances from unusual weather events or other unpredictable phenomena can significantly affect forest ecosystem structure and function. Determining the spatial extent of such disturbance events helps to evaluate the importance and impact of such disturbances. In the winter of 2005–2006, low snowfall and extremely cold temperatures resulted in the freezing of the upper layers of the soil to depths of 9 cm or more in many parts of the forest, resulting in ice-clogged stream channels and the deaths of many organisms that burrow into the soil during winter, including frogs and salamanders.

Forest floor disturbance due to the soil frost events has been postulated to increase the loss of nitrate from ecosystems.[15] To test this hypothesis during the soil-freezing event in 2005–2006, field surveys were

Red squirrel   Eastern chipmunk

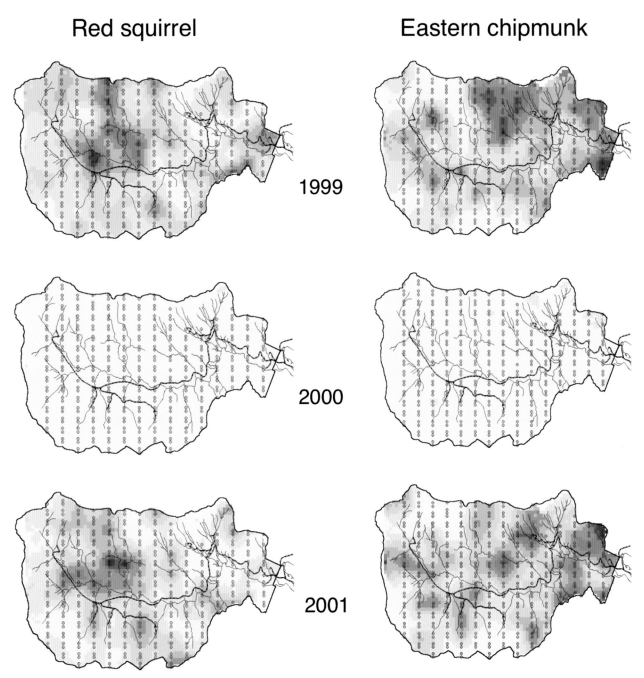

1999

2000

2001

Figure 15.7. Distribution and relative abundance of red squirrels and eastern chipmunks, 1999 to 2001. Both species are widely distributed, but they occur at higher densities (darker areas) in some parts of the forest and in some years. Annual differences reflect changes in population size related to seed production by forest trees, affecting overwinter survival and reproduction. (P. J. Doran and R. T. Holmes, unpublished)

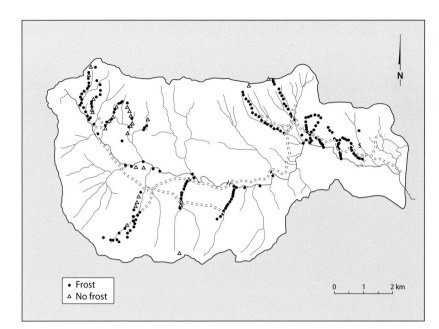

Figure 15.8. Spatial extent of soil frost at Hubbard Brook in late winter 2006. Double-dashed line represents gravel road. (Modified from Judd et al. 2011; reprinted with permission of Springer Science+Business Media)

Frost

No frost

0   1   2 km

conducted during the late winter along transects on both sides of the valley. The depth of the frozen layer was determined at 100-m intervals by digging small soil pits with an axe and by probing the soil with a steel rod at 10 randomly chosen locations near each sampling point. The results indicated that soil frost was widespread but not evenly distributed (fig. 15.8). Frost depth in the soil averaged 9 cm on the south-facing transects and 7 cm on the north-facing side of the valley. Overall, frost was found at 78% of the sites surveyed.[16] Contrary to predictions, nitrate concentrations in stream water did not increase in response to this massive soil frost disturbance, bringing into question the role of soil frost in generating widespread nitrate mobilization and loss in stream water.[17] These results also point to the need for further research on the role of soil frost in mobilizing nitrate.

SPATIAL DISTRIBUTION OF FISH IN HUBBARD BROOK STREAMS

Three species of fish inhabit the streams at Hubbard Brook: slimy sculpin, blacknose dace, and brook trout. In the early to mid-1960s, all three occurred in the mainstem of Hubbard Brook and the lower sections

of most tributary streams, with the upper limits of their distributions set by physical obstructions such as organic debris dams, waterfalls, and cascades along these smaller streams on steeper slopes.

A survey in 2001, however, found only brook trout in the mainstem of Hubbard Brook and its tributaries, and sculpin only in the lowest tributary, Norris Brook. This pattern was confirmed by a more intensive survey in the mid-2000s (fig. 15.9). The absence of sculpin and dace from streams in the upper portions of the valley in recent years is thought to be a result of the chronic and acute acidification of these streams during the early 1970s, when acidity peaked at Hubbard Brook.[18]

SPATIAL PATTERNS IN STREAMWATER CHEMISTRY

As we have discussed in Chapter 8, streamwater chemistry can be an important indicator and integrator of ecosystem processes. The quality of water leaving the system reflects the geologic substrate, soils, vegetation, and the biogeochemical interactions throughout the entire watershed. Examining the spatial variability in streamwater chemistry throughout the valley therefore provides a

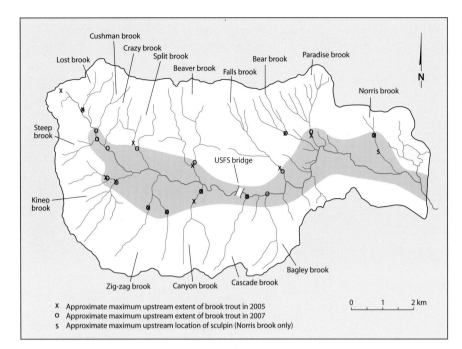

Figure 15.9. Distribution of fish in Hubbard Brook streams in late summer of 2005 and 2007 (green area). Fish do not occur in the upper reaches of the stream networks. (Modified from Warren et al. 2008)

way of assessing how these factors interact in forest ecosystems at larger scales.

In spring and fall 2001, water chemistry measurements were made in all of the major tributaries of Hubbard Brook. Water samples were taken at 100-m intervals from the point of first flowing water in the headwaters to the point where Hubbard Brook enters the Pemigewasset River (fig. 15.10). Streamwater chemistry varied by elevation, channel length, size, and type of drainage area (for example, dominated by in-stream seeps or not), reflecting the influence of soils, till, bedrock, vegetation, and hydrologic flow paths within the individual catchments. Spatial patterns for pH, calcium, nitrate, and sulfate were determined from samples collected in autumn. Relatively large upstream-downstream changes are evident for hydrogen ions, dissolved aluminum, and dissolved organic carbon concentrations in stream water, whereas small changes were found for sodium, chloride, and dissolved silica. The chemistry of the main Hubbard Brook changed remarkably little throughout its length in spite of the inputs from major tributaries. In the area of Mirror Lake, however, where there is

more human activity, higher levels of pollutants (for example, sodium chloride) were observed in the water of the main Hubbard Brook.[19] In a network analysis of these 2001 data, spatial patterns within the drainage network showed that some solutes had increasing homogeneity with downstream distance (dissolved organic carbon, sulfate, and aluminum), but others (nitrate, sodium) did not.[20]

This valley-wide study confirmed that the gauging weirs had been located in almost optimal positions for measuring change in streamwater chemistry within the headwater, gauged watersheds.[21]

IMPORTANCE OF SCALE IN ECOSYSTEM STUDIES

The scale at which ecological patterns and processes are examined is important and depends largely on the question(s) being asked, but also on the practicality of studying, quantifying, or even manipulating the system. If the biogeochemical studies had started with the entire valley as the unit of study, the research probably would have stopped in a year or two, because the investigators would have been discouraged by the lack of tractable findings, and any results would

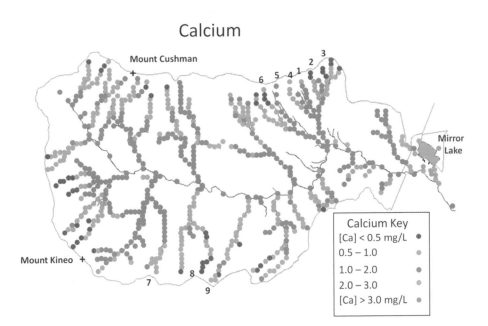

Streamwater pH

Mount Cushman

Mirror Lake

Mount Kineo

pH Key
pH < 4.5
pH 4.5 – 5.0
pH 5.0 – 5.5
pH 6.0 – 6.5
pH 5.5 – 6.0
pH 6.5 – 7.0
pH > 7.0

Calcium

Mount Cushman

Mirror Lake

Mount Kineo

Calcium Key
[Ca] < 0.5 mg/L
0.5 – 1.0
1.0 – 2.0
2.0 – 3.0
[Ca] > 3.0 mg/L

Figure 15.10. Streamwater chemistry (pH and calcium concentrations) at Hubbard Brook differs markedly in different parts of the valley. Data from surveys made in autumn 2001. Numbers 1 through 9 identify the experimental watersheds. (Modified from Likens and Buso 2006; reprinted by permission of Springer Science+Business Media)

have been vague at best. As it turns out, the size and placement of the small headwater watersheds were critical for understanding the impact of acid rain, as well as for providing a practical-sized unit for whole-watershed manipulations that produced quantitative results relevant to forest management. Indeed, what

we learned at the watershed scale led to new questions about patterns and processes at larger scales that we might not have considered otherwise. Furthermore, in hindsight, it is unlikely that in the 1960s we had the methods for quantifying ecosystem components at the forest-wide scale, but at the small-watershed scale, we did. What we learned at the watershed scale, particularly in terms of methods, allowed us to enlarge to the valley scale, but the converse would not have been true.

With larger plots or studies at broader scales, heterogeneity and diversity increase. This result was very evident in the bird distribution patterns on small plots versus forest-wide at Hubbard Brook. For instance, only 22 species of birds breed regularly on the 10-ha bird census plot near Watershed 6, but more than 66 occurred at the forest-wide scale.[22] The increased diversity and differences in distribution suggest new questions about patterns of habitat use by different species and the factors promoting diversity.

Similarly, expanding the scale of sampling for streamwater chemistry has led to new insights and new questions about water quality throughout the Hubbard Brook valley. It has been known for a long time that the streamwater chemistry of our long-term reference watershed (Watershed 6) differs with elevation from adjacent watersheds and from north-facing watersheds.[23] The recent network analysis

has opened many new questions about why the chemistry differs and the degree to which results from one stream can be generalized to others. Moreover, individual smaller watersheds with different water chemistry may be more appropriate for studying water-quality questions, such as treatment of acidity or water color. For purposes of understanding water quality exiting the entire valley in the main Hubbard Brook, sampling at a larger scale may be more relevant to management questions downstream. We have been monitoring water quality in the main Hubbard Brook at the base of the valley for almost four decades.[24] The chemistry at this sampling point gives an integrated value for streamwater chemistry for the entire valley and its contribution to the Pemigewasset River into which it flows.

These examples show how broadening the scale at which we study ecosystems can lead to new insights and important discoveries. Many of the environmental problems now extant require such analyses at regional or even larger scales.

16

HOW IS CLIMATE CHANGE AFFECTING THE FOREST ECOSYSTEM?

One of the most pressing environmental issues today is climate change and how it impacts the ecological systems that are critical in supporting life on earth, including humans. The topic of climate change is hotly debated across the United States and elsewhere, especially whether human activity is the major driver of this change. We and the vast majority of scientists believe that much of climate change is due to human activities, although some politicians and certain segments of the public reject this conclusion.[1] The extent and impact of climate change has many serious political and policy ramifications, not only for our country, region, and local communities but also worldwide.

Climate change has major consequences for the forest ecosystem at Hubbard Brook. Climate is a key driver of ecosystem functions and the services that they provide through effects on hydrological, biogeochemical, and biological processes.[2] Changes in climate are likely to alter nutrient flux and cycling, plant physiology, forest productivity, the presence and abundance of animal species, and other features of the forest ecosystem. With more than 50 years of high-quality meteorological and hydrological data from the intensively monitored and studied forest ecosystem at Hubbard Brook, researchers have documented these changing climatic patterns and the resultant changes in ecosystems, and have initiated studies of climate impacts on ecosystem functions. We illustrate here the directions of climate change recorded thus far, and provide some examples of how a changing climate is affecting the forest and associated aquatic ecosystems at Hubbard Brook.

METEOROLOGICAL TRENDS AT HUBBARD BROOK: 1950s TO THE PRESENT

Climate refers to the long-term average patterns of temperature, solar radiation, precipitation, and other meteorological variables in a region. At Hubbard Brook, these variables have been measured at meteorological stations operated by the U.S. Forest Service throughout the forest since the mid-1950s, using standardized and carefully calibrated instruments.[3] These measurements of key variables provide the basis for evaluating the effects of long-term climatic trends on ecosystem processes in our study system, which are as follows.

- Air and soil temperatures are warming (fig. 16.1). Although varying from year to year, the mean annual air temperatures, calculated by averaging the daily temperatures across each year, show an increasing trend between 1956 and 2012, for an increase of about 0.2°C per decade.[4] Air temperatures are about 1.2°C warmer today, on average, than in the 1950s. Air temperatures have increased during both the winter (December to February) and the summer (June to August) periods. Analyses of these patterns indicate that winter temperatures have generally warmed more than those during the summer months, and are more variable from year to year.[5]

  Soil temperatures are related to ambient air temperatures, although dampened and lagged with depth, and reach peak values in late summer. At Hubbard Brook, the mean soil temperature for the month of August at a depth of 61 centimeters shows a strong increasing trend over the 50 years of measurement. The fact that these mean late-summer soil temperatures at this depth have risen so strongly during the period of measurement clearly indicates the extent to which climate has warmed at Hubbard Brook.

- Precipitation is increasing (fig. 16.2A). Precipitation at Hubbard Brook has increased since the 1950s at a rate of about 5.3 cm of precipitation per decade. This trend means that annual precipitation increased by an average of about 25 cm (10 inches) over the 50-year period of the Study. Precipitation in summer has increased more than in the winter.[6] Both annual and winter precipitation amounts are projected to increase by 7–14% and 12–30%, respectively, by 2100.[7]

- Snowpack depth is decreasing (fig. 16.2B). Snow depth and water content have been measured

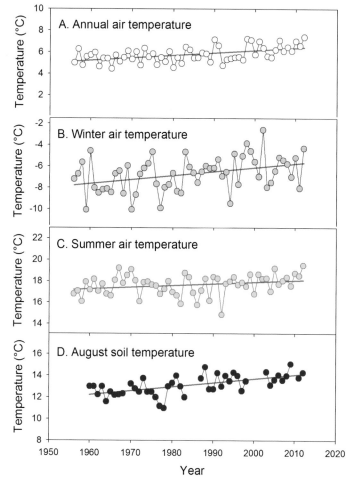

Figure 16.1. Trends in air and soil temperatures at Hubbard Brook over 50 years. All trends shown are statistically significant. Air temperature data are from weather station 1, located on the south-facing slope at an elevation of 1,364 m. Values shown for winter represent mean daily temperatures from December to February; those in summer from June to August. Soil temperature is measured nearby at a depth of 61 cm, with data shown indicating the mean soil temperatures for the month of August, the part of the year when soil temperatures reach maximal values. (A. S. Bailey et al. 2016)

weekly at fixed locations within the forest since the 1950s. The maximum snow depth recorded each year has declined at the rate of 4.8 cm/decade over the last 50 years, while the number of days with snow cover during the winter has declined by an average of 3.9 days/decade.[8] These changes are related to the generally

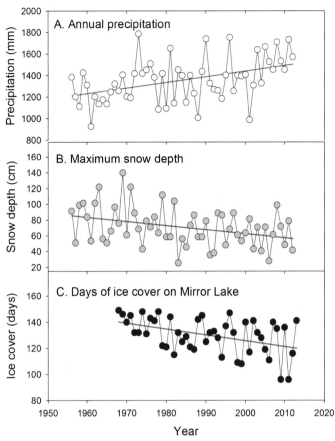

Figure 16.2. Trends in mean annual precipitation, maximum snow depth, and ice cover on Mirror Lake. All trends shown are statistically significant. Precipitation data are from weather station 1, and are calculated for the calendar year (January to December). (Precipitation and snow depth data from A. S. Bailey et al. 2016; ice cover data updated from Likens 2011)

warmer air temperatures in winter, the lower snowfall, and the greater frequency of thaws, sleet, and rain-on-snow events, which may lead to increased soil freezing.

- Duration of ice cover on Mirror Lake has been reduced (fig. 16.2C). The duration of ice cover is a good integrative measure of the heat budget for a lake (fig. 16.3). From 1968 to 1998, the period of ice cover on Mirror Lake declined at a significant rate of about −0.54 days per year.[9] In other words, by 1998, there were 17 fewer days of ice cover than in 1968. Between 1998 and 2009,

this trend did not continue steadily, even though ice-out in 2010 was the earliest date on record. In addition, the year-to-year variability in ice-cover duration increased significantly between 1998 and 2014. Over the entire period of record, from 1968 to 2014 (46 years), the change was about −0.48 days per year, or 22 fewer days of ice cover in 2014 than in the late 1960s. This change has come about primarily as a result of earlier breakup of the ice cover in spring, not later freeze-up in the autumn.[10]

These trends in climate recorded at Hubbard Brook are consistent with climate trends projected for the northeastern United States.[11] Climate change, however, is not just about increasing temperatures and precipitation, but also about the frequency and predictability of extreme weather events.[12] At Hubbard Brook, ice storms, heavy rain deluges, microbursts, and other extreme events have been documented over the past 50 years, which is too short a time to indicate whether these are increasing in frequency or not. In any case, the long-term record from Hubbard Brook will provide a basis for assessing the frequency of such events at some point in the future.

HOW DOES A CHANGING CLIMATE AFFECT THE FOREST ECOSYSTEM?

Climate potentially affects many components and processes of the forest ecosystem. Rising temperatures and increasing rainfall influence growth and production by plants, decomposition rates, and the flux and cycling of nutrients. In addition, these meteorological parameters influence the timing of many important biological phenomena, such as leaf expansion in spring, the length of the growing season, start and break from hibernation, reproduction of both plants and animals, and the arrival and departure times of migratory species. They also determine the suitability of the system for the establishment and survival of particular plant and animal species, and significant changes in climate may result in shifts in species distribution, both plant and

Figure 16.3. Ice melting on Mirror Lake in April. (Photo by D. C. Buso)

animal. Addressing the question of how the predicted climate-related changes affect the forest ecosystem is challenging but extremely important. We present here a few examples of the research at Hubbard Brook that bear on the effects of climate change.

Effects of warmer winters. One of the most obvious changes in climate at Hubbard Brook has been the warming of the winter period due not only to warmer temperatures but to more rain and sleet and, in general, more variability in weather patterns. One consequence is that the depth of the snowpack has been decreasing over time, and more rain events now occur during the winter, often leaving the ground and vegetation covered in ice rather than snow.[13]

This warming in the winter period has two major effects on nutrient dynamics in the system. One is to decrease the overwinter production of inorganic nitrogen when the soil is frozen.[14] The other is to increase the loss of nitrogen in stream water during the spring-to-summer transition and during the growing season due to lowered plant uptake.[15] The mechanisms underlying these effects have been determined at Hubbard Brook through a series of small-plot experiments. When snow is removed manually from the surface of these plots during the winter (fig. 16.4), inducing freezing of the upper soil layers, there is an increased loss of nitrate, phosphorus, and base cations (such as calcium

Figure 16.4. Experiments to test for the effects of warmer winter air temperatures involve shoveling snow to induce freezing in the upper layer of the soil. Depending on the experiment, snow is removed by shoveling within 48 hours of snowfall for the first five weeks of winter to mimic a later onset and accumulation of snow. A 3-cm snow layer is left on the ground during shoveling to avoid disturbing soil and litter and to maintain the albedo (surface reflecting quality) of a snow-covered forest floor. Researchers wear snowshoes while shoveling to reduce compaction (Templer et al. 2012). (Photo by A. Werner)

and magnesium) from the soil, due in large part to a reduction in the uptake of nutrients, particularly nitrogen, by plants.[16] The evidence thus far indicates that this reduced uptake is caused primarily by physical frost damage to the roots and to increased mortality of roots, especially the fine roots of forest trees.[17] Soil freezing, therefore, reduces the uptake of nitrogen by plants, making more nitrogen available to be lost by leaching. Snow removal had no measurable effect on nitrification processes or microbial biomass, suggesting that mineralization and other microbial processes are not significantly affected by soil freezing under those experimental conditions.[18] Furthermore, increased acidification of the soils due to soil freezing may also lead to reduced calcium:aluminum ratios in the foliage and potentially to reduced growth of twigs and foliage in the following season.[19]

These small plot-level experiments suggest how soil freezing may alter chemical processes in the soil and the uptake of elements by the plants, and illustrate how the effects of winter climate change on soil-nutrient-root relationships may have important consequences for ecosystem dynamics in future years. Whether or when these local effects might be large enough to alter whole-watershed nutrient budgets remains to be seen.

Soil freezing brought about by low snow cover also could affect soil-dwelling organisms, many of which are important decomposers, as well as their predators. For example, in another snow-removal experiment at Hubbard Brook, investigators found that the taxonomic richness and diversity (number and relative abundance of identifiable taxa) of soil arthropods were reduced by 30% and 22%,

Figure 16.5. A female moose surrounded by a sea of hobblebush and ferns. Hobblebush is an important food plant for moose, and the act of browsing affects nutrient dynamics in the forest ecosystem. (Photo by N. L. Rodenhouse)

respectively, in experimental (cleared, frozen soil) compared with reference (uncleared, unfrozen soil) plots.[20] The number of spiders was reduced by 57%, pseudoscorpions by 75%, hymenopterans by 57%, springtails by 24%, and larval flies by 33%. Thus, warmer winters with their diminishing snowpack will likely have significant effects on the soil arthropod communities, which in turn will influence the rates of litter decomposition, nutrient flux and cycling, food for other organisms such as birds, salamanders, and small mammals, and other ecosystem processes.

Snow depth and the duration of snow cover over the winter also influence browsing behavior and intensity of use by large mammals, such as moose (fig. 16.5).[21] This changed behavior can have unforeseen effects on other parts of the ecosystem. Low snow depth exposes understory plants, such as hobblebush and sugar maple and balsam fir saplings, to increased browsing. Experiments involving snow removal, increased browsing (simulated by mechanical cutting of vegetation), and fecal additions (with isotope-labeled nitrogen) indicate that activity of large herbivores coupled with changing climate via reductions in snow cover sets off a series of responses

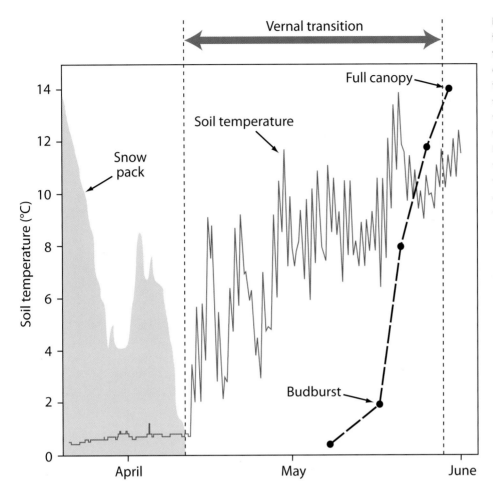

Figure 16.6. Spring phenology in the northern hardwood forest. The snowpack (shading on left) typically disappears in early to mid-April, leading to an abrupt increase in soil temperature (red line) as the sun passes through the leafless canopy and warms the forest floor. Budburst by the overstory trees (solid dots) begins in early May and the leaves are fully expanded by early June. The temporal relationship between these phenological events is changing over time, with consequences for nutrient cycling and flux and for other ecosystem dynamics. (Modified from Groffman et al. 2012; reprinted by permission of Oxford University Press)

in the system that result in an increased leaching of nitrogen (as nitrate) from the soil. The investigators found that any one of the experimental treatments alone did not result in significant alteration of nitrate leaching, but when combined (concurrent multiple stresses), balsam fir had the most significant response (highest loss of nitrate), followed by sugar maple and hobblebush. When plants are stressed by both herbivory (removal of foliage and twigs through browsing) and soil freezing (increased root mortality), as well as having additional nitrogen added in the form of feces, there is a lowered demand for nitrogen by the plants and potentially an increase in the loss of that nitrogen from the soil through leaching. This loss of nitrogen, if it were to occur at large spatial scales, would have many ecosystem consequences, including reductions in net primary productivity and impacts on water quality. This example illustrates the complexity of the interactions within the forest ecosystem, as well as the difficulties and challenges in unraveling the impacts of a changing climate.

Effects on the vernal transition. A critical period for the forest ecosystem is the transition from winter to spring (fig. 16.6). At this time, precipitation changes from mostly snow to rain, the snowpack melts, and days become longer, resulting in increased warming of the soil surface by solar radiation. Thawing of the soil, along with the increase in available soil water, promotes microbial activity, which increases the availability of nutrients, and leads to bud development and eventually leaf expansion, flowering, and increased photosynthetic activity of forest plants.[22]

Increasing precipitation, earlier snowmelt (wetter soils), and warmer soils—products of climate warming—influence biogeochemical dynamics by affecting nutrient uptake and loss. Such interactions in the hydrologic-nutrient cycle are important components of climate change effects, and are currently under intensive study at Hubbard Brook.[23]

Biological (microbial) activity in the soil increases when soil temperatures reach or exceed 4°C. This happens rapidly after the snow melts, with temperatures at the soil surface sometimes increasing by up to 8°C within a 48-hour period after snowmelt.[24] Nutrients released by microbial breakdown of organic matter become available quickly as the melt waters flush through the soil. Some of these nutrients are utilized by microbes, but others are taken up by perennial herbs, which suddenly begin to grow at this time (fig. 16.7). These spring ephemeral plants, whose growth is restricted to the period between spring snowmelt and summer canopy development, take up nitrogen, potassium, and other nutrients during this time. When their leaves die back in late May and June as the forest canopy becomes fully developed, these nutrients are returned to the soil, and become available for uptake by summer-growing plants or loss in stream water. In this way, it has been suggested that these spring ephemeral species contribute to a vernal dam, helping to reduce the loss of critical nutrients from the system during the spring thaw.[25] Plant roots also exude organic carbon in the spring, stimulating microbial uptake of nutrients in the soil.[26]

This important function of the spring ephemeral plants depends on tight synchrony with the timing of snowmelt and the availability of nutrients. Alterations in the timing of events during this vernal transition period, as might happen as the climate warms, would have important consequences. For instance, if warming temperatures were to affect the timing of melt and plant phenology differently, the belowground activity would become decoupled from onset of rapid plant nutrient and water intake, and lead to asynchrony in the timing of these events and a serious loss of nutrients from the system in stream water. At Hubbard Brook, the timing of snowmelt and the rapid increase in soil warming have advanced by about 14.3 days between 1956 and 2010, or 2.6 days per decade.[27] At the same time, the phenology of canopy closure (leaf expansion) has advanced only 6 days during that same period, or 1.1 days per decade. Such asynchrony in these events influences water and nutrient availability and decouples nutrient cycles, leading to increased loss of nutrients, potentially resulting in lower plant growth. Such changes will likely be important in how these forest ecosystems function as the climate becomes warmer.

Other phenological changes. In addition to the timing of events in early spring, meteorological conditions also affect many processes in the forest ecosystem. These include, but are not limited to, the onset and length of the seasons, leaf development by deciduous plants, leaf senescence in autumn, schedules and synchrony of emergence and mating activities of adult moths, the arrival and departure schedules of migratory animals, and the timing of reproduction and growth of both plants and animals.

At Hubbard Brook, not only is snowmelt happening earlier but also leaf fall in the autumn is occurring slightly later, resulting in a longer growing season for canopy trees (fig. 16.8). The earlier and warmer springs have resulted in earlier leaf-out by the deciduous trees. For example, budbreak by sugar maples has advanced at a rate of about 1.1 days per decade, or six days over the 50 years of the Study. Leaf-out in spring and longer growing seasons, along with the warmer temperatures, are likely to have a positive effect by increasing plant productivity and the rates of decomposition and nutrient cycling. In contrast, this warming effect will negatively affect the presence and distribution of some of the common plant species, like the conifers and sugar maple, whose ranges are predicted to shift northward with a warmer climate.[28] On the other hand, the warmer conditions and longer seasons may allow more southerly species, such as oaks and pines, to become established. The result will be a shift in the tree species composition of the forest, which will have effects on nutrient demands and flux,

Figure 16.7. Mats of trout lilies (and other spring ephemeral herbs) cover the forest floor in spring. These plants sprout, flower, and then the aboveground parts die during the vernal transition. (Photo by R. T. Holmes)

as well as the structure and dynamics of the forest food web.

Effects on bird distributions. Other components of the forest ecosystem are also likely to be affected by climate-caused phenological shifts. To take one example in detail, birds typically time their nesting to coincide with peak food availability for their nestlings. Studies in Europe have described a climate-induced mismatch between birds and the insects on which they feed, which has resulted in lower reproductive success for birds.[29]

No evidence exists thus far, however, for such a mismatch between birds and their insect food base at Hubbard Brook, based on long-term demographic data from the black-throated blue warbler population and its food supply.[30] Arrival times of this species, although varying from year to year, coincide with budburst of the dominant forest trees, while clutch initiation (laying of the first clutch of the season) is correlated with full canopy leaf expansion and with food abundance (caterpillar biomass) at time of egg-laying. Even though the arrival of spring is highly

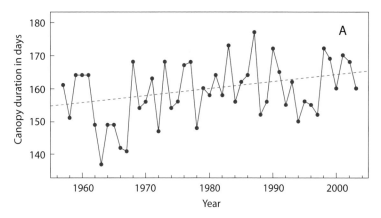

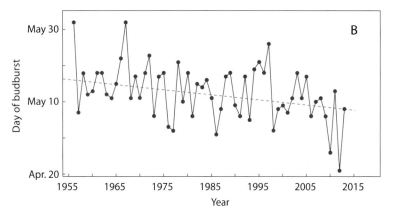

Figure 16.8. Duration of the deciduous forest canopy at Hubbard Brook has increased over the last 50 years, resulting from earlier leaf expansion in spring and later senescence in autumn (top). Canopy cover duration is calculated as the time between mid-spring (defined as when leaves have expanded to 50% of final length and obscure 50% of sky) and the midpoint of leaf fall in autumn (50% of leaves have fallen and leaves obscure 50% of sky). (Modified from Richardson et al., 2006; reprinted with permission of John Wiley & Sons, © 2006) Budburst by sugar maples, although highly variable, has advanced by an estimated 1.1 days per decade since 1956 (bottom). Day of budburst for this analysis was determined from field measurements of sugar maples near Weir 6, coupled with modeling of spring temperatures measured nearby. (Modified from Lany et al. 2015)

variable across years, the birds seem able thus far to adjust their breeding accordingly and thus maximize the number of young they produce. These warblers initiate breeding earlier in warmer springs and early breeders are more likely to attempt a second brood, both conditions resulting in the fledging of more young.[31] The question remains, however, whether this species and others at Hubbard Brook will be able to adjust to future changes in climate, which perhaps may be more extreme or unpredictable.

A frequently predicted consequence of a changing climate is that it will alter the distribution and abundance patterns of both plants and animals. The redistribution of tree species may be particularly important for birds, as their habitats are determined in large part by plant species and their associated insect faunas. Recent projections indicate that the breeding habitat for forest birds in eastern North America will

change the availability of suitable habitat for many species over the next 50 to 100 years.[32]

To test how climate change may affect bird populations in the Hubbard Brook valley, researchers have been examining the ecology of black-throated blue warblers along an elevation gradient.[33] Along this 600-m elevation gradient on the south-facing slopes of the valley, annual temperature decreases by an average of 2.3°C, and annual precipitation increases by 1.5 cm. Territories occupied by these warblers on the cooler and wetter upper slopes were found to be of higher quality, as evidenced by more food (insects) being present there, greater vegetation density (more places to find food), and by fewer predators, than those occupying sites lower in the valley. Also, warblers with high-quality territories at the higher elevations fledged more young per season and survived better (higher site return from year to year)

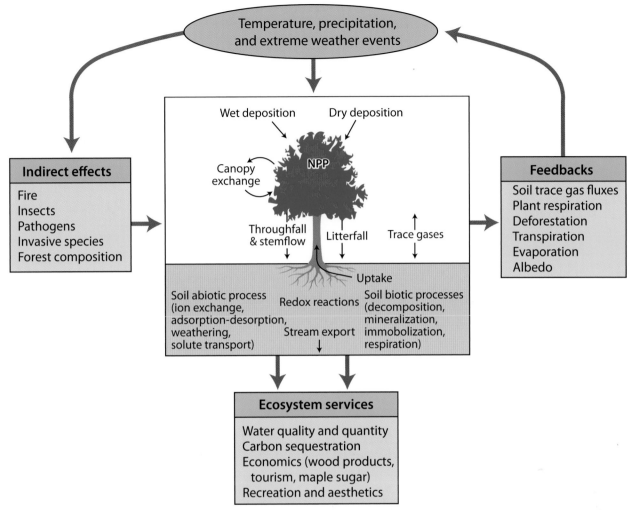

Figure 16.9. Summary of how temperature, precipitation, and other climate-related events affect ecosystem processes in northern hardwood forest ecosystems and the services they provide. Effects can be either indirect or direct and often influenced by feedbacks from other parts of the system. NPP = net primary productivity. (Modified from Campbell et al. 2009)

than did those living in poorer quality habitats at lower elevations. Whether or not birds are able to settle and occupy these higher quality sites at high elevations, however, depends on spring weather phenology. In warm springs, snow on the upper elevations melts sooner and the habitat is available for settlement when the birds arrive from their long migrations. In cold, late springs, the upper elevations are often still covered by snow when the birds begin to arrive, and in those years, more birds settle and become established at the lower elevations.

These findings illustrate how weather interacts with habitat quality for birds, which in turn affects bird settlement patterns, reproductive success, and survival. Because the upper elevational distribution of this warbler is set by the transition to boreal forest (at about 700 m elevation in the White Mountains), and because poor-quality habitat from the lower elevations is expected to expand upslope (and shift more northerly) as the climate warms, the hardwood habitats this species is best adapted to will be reduced, or at least be less suitable. Reductions in habitat quality will result in a shifting of the breeding range to the north, or for those remaining in the valley, lower fecundity and population density. The end result will be a population decline of this species at Hubbard Brook and probably in other parts of the northeastern United States.

∎

These are some of the indicators that climate change is occurring and having an effect on forest and associated aquatic ecosystems at Hubbard Brook. In some cases the effects are already obvious and measurable, and in others they are small or just beginning to occur. Some of these climate-induced changes directly affect ecosystem processes such as biogeochemical flux and cycling, soil processes, and plant physiology, while others act indirectly through changes in species composition, phenology, and hydrology (fig. 16.9). Continued long-term monitoring and study will be critical for describing and evaluating these changes and for predicting their future impact on the forest ecosystem.

PART 5

BROADER IMPACTS AND LOOKING TO THE FUTURE

Winter visitors on a tour of the Hubbard Brook forest.
(Photo by G. Wilson)

17

REACHING OUT
Hubbard Brook's Influence on Environmental Policy, Management, and Education

The research conducted at Hubbard Brook over the past 50 years has been extraordinarily productive and diverse, elucidating how forest and associated aquatic ecosystems work, how they respond to and are influenced by natural and anthropogenic disturbances, and how they provide important ecological services, such as clean air and water. The findings from the research are relevant in informing many national, regional, and local environmental issues, such as air pollution, water quality, and land management and conservation, and have been disseminated at many conferences and meetings, and published in a large number of scientific papers, books and reports (http://hubbardbrook.org/publications). They also appear in leading biology, ecology, and environmental science textbooks, and are used in many environmental science education programs in middle and secondary schools.[1] In addition, many university and postgraduate students have received training and experience in ecological research as participants in research activities.

We consider here these broader impacts of Hubbard Brook research, and the ways in which the research findings have been communicated to policy makers, natural resource managers, students, and the general public, including teacher professional development, curriculum development, training and research experiences for students and technicians, and other educational outreach activities.

CONTRIBUTIONS TO ENVIRONMENTAL POLICY AND NATURAL RESOURCES MANAGEMENT

Because of the increased understanding gained about how forest ecosystems work and how they are disrupted by disturbances such as atmospheric pollution and forest harvesting, Hubbard Brook research has many applications to policy and management issues. Here are three examples.

Acid rain and its impacts. The discovery of acid rain in 1963 at Hubbard Brook and its effects on forest and aquatic ecosystems has been one of the most significant findings of the Study. These results have played a major and influential role in the assessment

Figure 17.1. Gene E. Likens briefing President Ronald Reagan, the White House Cabinet, and members of the administration, September 1983. The briefing team was organized by William Ruckelshaus, the EPA administrator. (Photo from G. E. Likens)

of the scientific, economic, and political aspects of this major environmental problem, both in the United States and internationally, contributing importantly to a body of developing information that underpinned the passage of federal legislation, especially the amendments to the Clean Air Act in 1990. Moreover, the Study's high-quality, long-term data on precipitation and streamwater chemistry helped ward off aggressive attacks from various deniers and vested interests, and contributed to the development of federal government policy that led to regulations that resulted in less air pollution for the nation.[2] The long-term environmental monitoring at Hubbard Brook has since documented the subsequent declines in the deposition of sulfate and acidity, showing the effectiveness of these federal regulations.

Hubbard Brook scientists have published many scientific papers describing acid rain and its environmental effects.[3] They have also disseminated their findings through other media, including

op-ed pieces in leading newspapers, such as the *New York Times* and the *Wall Street Journal,* and popular magazines such as *Scientific American* and environmental journals.[4] Hubbard Brook scientists also deliver numerous lectures to professional and civic groups, and give interviews on local and national radio and television, which help spread the findings to larger audiences. They have provided information and input directly to policy makers and politicians by testifying before congressional committees and briefing high administrative officials, including President Ronald Reagan, his full Cabinet, the Environmental Protection Agency's administrator, and the president's science adviser (fig. 17.1). Scientists from Hubbard Brook have also served on technical subcommittees for the Environmental Protection Agency and other agencies, and committees formed by the White House and the Office of Science, Technology, and Policy. Similarly, information about the research findings has been communicated to local, state, and regional policy makers during visits to Hubbard Brook (fig. 17.2)

In a cost-benefit analysis of the Clean Air Act Amendments of 1990, which had a major focus on resolving acid rain in the United States, the Environmental Protection Agency estimated that for the period 1990 to 2020 the total costs of complying with these regulations would be about $0.4 billion per year, whereas the benefits are expected to be between $1.4 billion and $35 billion per year. These regulations have resulted in not only a cleaner environment but also a healthier one. By 2010, the 1990 amendments to the Clean Air Act are estimated to have prevented 160,000 deaths in the United States, primarily through reduction in atmospheric pollution. By 2020, the number of deaths prevented is estimated to be 230,000.[5]

Whole-watershed manipulations and forest harvesting. Another important topic on which Hubbard Brook data have been influential to policy development is in the evaluation of forest-cutting practices. The Study's whole-watershed manipulations involving different forest harvesting treatments have

Figure 17.2. Governor John Sununu of New Hampshire (second from left) touring a meteorological and precipitation sampling station at Hubbard Brook in the mid-1980s. Others in photo include Gene Likens (left), Herbert Bormann (orange shirt), Robert Pierce (second from right), and Richard Holmes (third from right). (Photo from G. E. Likens)

shown that the intensity and type of harvest can affect forest composition and growth rate, as well as streamwater quality and water yield. The long-term data from Hubbard Brook suggest that forest regrowth can be altered by different cutting practices, but that the forests eventually catch up in the accumulation of tree biomass after several decades.[6] But the more extreme practices of clear-cutting large blocks of forests or harvesting all trees at once result in marked increase in stream flow and loss of nutrients. More moderate practices, such as the progressive strip-cutting of Watershed 4, along with a protective buffer of living trees along the stream channel, reduced the loss of nitrates and other nutrients and promoted the establishment of desirable tree species such as yellow birch and sugar maple.[7]

When Watershed 2 was first deforested, the high nitrate concentrations in stream water leaving that watershed were alarming, in some cases exceeding U.S. Public Health standards by twofold. Similarly, high concentrations of nitrate were subsequently found in other commercial clear-cuts in the region,

including Watershed 101 at Hubbard Brook.[8] Modifications of forestry practices that attempt to reduce such loss of nutrients not only help to produce cleaner water but retain soil nutrients for future tree growth. These findings from the Study have important implications for sustainable harvests of northern hardwood forests.[9]

Decline in bird populations: implications for conservation. The results of the bird research at Hubbard Brook have implications for forest management and conservation practices. The research has shown that bird species respond differently to subtle changes in habitat (forest) structure and quality that are typical of forests undergoing natural succession. Some bird species have narrow preferences for forests of certain ages and characteristics, whereas others accept a wider range of forest conditions.[10] Management plans for creating or maintaining suitable habitat for a full spectrum of species therefore need to take into account which species are most likely to be affected and their particular preferences and requirements. Having stands of various ages is a good management strategy for conserving the widest range of bird species, including those that seem most specialized on forests of early and mid-successional stages.[11] This strategy can be achieved in part through natural disturbances leading to the presence of younger patches of vegetation (as in the shifting mosaic steady-state concept), but to support populations of threatened early successional bird species, maintaining a range of forest ages through carefully planned and well-thought-out forest harvesting can increase local populations.[12]

Other findings from the Hubbard Brook studies show that forest bird populations are strongly influenced by events and conditions in the breeding area, primarily through the factors that affect reproduction, such as weather, food availability, and nest predators. Plans for conservation and management must target high-quality breeding habitats that are crucial to the maintenance or enhancement of these populations. At the same time, winter habitat quality, again influenced by weather, food availability, and local population density during the nonbreeding season and almost certainly during migration also have significant effects on the survival of individuals and even on their reproductive performance once they return to the breeding areas. This finding implies that conservation and management plans need to take into account events throughout the annual cycle for migratory and probably most bird species. Information on the year-round population dynamics is therefore critical for the development of well-informed conservation or management plans.[13]

THE HUBBARD BROOK RESEARCH FOUNDATION'S SCIENCE LINKS PROGRAM

Results from research at Hubbard Brook have influenced public policy and helped to close the gap between ecosystem science and policy, especially with respect to air pollution regulations and forest management practices. The effort to make these findings known to a broader audience, including

policy makers and the general public, led in part to the establishment of the Hubbard Brook Research Foundation in 1993 (Box 17.1). One of the foundation's primary goals was to develop new initiatives that link ecosystem science to public policy.[14] An assessment by the foundation of the gap between ecosystem science and public policy indicated that ecosystem science was often underutilized and not well understood, and that effective communication between ecosystem scientists and policy makers was often lacking. To address both of these deficiencies, the foundation initiated Science Links, a program designed to integrate the findings from complex ecosystem science, as represented by results coming from the Hubbard Brook Ecosystem Study, with environmental policy. Supported initially by the Henry Luce Foundation, the Merck Family Fund, and other foundations, the goal was to synthesize and distill policy-relevant scientific data and to communicate that information to broader audiences, especially policy makers, stakeholders, and the general public.

Thus far, the foundation has completed four Science Links projects (fig. 17.3). Three of these focus explicitly on air pollution and its effects on the structure and function of forest and aquatic ecosystems in the northeastern United States. These deal specifically with increased emissions of sulfur and nitrogen oxides leading to acid rain ("Acid Rain Revisited"), dramatic changes in the nitrogen cycle ("Nitrogen Pollution: From Sources to the Sea"), and mercury emissions, deposition, and effects ("Mercury Matters"). The fourth ("Carbon and Communities") deals with carbon dioxide emissions and steps that might be taken to reduce them in various land-use settings.

The production of a Science Links project involves many steps. Topics to be addressed are chosen by one or more Hubbard Brook scientists, working in concert with foundation staff and trustees. Topics are based on pressing environmental issues of regional or national scope in need of synthesis and explanation and, in most cases, to which Hubbard Brook data can contribute. A project team is then assembled, usually consisting of 10 to 12 scientists,

Box 17.1. The Hubbard Brook Research Foundation

An unusual feature of the Hubbard Brook Ecosystem Study was the creation of the Hubbard Brook Research Foundation, a nonprofit 501(c)(3) organization whose mission is to support and sustain the Study by owning and managing housing and laboratory facilities for scientists and their research teams; to bridge the gap between ecosystem science and public policy by enhancing the exchange of information among scientists, policy makers, and land managers; and to foster public understanding of ecosystems and their importance to society by

Pleasant View Farm, which has provided housing and laboratory facilities at Hubbard Brook for researchers and students since the mid-1960s. (Photo, circa 1978, by S. P. Hamburg)

providing educational programs and other outreach activities. Established in 1993, the foundation is governed by a board of trustees, about half of whom are scientists, mainly from the Study, and the other half interested and committed supporters of ecosystem research from the community and the region. A citizen advisory council also provides input to the policy and education outreach processes, and an executive director and staff carry out the functions and activities of the foundation.

The foundation supports the scientific community at Hubbard Brook in numerous ways, from providing physical facilities for field researchers to publicizing the results of studies. The living and work facilities have been particularly crucial to the development and maintenance of the Study. In the early years, most researchers lived and worked together at Pleasant View Farm, a nearly 200-year-old farmhouse, which once served as an inn and stagecoach stop. Pleasant View provided a melting pot that promoted discussions, collaborations, and long-lasting friendships among the residents, and significantly enhanced the research effort. Today, these opportunities have been greatly expanded by the addition of a group of cottages and study facilities for researchers, located on Mirror Lake between Pleasant View and the forest itself.

As the educational and outreach arm of the Study, the Hubbard Brook Research Foundation links the environmental science results from Hubbard Brook research to public policy, and promotes environmental literacy in both formal and informal science education. Major contributions include creating and distributing public school environmental education curricula, publishing summary reports of studies of immediate public importance, and maintaining a website making available these as well as a database of all the scientific publications relating to the Study (http://hubbardbrookfoundation.org). Furthermore, foundation personnel arrange visits to Hubbard Brook for public officials and influential business leaders, organize high-level seminars on special topics, and manage informal events that facilitate interactions among scientific researchers and between scientists and the public.

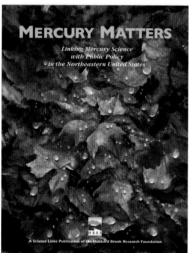

Figure 17.3. Four Science Links reports produced by the Hubbard Brook Research Foundation. (Reproduced by permission of the Hubbard Brook Research Foundation)

some of whom conduct research at Hubbard Brook and some of whom do not, and four to six policy advisers chosen from relevant state and federal agencies. The team reviews and discusses the issue under consideration, and examines and refines the question(s) being addressed to ensure their relevance and appropriateness and to decide how best to synthesize the available information. A postdoctoral fellow is usually engaged to coordinate data collection and management and to perform analyses including modeling to assess the consequences of alternative policy scenarios. Long-term data, often from Hubbard Brook, form a key part of most of these projects. The team meets periodically to measure progress, refine the questions as needed, review the analyses and models, and eventually to arrive at a consensus of the results. This process typically takes three to four years to complete. Funding of the project is entirely from private foundations, with seed money provided by the U.S. Forest Service and the foundation.

All of this work and deliberation results in a Science Links report, one or more peer-reviewed publications, often in *BioScience, Environment,* or other environmentally oriented journals, fact sheets, and other summary documents, all based on the syntheses and conclusions from the Science Links team. The Science Links reports are written to be understandable by a wide range of readers, with conceptual diagrams illustrating basic principles as well as research results. These reports are distributed widely and available for downloading from the foundation's website (http://hubbardbrookfoundation.org). The reports and publications review the state of the science, and the likely responses and unintended consequences of multiple scenarios. There is no advocacy for specific policy outcomes—only a synthesis of the science and a clear consideration of the policy options.

The work doesn't stop there, however. Once the reports are complete and are in the process of being published, the foundation staff and team scientists meet with appropriate audiences to disseminate and discuss their findings.[15] These meetings have included congressional briefings and testimony, and presentations to state and federal agencies, natural resource managers, science educators, and the scientific community. Press releases are distributed for local, state, and national media, and press conferences are often held. The materials produced are distributed to the public and especially to teachers. For the "Acid Rain Revisited" project, a teacher's guide was also developed ("Exploring Acid Rain: A Curriculum Guide and Resources for Teachers of Grades 7–12"), which is available on the foundation's website (http://hubbardbrookfoundation.org/exploring-acid-rain-a-curriculum-guide).

Although producing a Science Links project is a demanding and time-consuming process, it is an effective one. The first three projects have resulted in three core and nine peer-reviewed publications, more than 100 policy briefings, and nine documents for nontechnical audiences.[16] The fourth project, on carbon use, takes a practical approach, summarizing information for county planners, municipal agencies, and local governments in making land-use decisions and other policies intended to reduce carbon emissions. The publications are frequently cited, and are in demand from a wide variety of potential users.[17] Thus, the Science Links program serves an extremely important role in the Study by communicating important research results and their implications for policy makers, land managers, and the general public,

and helps to bridge the gap between ecosystem science and public policy.

EDUCATIONAL OUTREACH FROM HUBBARD BROOK

Although Science Links has become an important and effective way to reach policy makers and other audiences, Hubbard Brook scientists and their graduate students and postdoctoral fellows and especially the staff of the foundation are also actively involved in science and environmental education. Some examples of these educational outreach activities and learning opportunities promoted by the Hubbard Brook community are as follows.

The Hubbard Brook Roundtable. The roundtable, initiated in 2006, engages leaders in forest science, governmental agencies, the timber industry, and other interested parties in a process of facilitated dialogue to connect research findings about forest ecosystems with public policy and management strategies. Topics deal with environmental issues of local, regional, and national concern, such as ecological forest harvest, water availability and use, and the local effects of climate change. Outputs of Hubbard Brook Roundtables have included white papers co-written by foundation staff and roundtable participants, op-ed pieces published in local, state, and national newspapers, media coverage, and scientific publications.

Forest Science Dialogues. The Dialogues program is an informal science education venture supported by the National Science Foundation's Advancing Informal STEM Learning (AISL) program. Forest Science Dialogues began in 2014 with a goal of expanding the Hubbard Brook Roundtable model to bring Hubbard Brook scientists together with stakeholders and community members from the region for action-oriented dialogue about social, economic, and policy-related concerns in the rural regions of northern New England. The objective is to open and sustain new channels for communication between scientists and nonscientists, for the sharing of information, knowledge, perspectives, and ideas in multiple

Figure 17.4. Jacquelyn Wilson, science educator with the Hubbard Brook Research Foundation, describing findings from the Study to a group of middle-school students. (Photo by R. L. Pinsonneault)

Figure 17.5. Student group from Dartmouth College on a tour of the recently harvested Watershed 5, led by Wayne Martin (second from left), U.S. Forest Service. (Photo by N. L. Rodenhouse)

directions: local citizens learning from Hubbard Brook scientists and Hubbard Brook scientists learning from local citizens.

The Environmental Literacy Program (ELP). This program is a cooperative effort between the U.S. Forest Service and the Schoolyard LTER program of the National Science Foundation, designed to support formal science education through teacher professional development, the development of data-based teaching resources, on-site field trips for visiting schools and the general public, and cooperation with other regional groups engaged in science education (fig. 17.4). As a member of the New Hampshire Education and Environment Team (NHEET), the foundation supports education for grades K through 8 by providing teacher professional development that focuses on science process skills and ecological literacy. The foundation's educational efforts contribute to science education in grades 7–12 through the development of data-based lessons and also teacher professional development. Foundation

staff members, in consultation and collaboration with Hubbard Brook scientists, work with local teachers to develop classroom resources using results from Hubbard Brook research.

Research Experience for Teachers (RET). This program offers opportunities for teachers to gain research experience during the summer, working in collaboration with ongoing research teams. Scientists and foundation staff work with these and other local teachers during the school year to develop and test new materials to aid in science education.

Research Experience for Undergraduates (REU). Dozens of students each year are involved in research activities at Hubbard Brook, either as field assistants working as part of the research teams or through more organized Undergraduate Research Participation and Research Experience for Undergraduate programs supported by the National Science Foundation. Students come from many different colleges and universities, and work closely with Hubbard Brook scientists during the course of a summer, participate in seminars and discussions, and frequently develop their research into an undergraduate honors thesis

Figure 17.6. Scientists, students, and visitors at the 2003 Hubbard Brook Cooperators' Meeting, an annual event in which researchers present their ideas and results. The Robert S. Pierce Ecosystem Laboratory of the U.S. Forest Service is in the background. (Photo from Northern Research Station, U.S. Forest Service)

at their home institutions. Research for more than 75 undergraduate honors theses has been completed on studies conducted at Hubbard Brook, and the bird research team alone has involved more than 500 students as field assistants and interns during the past 47 years.

Tours, Lectures, and Discussions. Scientists, foundation staff, and U.S. Forest Service personnel are actively engaged in educational and outreach activities. They conduct tours of the Hubbard Brook forest for student groups, legislators, and the general public (fig. 17.5), in which past and ongoing research activities are demonstrated or discussed. Talks, both formal and informal, are given to local schools, community groups, and other special events. For

those farther away, there is an online walking tour of the forest (http://www.hubbardbrook.org/w6_tour /w6-welcome.htm).

The Hubbard Brook Cooperators' Meeting. This is an annual conference involving scientists from the Study, their students, postdocs, technicians, field assistants, and visiting scientists, held to foster exchange of ideas and results (fig. 17.6). Officials from nonprofit organizations, state and federal agencies, educators, international visitors, as well as interested members of the public also attend these conferences.

■

Another important outcome of the Study has been to provide experience and professional training to hundreds of students and technical support staff over the past 50 years. More than 100 Ph.D. students, dozens of postdoctoral fellows associated with Hubbard Brook scientists, and many technicians have participated in the research and have contributed in important ways to the discoveries made. Many of these individuals have continued in professional activities and now occupy prominent positions in academia, government, nonprofit organizations, and other agencies throughout the country and abroad.

Through all of these outreach and educational activities, the results of Hubbard Brook research are made available for use by the wider scientific community, policy makers, land managers, and the interested general public. Perhaps most important, generations of students have been exposed to Hubbard Brook, its researchers, their unique way of thinking about ecosystems, and the findings from the Study that help to understand and manage the natural world better.

18

A LOOK AHEAD
The Forest Ecosystem in the Future

The research conducted at Hubbard Brook has provided a treasure trove of information about the northern hardwood forest and associated aquatic ecosystems, which provides a solid foundation for predicting what might happen in the years ahead. Even with our knowledge from these long-term studies, however, our crystal ball is no clearer than anyone else's when it comes to making predictions. There are a great many uncertainties, but the ecosystem approach and ecosystem thinking provide a clearer window for seeing into some predictions about the future.

Forest ecosystems generally are highly resilient to change and have great recuperative capacity after disturbance, but in the 50 years of our Study, both the environment and the ecosystems at Hubbard Brook have changed markedly. We would expect further change in the next 50 years. Although our predictions may be speculative, we think it is important not to be caught totally unaware if the events we describe were to occur. This effort is somewhat similar to disaster planning, where being ready and informed might alleviate the worst of possible outcomes, yet also provide important opportunities for mitigation and additional learning.[1] The long-term data from Hubbard Brook also provide a good baseline against which we can judge the impact of changes that might occur in the future to the ecosystems at Hubbard Brook and more generally to forested regions across the northeastern United States and southeastern Canada.

What are some of the major issues for consideration in the future? What will the vegetation, animals, biogeochemistry, and hydrology of Hubbard Brook and nearby areas be like in 20 years, or 50 years? What will be the ecosystem effects resulting from diminishing acidity in precipitation, from climate change, or from invasive species and disease (ticks and Lyme disease, emerald ash borer, hemlock woolly adelgid, gypsy moth), and others we haven't even thought about?

SOME PREDICTIONS

Air quality and atmospheric pollution. The acidity of precipitation is currently declining, but if the federal controls established over the past several decades and switching to cleaner fuels were to be reduced, this situation could reverse quickly. Because of the forest's increased sensitivity to acid inputs following the degraded buffering capacity of soils due to leaching of base cations, such as calcium and magnesium, over the past six decades, forest health could then also decline rapidly and markedly if acid inputs were to increase.

Because of tighter air quality controls, atmospheric deposition of nitrate and calcium also are declining at Hubbard Brook, which might be expected to continue in the future. As important nutrients for the forest ecosystem, major decreases in these inputs also could affect the composition, quality, and vigor of the vegetation, and decreased calcium inputs could further affect the soils' buffering capacity. The future of air quality is uncertain because aspects of it are politically driven, especially regarding whether federal regulations will be maintained, increased, or decreased. Indeed, more combustion of coal and oil in the United States could result in increased emissions of sulfur and nitrogen oxides if not accompanied by equivalent environmental controls and would, in turn, slow the recovery from acidification and shift the pH of precipitation and stream water to much lower and less desirable values.

With the overall depletion of nutrients in the ecosystem and declining inputs from the atmosphere, where will the nutrients come from to fuel growth of organisms in the future? This problem could become very serious for forest growth, aquatic organism survival, and carbon sequestration in the next 20 years.

Using information from the long-term study of Mirror Lake, including paleoreconstruction of historical changes in its catchment, we can make predictions about potential changes in surface water chemistry in the Hubbard Brook valley.[2] The chemistry of Mirror Lake was characterized by calcium bicarbonate at a pH between 7 and 7.5 for more than 13,000 years, from the lake's formation 14,000 years ago to about 200 years ago. About 200 years ago, increased concentrations of nitrate and phosphate entered the lake in drainage water from human activity, such as limited agriculture in the catchment. Some 65 years ago, the water in Mirror Lake changed to become dominated by calcium and sulfate at a pH of 6 to 7 from the inputs of acid rain, and about 35 years ago, to water characterized by calcium, sodium sulfate, and chloride at a pH of 6.4 to 6.5 from the combined pollution of acid rain and road salt added to nearby Interstate Highway 93 and access roads around the lake.[3] Headwater streams have experienced this same general rapid change in chemistry, but they are much less affected by road salt than Mirror Lake and much more acidic currently than water in Mirror Lake. Headwater streams are decreasing in acidity now, in accordance with the decline in acidity of precipitation, and are expected to follow this trend in the future, assuming that emissions continue to decline.

What will be the chemistry of these surface waters in 20 or 50 years? We don't know, but we can make some predictions.[4] The chemistry of precipitation and stream water are on robust trajectories to become extremely dilute in only a few years, so in 20 years it is likely that headwater streams will be even more nutrient poor and less acidic than was the case before the Industrial Revolution. Water in the streams therefore should continue to look clean and pure (fig. 18.1). Looks, however, may be deceiving if they become too dilute.

Precipitation has become less acidic in recent decades, with the trend actually accelerating over the past 15 years in response to reductions in emissions of sulfur and nitrogen oxides. Currently precipitation values appear to be headed toward a pH of 5.5 by 2020. Surprisingly, streamwater pH is responding in parallel, rising steeply, also with a high correlation coefficient for the trend. This upward trajectory is expected to cross above an average pH value of 5.5 within in the next few years.[5] Such high average pH values probably have not occurred in stream water at

Figure 18.1. Based on projections from trends measured at Hubbard Brook over the past 50 years, stream water is predicted in the future to have lower nutrient content and a higher pH than in the last decade—that is, to be more like distilled water. (Photo by R. T. Holmes)

Hubbard Brook since about 1950.[6] Even so, meltwater events in the headwater streams can still be expected to have much lower pH values and higher toxic aluminum concentrations that would be harmful to aquatic organisms.[7] These trends raise the question of what the *maximum* pH of stream water in the gauged watersheds at Hubbard Brook will be in the next 20 years, as acid rain continues to decline. This question is not only of intellectual interest: pH is important in the mobilization of aluminum (which is toxic in the dissolved form), weathering of minerals, and the welfare of aquatic organisms.

It is highly relevant to point out that on June 25–26,

2014, 12.2 cm of rain fell at Hubbard Brook within an 8-hour period, and was the most dilute precipitation event in the Study's history. Rain falling that day had a specific conductance of 2.4 μS/cm and a pH of 5.2, already exceeding the predictions that we had made previously, and also exceeded the levels we had established for specific conductance before the time of the Industrial Revolution. This could be a harbinger of conditions in the future. Such dilute waters (precipitation and stream water), if they were to occur on a consistent basis, would present serious ecological problems—for example, because of osmotic and nutrient constraints for aquatic organisms, nutrient availability concerns for terrestrial organisms, and relationships and reactions for biogeochemical dynamics within the ecosystem. Streamwater chemistry has been relatively stable for thousands of years at Hubbard Brook, but now is poised to change relatively quickly and markedly to values not occurring here previously. The full ecological and biogeochemical consequences of such changes are not clear.

Temperature and climate. Warmer and potentially wetter climate is predicted for New England in the future, but with enhanced hydrologic extremes, for example, more frequent drought and extended low stream flow periods in summer.[8] Normally soils do not freeze during the winter at Hubbard Brook because of a deep snow cover and a thick humus layer. Occasionally, because of a thin or nonexistent snow cover (for example, following a January thaw), soils may freeze. Soil frost can mobilize significant nitrate loss in stream water from the forest. If soil frost were to become less frequent and less intense in response to warmer winters, would nitrate export in stream water continue to decrease on a long-term basis? Or, if snow cover were to become less with climate change, as is expected, would soils freeze more frequently and nitrate and hydrogen ion losses in stream water increase?[9]

Hydrology. If evapotranspiration were to decline significantly as a result of increased temperature, atmospheric carbon dioxide, and change in dominant tree species, then annual stream flow would increase, resulting in increased nutrient export, increased erosion and reduced effectiveness of in-stream processing of nutrients in reducing nutrient export. The increased trend in precipitation amount would exacerbate this response. Would the strong relationship between annual precipitation and stream flow deteriorate with this scenario, and what would be the implications for downstream water quantity and quality?

Vegetation. In response to climate changes, we would predict a surge in the growth of red spruce, because of increasing temperatures, especially during

Figure 18.2. Red spruce trees might improve in health following the decline in acid precipitation. (Photo by W. S. Schwenk)

the winter period, increasing precipitation amount, and less acid precipitation (fig. 18.2). Decreasing acidity of precipitation would reduce the leaching of calcium from needles, which provides protection from extreme cold, and combined with increasing temperature, particularly during winter, would result in *less* winter damage to red spruce. This response already seems to be happening in some areas of the northeastern United States.[10]

Ultimately, however, we might expect decline and then loss of red spruce and balsam fir from the valley, because rising temperatures would exceed the tolerance levels for these species. There also could be a loss of sugar maple, except at higher elevations where pockets of suitable soil exist, because of increased temperature; American beech, ash, and hemlock could be greatly diminished because of invasive pests and pathogens (mature beech already has been devastated from beech bark disease, and the hemlock adelgid is on the horizon). White pine, red and white oak, vines, and various exotics might replace these current dominant species, but yellow birch probably would persist.

Such changes could have large effects on the

hydrology and biogeochemistry of the system. For example, rates of nitrification of sugar maple leaves on the forest floor is much higher than for other tree species, favoring faster nitrogen cycling and larger loss of nitrogen in stream water.[11]

Animals. As the forest ecosystem responds to climate change, the animal species are also likely to be affected. Warming temperatures, changing precipitation regimes, and greater variability in weather will affect many species through changes to their survival or reproduction. Birds that reside year-round in the forest, for example, are likely to benefit from climate warming, especially warmer winters.[12] Such species as the white-breasted nuthatch, downy and hairy woodpeckers, and black-capped chickadee, whose numbers are limited mostly by winter conditions, might be expected to increase. Similarly, birds whose northern range limits are set by winter temperatures, such as tufted titmouse and northern cardinal, should increase as the climate warms. Neither of the latter species is found regularly in the forest at Hubbard Brook today, but they might be expected to be there in the not too distant future.

Breeding ranges and therefore local abundances of some species may shift or contract as climate changes.[13] For instance, model projections for black-throated blue warblers indicate marked contraction of suitable habitat under high anthropogenic carbon dioxide emission scenarios.[14] Similarly, the Bicknell's thrush, which in New England is restricted to high-elevation spruce-fir forests and occurs only in very small numbers along the ridge tops at Hubbard Brook, is particularly vulnerable to a warmer climate, especially if the balsam fir and red spruce were to disappear. It has been estimated that a further increase of as little as 1°C will reduce suitable habitat for this species in the New England area by more than half.[15] Given current rising temperature trends, it is unlikely that the very small numbers of this species that presently occur along the ridgelines of the Hubbard Brook valley will persist for long. Likewise, blackburnian and blackpoll warblers, species that are closely associated with stands of red spruce and

balsam fir, will likely decline in the valley as those tree species retreat northward, while more southerly species, such as red-bellied woodpeckers and Carolina wrens, might be expected to increase.

Not only will shifts in the distributions of major tree species and perhaps entire plant communities affect the occurrence and viability of birds and other animals in the forest, but some tree species may be greatly reduced or even entirely eliminated through the actions of invasive insects, such as the hemlock woolly adelgid or the emerald ash borer. The loss of hemlock might be expected to affect the occurrence and numbers of various animal species, such as black-throated green warblers and perhaps blue-headed vireos.[16]

Warmer winters with less snow cover and a greater frequency and extent of soil freezing events are likely to have negative impacts on hibernating animals, such as the woodland jumping mouse, most amphibians, and many overwintering insects (all Lepidopteran larvae, for example). Increasing temperatures in both summer and winter may lead to the disappearance of moose from the valley and perhaps to an increase in the deer population. The impact of such a change on browsing of vegetation like hobblebush, particularly at high elevations, could be very large.[17] Also, the predicted increased frequency of droughts that would result in lower stream flows and drier upland conditions, particularly during summer, could have serious survival consequences for salamanders and other amphibians.[18] Moreover, changes in plant species and their associated insects may influence habitat quality for salamanders, black-throated blue warblers, and other songbirds by influencing food availability and predator numbers.

The tiny black-legged or deer ticks and the somewhat larger dog ticks have been absent from Hubbard Brook, but they are already starting to appear at lower elevations in the forest. The causes for this invasion may be related to climate change, allowing ticks to expand their range,[19] or possibly to their increased transport from highly infected areas on pets by second-home owners in the region. Whatever the cause, these ticks are well-known vectors for transmitting disease (including Lyme disease) to humans. The presence of ticks would greatly change the experience of walking through the understory of the forest and on various animal-to-animal interactions. The much larger ticks found on moose already are a major detrimental factor to the health and survival of this large herbivore. Current evidence suggests that these parasitic ticks survive better in warmer winters, and thus their impact on moose is increasing with climate change.[20]

Understanding how animals, including humans, will respond to these future environmental changes remains a major challenge for scientists at Hubbard Brook and elsewhere.

SOME LARGE-SCALE IMPLICATIONS OF THESE CHANGES

Water quality. Hubbard Brook is a tributary to the Pemigewasset River (2,644 km² area), which in turn drains into the Merrimack River (12,200 km² area), a major source of drinking water in an area of New England with nearly 200 communities and over 2 million human inhabitants.

There have been unsuccessful attempts to reintroduce Atlantic salmon into the Merrimack River watershed, most recently in 2012. This species requires high-quality headwater streams, which includes some that are found in the upper Pemigewasset River system, including Hubbard Brook. If such reintroductions were to be successful in the future, it would be essential to evaluate and manage the impacts of disturbances on water quality in these smaller headwater catchments to understand and protect the larger system for maintaining salmon habitat. Because they are relatively small in area and chemically dilute, headwater streams are among the most sensitive of aquatic resources to be affected by pollutants and climate change, and these changes are reflected downstream. How might the greater frequency and intensity of extreme disturbances (wind, drought, floods, ice storms) resulting from

climate change affect the quantity and quality of water in this larger drainage system in the future?

Acid deposition has thoroughly permeated the Hubbard Brook valley (and all of New England) since the mid-1950s. The watershed-ecosystems at Hubbard Brook now have maximum sensitivity to acid deposition because of reduced buffering provided by the native geologic substrate and from loss of base cations in the soil during previous decades. Nevertheless, the acidity of precipitation is declining, and streams are now recovering from acidic deposition and should have much lower concentrations, including acidity, within two or three decades if deposition rates continue to decline. The full impact of strong acid anions, such as sulfate and nitrate, depends on the availability of neutralizing base cations in the substrate, and on hydrologic pathways through the soil. Larger, more complicated watersheds, with longer flow paths for water, probably will be slower to off-load acid sulfate and nitrate that have accumulated from atmospheric deposition, even with reduced deposition now occurring.[21]

These findings have important ramifications for setting and maintaining water-quality standards downstream, including the requirements for young salmonids in the headwaters. Human demand for high-quality water almost certainly will increase in the coming decades, which must be an integral part of these scenarios and will require careful planning, management, and monitoring.

Timber harvest. Whether there will be a resurgence of timber harvesting for fuelwood on a massive scale in the White Mountains and surrounding areas is unknown, but given recent projections concerning the increasing size of human populations worldwide, and the availability and economy of fossil fuels and other energy sources to meet their demands, increased fuelwood harvesting is a distinct possibility.[22] A return to earlier conditions in New England when forests were overharvested is not desirable. That prospect was sketched out graphically by Dr. Seuss and the Lorax, with his beloved Truffula trees being cut and removed while the skies filled with ugly pollution.[23] The scenario projected in *The Lorax* certainly must be avoided, but it provides everyone, the public as well as foresters and land managers, a vivid reminder of what could happen without careful protection, conservation, and management of forest ecosystems. Forests today need to be kept or managed as functioning ecosystems that provide fresh air, clean water, habitat, and other ecosystem services as well as forest products.

Forest and associated aquatic ecosystems provide many services on which humans depend, such as high-quality water, air filtration, local climate modification, aesthetics, recreation, wood products, and food. All of these services are fueled ultimately from the energy of solar radiation. Finding ways to avoid or mitigate the more negative predictions (mostly human caused) given here is a major goal in protecting these vital ecosystems that provide so many of the life support systems for humans.

The beauty we see in the complexity of a forest ecosystem and the vital services these natural ecosystems provide must be the guiding principle in managing the future for what we want to see in 50 years when we look again into that crystal ball. F. H. Bormann and Likens in 1977 aptly summed up the importance of understanding how ecosystems function:

To the untrained eye, forests appear static; in reality, they are sites of intense activity. Each year, millions of gallons of precipitation, trillions of calories of energy, and vast tonnages of gases flow into each square mile of forest. Thousands of plant, animal, and microbe species use these elements to live and reproduce, regulating the flow of water, energy, and nutrients through the ecosystem. Forests, with their complement of animal and plant species, provide immensely important benefits to humans—modifying stream flow and erosion, filtering the air and water that constantly flow through the landscape, and releasing sediment-free water to underground and surface water supplies. Stands of trees also modify local climate by moderating temperatures and humidity. . . . Although arguments for the preservation of natural land in regions undergoing development are most often based on the need for recreation areas, wildlife preserves, and parks, a more basic argument exists, namely, the biologic and economic health of [humans].[24]

EPILOGUE

Step into the Forest—2065

SPRING

As you approach the forest on this sultry spring morning, heavy black storm clouds are gathering to the northwest. It is unseasonably warm, and there is a smell of rain in the air. You stop to tuck pant legs into your socks and apply some insect repellent with DEET to your pant legs, socks, and hiking boots, and maybe the sleeves of your long-sleeved shirt, to help ward off disease-bearing ticks lurking on the underbrush. You notice a low-flying unmanned drone overhead collecting data on the vital characteristics of the forest, such as plant growth and chemistry, soil moisture and temperature, streamwater chemistry, the locations and movements of birds, mammals, and invertebrates, and audio recordings of their calls. The drone is also downloading data collected electronically and remotely at several locations in this "smart" forest. You send a signal to the drone that you are about to enter the forest and turn on your personalized electronic internal health-monitoring system.

You step into the forest and let your human senses take over. There are many standing dead and recently fallen sugar maples, ash, and hemlocks—victims of climate change, invasive pests, nutrient depletion, and violent storms. The brook is running full and fast, but not overflowing its banks. If you look carefully you may see a caddisfly in its case or other stream invertebrates, but they are rare because the stream has been depleted of nutrients. The running water makes a loud roar as it tumbles over large boulders in the streambed and rushes through the weirs.

The earthy smells of decay and volatile organic compounds fill your nostrils as you make your way through the thick understory of hobblebush and beech saplings. Signs of deer browsing on the understory plants are common, because the deer herd is quite large. Stand quietly and you might see a small fawn or two. Spring herbaceous ephemerals, like spring beauty, red trillium, and trout lily, are common, providing points of color and beauty on the surface of the otherwise drab forest floor. As you entered the forest, you probably noticed invasive garlic mustard plants growing in profusion along the edge of the road.

Floods might increase with increasing precipitation and more intense storms. (Photo by D. C. Buso)

SUMMER

The heat and humidity are oppressive, but it is cooler inside the forest under the shade of the canopy. It feels like conditions are right for a strong thunderstorm, which is common in summer. Mosquitoes and black flies seem to like these conditions, and their swarms are a nuisance even with a head net. Bats have been decimated by disease and are no longer here to feed on large numbers of mosquitoes and other flying insects. Falling frass is obvious, by both sound and sight, because of the large numbers of defoliating caterpillars feeding high in the canopy.

Pin cherry and red raspberry are growing in the larger openings in the forest. Try a raspberry—they are very tasty. The bears think so too! If you stay on the network of raised boardwalks, which has been provided by the Forest Service to accommodate visitors and scientists, the walking is easier and less damaging to the forest floor. Scratch away some of the thin forest floor material and look for invasive earthworms, which are now abundant in the soil. These earthworms are actively consuming the surface organic matter, recycling nutrients, and changing the species composition of forest floor plants and shrubs.

The weather has been sunny and hot. Now it looks like rain, however, so be prepared. The rain would be welcome, since drought conditions occurred earlier in the summer and stream flow has been minimal for most of the season.

AUTUMN

This season of transition is a beautiful time to walk into the forest. The biting insects are mostly gone, temperature and humidity are lower, and the color season is greatly extended compared with earlier years, lasting well into November before the leaves fall from the deciduous trees. Colors aren't as vibrant now without the sugar maples, but the oaks and pin cherries provide fall food for the small mammals of the forest that are getting ready to hibernate. You see a large patch of genetically modified sugar maples that has been planted by the Forest Service as a test of their hardiness in this new climate. It is said that the maple

Biting insects, such as mosquitoes and black flies, are abundant, and you slip on a head net for protection. There have been recent reports of major West Nile virus outbreaks in the area. This virus is carried by mosquitoes and infects birds and humans and other mammals. Besides the usual chickadees and warblers, you might see or hear red-bellied woodpeckers, Carolina wrens, cardinals, or titmice, species that would not have been seen in the forest 50 years ago. You might even see flocks of crows in forest gaps or along the forest edges.

The snow has melted and the forest canopy has already leafed out, so the light is relatively dim inside the forest. There are large openings, however, where mature trees have died or blown over, leaving gaps in the canopy where sunlight can penetrate and form large patches of brilliant light on the forest floor on a cloudless day. Around you are yellow birch and oak trees and many vines, such as woodbine, wild grape, poison ivy, Virginia creeper, and multiflora rose, which cover many of the trees and saplings. These vines weren't seen in the forest a few decades ago.

syrup they produce is not as good as from the original trees. The leaves of woodbine and poison ivy are a brilliant scarlet.

As you walk through the forest, the patchwork of downed trees and tangled brush will be obvious and difficult to traverse in places, as all this debris has played havoc with the boardwalks. Many trees were knocked over by recent hurricanes, which are now much more common in the area. If you look carefully, you will see other species of invasive vines appearing on the edges of the disturbed areas. Ferns cover the forest floor beneath smaller openings in the canopy.

Listen carefully and you can hear the brook gurgling along, carrying its load, albeit a much smaller load, of dissolved nutrients downstream.

WINTER

Notice that the snowpack is thin, with many open patches of bare ground, and if you check, these will probably be frozen several centimeters deep. It would be good if you brought a poncho, because rain is much more common now even on a winter's day.

No one uses snowshoes anymore, so the walking is difficult over the slippery patches of snow and open ground. But the forest is alive with overwintering birds, such as chickadees, robins, crows, woodpeckers, and seed-eating finches, including goldfinches, siskins, and grosbeaks. If you are lucky, you might be able to witness a barred owl catch a mouse, red squirrel, or snowshoe hare. Owls are common now because of the increased numbers of small mammals in the forest. What might look like patches of plowed ground are really the result of rooting by European wild boar, which have invaded the forest.

■

The scenarios depicted here are based on current projections of warming temperatures in the continued absence of effective controls on emissions of carbon dioxide to the atmosphere. Things could, of course, turn out quite differently if governments and international organizations manage to take meaningful action to limit greenhouse gases. The "story" of the ecosystem unfolds carefully and deliberately, and as Aldo Leopold has said, "to hear even a few notes . . . of the song of a river . . . you must first live here for a long time and you must know the speech of hills and rivers."[1]

APPENDIX 1. SCIENTIFIC UNITS

CONVERSIONS AND ABBREVIATIONS

Some useful definitions and conversions of scientific units.

LENGTH AND AREA
millimeter (mm) = 0.04 inches (in)
centimeter (cm) = 0.4 inches (in)
meter (m) = 3.3 feet (ft)
kilometer (km) = 0.6 miles (mi)
square kilometer (km²) = 0.4 square miles (mi²)
hectare (ha) = 10,000 m² = 2.5 acres (ac)

MASS AND WEIGHT
gram (g) = 0.035 ounces (oz)
kilogram (kg) = 2.2 pounds (lb)
tonne (1,000 kg) = 1.1 short tons (or 2,204 pounds)

VOLUME
liter (L) = 1.06 quarts (qt)
cubic meter (m³) = 35 cubic feet (ft³)

OTHER
gram equivalent (eq) weight = weight of a substance that will react with 1 gram of hydrogen (e.g., Ca^{2+} = 20.0 mg eq/L)
mole (mol) = 1 gram-molecular weight of a substance (e.g., 1 mole calcium = 40.08 g)
molar concentration = number of gram-moles of solute per liter of solution
temperature, degrees Celsius (°C) to degrees Fahrenheit (°F) = (9/5 °C) + 32
calorie = amount of heat required to raise 1 gram of water 1 degree Celsius
electrical conductivity, measured in microsiemens/ centimeter (μS/cm)

The metric system uses modifiers in multiples of ten.
micro (μ) = 1/1,000,000
milli (m) = 1/1,000
centi (c) = 1/100
kilo (k) = 1,000
Mega (M) = 1,000,000
Tera (T) = 1,000,000,000,000

APPENDIX 2. SCIENTIFIC NAMES AND LISTS OF SELECTED ORGANISMS

This appendix contains the scientific names of organisms mentioned in the text of this book, along with lists, organized by life form or ecological characteristics, of the more common species. For more complete taxonomic lists of organisms, including those in and around Mirror Lake, see Holmes and Likens (1999) and www.hubbardbrook.org/data/SpeciesLists.

Plants

TREES

Deciduous Trees

American beech	*Fagus grandifolia*
American mountain ash	*Sorbus americana*
Bigtooth aspen	*Populus grandidentata*
Black cherry	*Prunus serotina*
Gray birch	*Betula populifolia*
Mountain maple	*Acer spicatum*
Paper (white) birch	*Betula papyrifera*
Pin cherry	*Prunus pensylvanica*
Quaking aspen	*Populus tremuloides*
Red maple	*Acer rubrum*
Red oak	*Quercus rubra*
Striped maple	*Acer pensylvanicum*
Sugar maple	*Acer saccharum*
White ash	*Fraxinus americana*
Yellow birch	*Betula alleghaniensis*

Coniferous Trees

Balsam fir	*Abies balsamea*
Eastern hemlock	*Tsuga canadensis*
Eastern white pine	*Pinus strobus*
Red spruce	*Picea rubens*

SHRUBS

Alternate-leaved dogwood	*Cornus alternifolia*
Elderberry	*Sambucus* spp.
Fly honeysuckle	*Lonicera canadensis*
Hobblebush	*Viburnum lantanoides* (formerly *V. alnifolium*)
Raspberry	*Rubus* spp.

SPRING EPHEMERALS OF THE FOREST FLOOR

Bunchberry	*Cornus canadensis*
Dutchman's breeches	*Dicentra cucullaria*
Dwarf ginseng	*Panax trifolius*
Foamflower	*Tiarella cordifolia*
Jack-in-the-pulpit	*Arisaema triphyllum*
Marsh blue violet	*Viola cucullata*
Painted trillium	*Trillium undulatum*
Red trillium	*Trillium erectum*
Spring beauty	*Claytonia caroliniana*
Squirrel corn	*Dicentra canadensis*
Starflower	*Trientalis borealis*
Trout lily	*Erythronium americanum*
Yellow violet	*Viola rotundifolia*

SUMMER-GREEN HERBACEOUS PERENNIALS OF THE FOREST FLOOR

Bladder sedge	*Carex intumescens*
Blue-bead lily	*Clintonia borealis*
Bristly clubmoss	*Lycopodium annotinum*
Canada mayflower	*Maianthemum canadense*
False solomon's seal	*Smilacina racemosa*
Hay-scented fern	*Dennstaedtia punctilobula*
Indian cucumber root	*Medeola virginiana*
Indian pipe	*Monotropa uniflora*
Lady fern	*Athyrium filix-femina*
Large round-leaved orchid	*Platanthera orbiculata*
New York fern	*Thelypteris* [formerly *Dryopteris*] *noveboracensis*
Pink wood sorrel	*Oxalis montana*
Rose twisted-stalk	*Streptopus lanceolatus*
Running pine	*Lycopodium clavatum*
Sharp-leaved aster	*Aster acuminatus*

Shining clubmoss	*Huperzia* [formerly *Lycopodium*] *lucidula*
Wild oats	*Uvularia sessifolia*
Wild sarsaparilla	*Aralia nudicaulis*
Wood fern	*Dryopteris carthusiana* [formerly *D. spinulosa*]

Invertebrates

Only species mentioned in the text are listed here. Lists for others can be found in Holmes and Likens (1999) and at hubbardbrook.org/data/SpeciesLists/.

American idia moth	*Idia americalis*
Beech scale	*Cryptococcus fagisuga*
Canadian swallowtail	*Papilio canadensis*
Little white lichen moth	*Clemensia albata*
Saddled prominent	*Heterocampa guttivitta*

Fish

Atlantic salmon	*Salmo salar*
Blacknose dace	*Rhinichthys atratulus*
Brook trout	*Salvelinus fontinalis*
Slimy sculpin	*Cottus cognatus*

Amphibians

FROGS AND TOADS

American toad	*Anaxyrus* (formerly *Bufo*) *americanus*
Gray tree frog	*Hyla versicolor*
Green frog	*Lithobates clamitans*
Northern leopard frog	*Lithobates pipiens*
Pickerel frog	*Lithobates palustris*
Spring peeper	*Hyla crucifer*
Wood frog	*Lithobates sylvaticus*

SALAMANDERS

Dusky salamander	*Desmognathus fuscus*
Red-backed salamander	*Plethodon cinereus*
Red-spotted newt	*Notopthalmus viridescens*
Spotted salamander	*Ambystoma maculatum*
Spring salamander	*Gyrinophilus porphyriticus*
Two-lined salamander	*Eurycea bislineata*

Reptiles

| Eastern garter snake | *Thamnophis sirtalis* |

Birds, Grouped by Residence Status

PERMANENTLY RESIDENT SPECIES

Barred owl	*Strix varia*
Black-capped chickadee	*Poecile atricapillus*
Blue jay	*Cyanocitta cristata*
Downy woodpecker	*Picoides pubescens*
Hairy woodpecker	*Picoides villosus*
Pileated woodpecker	*Dryocopus pileatus*
Red-breasted nuthatch	*Sitta canadensis*
Ruffed grouse	*Bonasa umbellus*
Saw-whet owl	*Aegolius acadacus*
White-breasted nuthatch	*Sitta carolinensis*
Wild turkey	*Melagris gallopavo*

FOREST-BREEDING MIGRATORY SPECIES

Black and white warbler	*Mniotilta varia*
Blackburnian warbler	*Setophaga fusca*
Blackpoll warbler	*Setophaga striata*
Black-throated blue warbler	*Setaphaga caerulescens*
Black-throated green warbler	*Setophaga virens*
Blue-headed vireo	*Vireo solitarius*
Broad-winged hawk	*Buteo platypteris*
Brown creeper	*Certhia familiaris*
Canada warbler	*Cardellina canadensis*
Chimney swift	*Chaetura pelgica*
Dark-eyed junco	*Junco hyemalis*
Eastern wood pewee	*Contopus virens*
Golden-crowned kinglet	*Regulus satrapa*
Hermit thrush	*Catharus guttatus*
Magnolia warbler	*Setophaga magnolia*
Ovenbird	*Seiurus aurocapilla*
Red-eyed vireo	*Vireo olivaceus*
Red-tailed hawk	*Buteo jamaicensis*
Rose-breasted grosbeak	*Pheucticus ludovicianus*
Ruby-throated hummingbird	*Archilochus colubris*
Scarlet tanager	*Piranga olivacea*

Sharp-shinned hawk	*Accipiter striatus*
Swainson's thrush	*Catharus ustulatus*
Winter wren	*Troglodytes hiemalis*
Yellow-bellied sapsucker	*Sphyrapicus varius*
Yellow-rumped warbler	*Setophaga coronata*

SUCCESSIONAL OR DISTURBED-SITE-BREEDING MIGRATORY SPECIES

Chestnut-sided warbler	*Setophaga pensylvanica*
Common yellowthroat	*Geothlypis trichas*
Eastern bluebird	*Sialia sialis*
Indigo bunting	*Passerina cyanea*
Mourning warbler	*Oporornis philadelphia*
Rufous-sided towhee	*Pipilo erythrophthalmus*
Song sparrow	*Melospiza melodia*
White-throated sparrow	*Zonotrichia albicolis*

WINTER RESIDENTS (mostly non-breeding, numbers highly variable from year to year)

American goldfinch	*Carduelis tristis*
Common redpoll	*Carduelis flammea*
Evening grosbeak	*Coccothraustes vespertinus*
Pine grosbeak	*Pinicola enucleator*
Pine siskin	*Carduelis pinus*
Red crossbill	*Loxia curvirostra*
White-winged crossbill	*Loxia leucoptera*

Mammals, Grouped by Order or Family and Feeding Habits

SHREWS AND MOLES (litter insectivores)

Short-tailed shrew	*Blarina brevicauda*
Pygmy shrew[a]	*Microsorex hoyi*
Masked shrew	*Sorex cinereus*
Hairy-tailed mole	*Paascalops breweri*

BATS (aerial insectivores)

Little brown bat[b]	*Myotis lucifugus*
Large brown bat[b]	*Eptesicus fuscus*
Red bat[b]	*Lasiurus borealis*
Hoary bat[b]	*Lasiurus cinereus*

CARNIVORES

Coyote	*Canas latrans*
Red fox	*Vulpes fulva*
Bobcat	*Lynx rufus*
Fisher	*Martes pennanti*
American marten[a]	*Martes americana*
Long-tailed weasel	*Mustela frenata*
Short-tailed weasel[a]	*Mustela erminea*
Mink	*Mustela vison*
Striped skunk	*Mephitis mephitis*
Raccoon	*Procyon lotor*

BEAR (large omnivore)

Black bear	*Ursus americanus*

RODENTS (seed, fruit, and insect feeders)

Deer mouse	*Peromyscus maniculatus*
Northern flying squirrel	*Glaucomys sabrinus*
Southern flying squirrel[a]	*Glaucomys volans*
Eastern gray squirrel	*Sciurus carolinensis*
Eastern chipmunk	*Tamias striatus*
American red squirrel	*Tamiasciurus hudsonicus*

RODENTS (fungus, leaf, bud, and insect feeders)

Woodland jumping mouse	*Napaeozapus insignis*
Boreal red-backed vole	*Clethrionomys gapperi*

BEAVER AND PORCUPINE (bark and wood chewers)

North American beaver	*Castor canadensis*
North American porcupine	*Erethizon dorsatum*

HARE (browser/grazer)

Snowshoe hare	*Lepus americanus*

UNGULATES (browsers)

Moose	*Alces alces*
White-tailed deer	*Odocoileus virginianus*

Notes
[a] Rare or very infrequently encountered
[b] Status and distribution unknown or poorly known

NOTES

Chapter 1. Ecosystem and Ecological Studies at Hubbard Brook

1 Likens 1992.
2 Likens 1985a; Likens 1992.
3 Bormann and Likens 1967.
4 Likens et al. 1977.
5 Holmes 2011.
6 Lindenmayer and Likens 2009.
7 Rothamsted Research: http://rothamsted.ac.uk.
8 Reasons for the longevity and productivity of our Study are essentially the same as for Rothamsted, except for stable funding. Rothamsted has an endowment, whereas Hubbard Brook research, except for federal funding for the Forest Service operations, has depended on short-term grants from many sources to individuals or groups of investigators, with new proposals submitted and acted upon at 1- to 6-year intervals.
9 Bormann and Likens 1967.

Chapter 2. The Small Watershed-Ecosystem Approach

1 Hoover 1944; Hewlett 1971.
2 Bormann and Likens 1967.
3 Likens et al. 1977; Likens and Bormann 1972.
4 Bormann and Likens 1979: page vi.
5 Sopper and Lull 1965, 1970; Federer 1969; Hornbeck et al. 2001; McGuire and Likens 2011.
6 Sopper and Lull 1965, 1970; Likens 2013.
7 This conclusion was based on early geological work in the valley, hydrologic comparisons between adjacent watersheds, a budgetary analysis of the conservative anion chloride, and an analysis of the long-term stream-water chemistry records, particularly at low flows (Likens 2013).
8 Likens 1985c.
9 Likens 1985a; Winter and Likens 2009.
10 Likens 2004.

Chapter 3. Physical Setting and Climate

1 Bormann and Likens 1979.
2 Bormann and Likens 1979; R. E. Bormann et al. 1985.
3 Likens 2013.
4 A. S. Bailey et al. 2003.
5 Bormann et al. 1974.
6 Likens and Bormann 1974b.
7 Bilby and Likens 1980; Steinhart et al. 2000.
8 A. S. Bailey et al. 2003.
9 A. S. Bailey et al. 2003; Likens and Bormann 1974b.

Chapter 4. The Forest

1 Likens and Davis 1975; Davis 1985.
2 Davis 1985.
3 Davis 1985.
4 Davis 1985.
5 Davis 1985; R. E. Bormann et al. 1985.
6 Davis 1985.
7 Cogbill 1989.
8 Vadeboncoeur et al. 2012.
9 Chittenden 1905; Hamburg and Cogbill 1988.
10 Mann 2005.
11 Davis 1985; S. P. Hamburg, personal communication.
12 Davis 1985.
13 Bormann and Likens 1979.
14 Hamburg 1984.
15 Cogbill 1989.
16 Cogbill 1989.
17 Cogbill 1989; D. C. Buso, personal communication.
18 Federer 1969; Hornbeck 2001.
19 Cogbill 1989; D. C. Buso, personal communication.
20 Merrens and Peart 1992; see Chapters 11 and 12.
21 Schwarz et al. 2003.
22 Peart et al. 1992.
23 Bofinger 2001.
24 Rhoads et al. 2002.
25 Rhoads et al. 2002; Weeks et al. 2009.
26 Rhoads et al. 2002.
27 Weeks et al. 2009.
28 Amphibians (two-lined salamander, green frog) were found frozen in ice along the streambeds in late March 2006 (D. C. Buso, personal communication). This occurred after a long cold period with little snow.
29 Groffman et al. 2001b; Judd et al. 2011.
30 Gosz et al. 1978; R. T. Holmes, personal observation.
31 Gosz et al. 1978; N. K. Lany, personal communication.
32 Gosz et al. 1972.
33 See chapters 8 and 11.
34 Gosz et al. 1972; Likens 2013.
35 Lovett et al. 2006.
36 Houston 1994.
37 Van Doorn 2014.

38 Schwarz et al. 2003.

39 Bormann et al. 1970; Schwarz et al. 2003; Van Doorn et al. 2011.

40 Bormann et al. 1970.

41 Siccama et al. 2007.

Chapter 5. A Rich Array of Organisms and Their Interactions

1 Holmes and Likens 1999.

2 Bormann et al. 1970; Van Doorn et al. 2011.

3 Kelly 1994.

4 Janzen 1971.

5 Moreira et al. 2014.

6 Cleavitt et al. 2008b.

7 Cleavitt et al. 2014.

8 Forcier 1975; Schwarz et al. 2003.

9 Siccama et al. 1970.

10 Muller 1978.

11 Bidartondo et al. 2002.

12 Holmes and Likens 1999.

13 Taylor et al. 2014.

14 C. Wood, M. Fisk, and T. Fahey, personal communication.

15 M. Fisk, personal communication.

16 Cleavitt et al. 2011, 2014.

17 Cleavitt et al. 2011; N. L. Cleavitt, personal communication.

18 L. M. Christenson, personal communication.

19 Sridevi et al. 2012.

20 Fisk et al. 2010.

21 Stange et al. 2011; N. K. Lany and M. P. Ayres, personal communication.

22 Thornton 1974.

23 Skeldon et al. 2007.

24 Skeldon et al. 2007.

25 Warren et al. 2008.

26 Lowe et al. 2006; Greene et al. 2008.

27 Burton and Likens 1975a.

28 G. E. Likens, unpublished.

29 Holmes and Likens 1999; see updated list at www .hubbardbrook.org.

30 Holmes and Sturges 1975; Holmes et al. 1986; Holmes and Sherry 2001.

31 Holmes and Sturges 1975.

32 Holmes and Sturges 1975.

33 Holmes 2011.

34 Holmes et al. 1986; Holmes 1990.

35 Holmes et al. 1979; Robinson and Holmes 1982, 1984.

36 Holmes 2011.

37 Potter 1974.

38 Conrod 2008.

39 Groffman et al. 2012; L. M. Christenson, personal communication.

40 Samuel 2007; Rodenhouse et al. 2009; Morse 2015.

41 Pletscher et al. 1989; D. C. Buso, personal communication.

42 Whitlaw and Lankster 1994.

Chapter 6. How Is Energy Transformed?

1 Odum 1959; Gosz et al. 1978.

2 Fahey et al. 2005.

3 Whittaker et al. 1974; Siccama et al. 2007; Campbell et al. 2013.

4 Battles et al. 2014.

5 Fahey et al. 2005.

6 Burton and Likens 1975a.

7 Holmes and Sturges 1975.

8 Burton and Likens 1975a; Holmes and Sturges 1973, 1975.

9 Material in the following section from Gosz et al. 1978.

10 Phillips and Fahey 2005.

Chapter 7. Hydrology

1 Likens 2013.

2 Likens 2013.

3 Likens 2013.

4 Bormann and Likens 1967; Likens 2013.

5 McGuire and Likens 2011.

Chapter 8. Biogeochemistry

1 Likens 2013.

2 Likens 2002; Likens and Buso 2012, 2015; Likens and La-Baugh 2009.

3 Siccama and Smith 1978; Siccama et al. 1980; Smith and Siccama 1981; C. E. Johnson et al. 1995.

4 Siccama and Smith 1978.

5 Driscoll et al. 2007a, b; Dittman et al. 2010.

6 Butler et al. 2008.

7 Driscoll et al. 2007a.

8 Dittman et al. 2010.

9 Demers et al. 2010.

10 Ollinger et al. 2002a; Hong et al. 2006.

11 Hong et al. 2006.

12 Aber et al. 2002; Ollinger et al. 2002a.

13 Bormann et al. 1969, 1974.

14 Likens 2013.

15 N. M. Johnson et al. 1969.

16 Bormann et al. 1974.

17 Likens 2013.

18 Hunt 1967.

19 Fisher and Likens 1972; Bormann et al. 1969.

20 Bormann et al. 1969.

21 Bilby and Likens 1980.

22 Goodale et al. 2003; Bernhardt et al. 2005; Bernal et al. 2012; Likens 2013.
23 Likens 2010.
24 Lovett et al. 1992; Likens 2013.
25 Mitchell et al. 2011.
26 Likens 2013.
27 Likens et al. 1994; Likens 2013.
28 Likens et al. 1998; Likens 2013.
29 Likens et al. 1998.
30 Likens et al. 1998.
31 Burton and Likens 1975a.
32 Sturges et al. 1974.
33 Yanai et al. 2014.
34 Likens 2013.
35 Likens et al. 2002; Likens 2013.
36 Likens et al. 2002.
37 Likens et al. 2002.
38 Sturges et al. 1974; Likens 2013.
39 Likens et al. 2002.
40 Likens et al. 2002.
41 Likens et al. 2002; Likens 2013.
42 Likens 2013.
43 Likens 2013.
44 Lovett et al. 2005.
45 Lovett et al. 2005; Svensson et al. 2012.
46 Likens 2013.
47 Likens 2013.
48 Steinhart et al. 2000; Wexler et al. 2014.
49 Likens 2013.
50 Hornbeck et al. 1997, 2001; see also Fahey et al. 2015.
51 Likens and Buso 2012.

Chapter 9. The Discovery of Acid Rain at Hubbard Brook

1 Likens et al. 1972.
2 Likens and Bormann 1974a; Likens 1989.
3 Galloway et al. 1982; Likens et al. 1987; Keene et al. 2015.
4 Likens 1984b.
5 Draxler and Hess 1998.
6 Stunder 1996.
7 Likens et al. 2005.
8 Likens 2010.
9 Likens 1992; McNeil and Culcasi 2014; Likens et al. 1972.
10 See www.un.org/apps/news/story.asp?NewsID=20274&Cr=cartoon&Cr1=.
11 Likens 1992.
12 Cogbill and Likens 1974.
13 McNeil and Culcasi 2014.
14 Likens et al. 1979.

Chapter 10. The Consequences of Acid Rain and Other Air Pollutants

1 Likens and Butler 2014; see also www.epa.gov/acidrain/effects/index.html; Weathers et al. 2007.
2 Galloway and Likens 1981; Likens and Lambert 1998.
3 Likens and Lambert 1998.
4 Butler et al. 2011.
5 Likens et al. 1996, 1998.
6 Bormann and Likens 1979.
7 Rastetter et al. 2013.
8 Brady and Weil 2007.
9 Lawrence et al. 1995.
10 Likens et al. 1998; Likens 2013.
11 Hawley et al. 2006; Kosiba et al. 2013.
12 Weathers et al. 2006; Likens and Bailey 2014.
13 Horsley et al. 2002; Bailey et al. 2004, 2005; Hallett et al. 2006; Long et al. 2009; Moore et al. 2012; Likens and Bailey 2014; Halman et al. 2015.
14 N. M. Johnson et al. 1981; Driscoll 1984a; Edzwall 2014.
15 Cho et al. 2012.
16 Cho et al. 2010.
17 Battles et al. 2014.
18 Green et al. 2013.
19 T. J. Fahey, personal communication.
20 Likens and Buso 2012.
21 Likens and Buso 2012.
22 Graveland et al. 1994.
23 Taliaferro et al. 2001.
24 Pabian and Brittingham 2007.
25 Pabian and Brittingham 2011.

Chapter 11. The Effects of Forest Harvesting and Other Disturbances

1 Likens 1985c.
2 Campbell et al. 2013.
3 Reiners 1992.
4 Likens et al. 1970.
5 Likens et al. 1970.
6 Likens and Bormann 1974c.
7 Pierce et al. 1972.
8 Martin and Pierce 1980; Martin et al. 1986.
9 Likens and Bormann 1974c.
10 C. E. Johnson et al. 2014.
11 Pardo et al. 2002.
12 Likens and Buso 2016.
13 Lindenmayer and Likens 2010.
14 Likens et al. 1978; Vadeboncoeur et al. 2014.

Chapter 12. How Does the Forest Ecosystem Recover After Harvesting and Other Disturbances

1 Likens et al. 1970.
2 Marks 1974; Likens et al. 1978; Bormann and Likens 1979; Hornbeck et al. 1987; Martin and Hornbeck 1989, 1990; Reiners 1992; Mou et al. 1993; Tierney and Fahey 1998; Reiners et al. 2012; Campbell et al. 2013; Battles et al. 2014.
3 Likens et al. 1978; Bormann and Likens 1979; Vadeboncoeur et al. 2014.
4 Bormann and Likens 1979.
5 Marks 1974; Tierney and Fahey 1998.
6 Marks 1974.
7 Marks and Bormann 1972.
8 Marks 1974.
9 Marks 1974.
10 Tierney and Fahey 1998.
11 Likens and Bormann 1974c; White 1974; Federer et al. 1989; Likens et al. 1996, 1998, 2002.
12 Bormann and Likens 1979; Hornbeck et al. 1987.
13 Likens et al. 1978; Vadeboncoeur et al. 2014.
14 Likens et al. 1978.
15 Franklin and Johnson 2014.

Chapter 13. How Stream Ecosystems Are Integrated with Their Watersheds

1 Fisher and Likens 1972.
2 Mayer and Likens 1987.
3 Fisher and Likens 1972.
4 R. J. Hall et al. 1980.
5 R. O. Hall et al. 2001.
6 Likens et al. 2004a.
7 Likens et al. 2004a.
8 Likens and Bilby 1982.
9 Bilby 1979.
10 Bilby and Likens 1980.
11 Warren et al. 2009.
12 Likens and Bilby 1982.
13 Marks 1974.
14 Likens and Bilby 1982.
15 Macneale et al. 2005.
16 Macneale et al. 2005.
17 Lowe et al. 2008.

Chapter 14. What Causes Population Change in Forest Birds?

1 Holmes and Sturges 1973, 1975; Sturges et al. 1974.
2 Terborgh 1989; Robbins et al. 1989.
3 Holmes and Sherry 2001; R. T. Holmes, unpublished data.
4 Holmes 2007, 2011.

5 Holmes and Sherry 2001; Hunt 1996.
6 Holmes et al. 1986.
7 Sherry and Holmes 1992; Holmes 2011.
8 Sillett and Holmes 2005.
9 R. T. Holmes, unpublished; Taliaferro 2001.
10 Rodenhouse and Holmes 1992.
11 Nagy and Holmes 2005.
12 Sillett et al. 2000; Townsend et al. 2015.
13 Sillett et al. 2004.
14 Sillett and Holmes 2005.
15 McPeek et al. 2001; Rodenhouse et al. 2003; Sillett and Holmes 2005.
16 Holmes and Sherry 1992.
17 Rubenstein et al. 2002; Norris et al. 2006.
18 Holmes et al. 1989; M. D. Johnson et al. 2006.
19 Holmes et al. 1989; Marra 2000.
20 Marra et al. 1998; Marra and Holmes 2001.
21 Holmes et al. 1989.
22 Sherry and Holmes 1996; M. D. Johnson et al. 2006.
23 Lovette and Holmes 1995.
24 Marra et al. 1998.
25 Norris et al. 2006; Reudink et al. 2009.
26 Sillett and Holmes 2002.
27 Faaborg et al. 2010a, b.
28 M. D. Johnson et al. 2006; Wilson et al. 2011; Marra et al. 2015.
29 Rubenstein et al. 2002.

Chapter 15. Scaling Up

1 Bormann and Likens 1967, 1979.
2 Because Hubbard Brook was initially selected as a research site based on its suitability for hydrologic studies in New England, not all aspects of the local ecosystems (such as geology, flora, or fauna) are necessarily representative of the broader region (Fahey et al. 2015). Nevertheless, extrapolating results from the Study to larger scales can provide insights into ecological processes and how they operate across the broader forest landscape (e.g., Marra et al. 1998; Jones et al. 2003; Keene et al. 2015).
3 Bormann et al. 1970; J. J. Battles, personal communication.
4 Schwarz et al. 2003; Van Doorn et al. 2011.
5 Van Doorn 2014; J. J. Battles, personal communication.
6 Fahey et al. 2005.
7 Ollinger et al. 2008.
8 Ollinger et al. 2002b.
9 Doran and Holmes 2005; Goetz et al. 2010; Swatantran et al. 2012; Whitehurst et al. 2013.
10 Whitehurst et al. 2013.
11 Gillen et al. 2015.

12 Doran and Holmes 2005; Goetz et al. 2010; Swatantran et al. 2012.

13 Doran 2003; S. J. K. Frey and R. T. Holmes, unpublished.

14 P. J. Doran and R. T. Holmes, unpublished.

15 Judd et al. 2011; Likens 2013.

16 Judd et al. 2011.

17 Groffman et al. 2001a, b; Fitzhugh et al. 2001; Likens 2013; Judd et al. 2011.

18 Warren et al. 2008.

19 Likens and Buso 2006.

20 McGuire et al. 2014.

21 Likens and Buso 2006.

22 S. J. K. Frey and R. T. Holmes, unpublished.

23 Likens 1999, 2013; C. E. Johnson et al. 2000; Likens et al. 2002; Palmer et al. 2005.

24 Likens et al. 2002.

Chapter 16. How Is Climate Change Affecting the Forest Ecosystem?

1 Molina et al. 2014; International Panel on Climate Change 2014.

2 Campbell et al. 2009; Groffman et al. 2012; Rustad et al. 2012.

3 A. S. Bailey et al. 2003; Campbell et al. 2007a.

4 Groffman et al. 2012.

5 Campbell et al. 2010; Hamburg et al. 2013.

6 A. S. Bailey et al. 2016.

7 Campbell et al. 2009.

8 Campbell et al. 2010.

9 Likens 2011.

10 Likens 2011.

11 Hayhoe et al. 2007, 2008.

12 International Panel on Climage Change 2012.

13 Likens 2013.

14 Judd et al. 2007.

15 Fitzhugh et al. 2001.

16 Groffman et al. 2001a, b; Fitzhugh et al. 2003; Comerford et al. 2013; Campbell et al. 2014.

17 Tierney et al. 2001; Cleavitt et al. 2008a; Comerford et al. 2013; Campbell et al. 2014.

18 Groffman et al. 2001b.

19 Comerford et al. 2013.

20 Templer et al. 2012.

21 Christenson et al. 2014.

22 Groffman et al. 2012.

23 Durán et al. 2014; Campbell et al. 2014.

24 Groffman et al. 2012.

25 Muller and Bormann 1976; Muller 1978.

26 Zak et al. 1990.

27 Groffman et al. 2012.

28 Iverson et al. 2008.

29 Both et al. 2005; Visser et al. 2006.

30 Lany et al. 2015.

31 Townsend et al. 2013.

32 Rodenhouse et al. 2009; Matthews et al. 2011; North American Bird Conservation Initiative 2014.

33 Rodenhouse et al. 2009; Holmes 2011.

Chapter 17. Reaching Out

1 For example, see *Campbell Biology* (Reece et al. 2013); *Biology* (Raven et al. 2014); *The Ecological World View* (Krebs and Elwood 2008); *Essentials of Environmental Science* (Friedland et al. 2012).

2 Likens 1992, 2010.

3 For example, Likens et al. 1972; Cogbill and Likens 1974; Likens and Bormann 1974a; Galloway et al. 1976; N. M. Johnson et al. 1981; Likens et al. 1987; Likens 2010.

4 Likens et al. 1972, 1979; Likens 1974, 1984a, b, 1987.

5 Environmental Protection Agency 2011.

6 Reiners et al. 2012; Battles et al. 2014; Campbell et al. 2013.

7 Martin and Hornbeck, 1989, 1990.

8 Likens and Bormann 1974a; Hornbeck et al. 1997.

9 Likens et al. 1978, Vadeboncoeur et al. 2014.

10 Holmes 2011.

11 Costello et al. 2000.

12 DeGraff and Yamasaki 2003.

13 Faaborg 2010b; La Sorte et al. 2015; see also the Migratory Connectivity Project at www.migratoryconnectivity project.org.

14 See http://hubbardbrookfoundation.org; Driscoll et al. 2011.

15 Driscoll et al. 2011.

16 Driscoll et al. 2011.

17 Driscoll et al. 2011.

Chapter 18. A Look Ahead

1 Lindenmayer and Likens 2009; Lindenmayer et al. 2010a, b.

2 Buso et al. 2009; Likens and LaBaugh 2009.

3 Buso et al. 2009; Likens and LaBaugh 2009; Likens and Buso 2010.

4 Likens and Buso 2012.

5 Likens and Buso 2012.

6 Likens and Buso 2016.

7 Demers et al. 2010.

8 Union of Concerned Scientists 2006, Hayhoe et al. 2007.

9 Groffman et al. 1999, 2001b.

10 Kosiba et al. 2013; P. G. Schabeg, personal communication.

11 Lovett et al. 2002; Lovett and Mitchell 2004; Christenson et al. 2010.
12 Rodenhouse 2009; Matthews et al. 2011.
13 North American Bird Conservation Initiative 2014.
14 Matthews et al. 2011.
15 Rodenhouse et al. 2008, 2009.
16 Tingely et al. 2002.
17 Christenson et al. 2014.
18 Rodenhouse et al. 2009.

19 Cary Institute of Ecosystem Studies 2015.
20 Samuel 2007; Rodenhouse et al. 2009.
21 Mitchell and Likens 2011.
22 Vadebonocoeur et al. 2014.
23 Seuss 1971.
24 Bormann and Likens 1977.

Epilogue
1 Leopold 1949.

BIBLIOGRAPHY

Aber, J. D., S. V. Ollinger, C. T. Driscoll, G. E. Likens, R. T. Holmes, R. J. Freuder, and C. L. Goodale. 2002. Inorganic nitrogen losses from a forested ecosystem in response to physical, chemical, biotic, and climatic perturbations. *Ecosystems* 5:648–658.

Bailey, A. S., J. W. Hornbeck, J. L. Campbell, and C. Eagar. 2003. Hydrometerological database for Hubbard Brook Experimental Forest: 1955–2000. U.S. Forest Service, General Technical Report NE-305. 36 pp.

Bailey, A. S., J. L. Campbell, M. B. Green, and L. Rustad. 2016. Long-term trends in foundation datasets at Hubbard Brook Experimental Forest, Woodstock, New Hampshire, USA. U.S. Forest Service, General Technical Report.

Bailey, S. W., B. Mayer, and M. J. Mitchell. 2004. Evidence for influence of mineral weathering on stream water sulfate in Vermont and New Hampshire (USA). *Hydrological Processes* 18:1639–1653.

Bailey, S. W., S. B. Horsley, and R. P. Long. 2005. Thirty years of change in forest soils of the Allegheny Plateau, Pennsylvania. *Soil Science Society of America Journal* 69:681–690.

Battles, J. J., T. J. Fahey, C. Driscoll, J. Blum, and C. E. Johnson. 2014. Restoring soil calcium reverses forest decline. *Environmental Science and Technology Letters* 1:15–19.

Bernal, S., L. O. Hedin, G. E. Likens, S. Gerber, and D. C. Buso. 2012. Complex response of the forest nitrogen cycle to climate change. *Proceedings of the National Academy of Sciences, USA* 109:3406–3411.

Bernhardt, E. S. 2001. Nutrient demand and nitrogen processing in streams of the Hubbard Brook Experimental Forest. Ph.D. dissertation, Cornell University. Ithaca, N.Y. 185 pp.

Bernhardt, E. S., and W. H. McDowell. 2008. Twenty years apart: Comparisons of DOM uptake during leaf leachate releases to Hubbard Brook valley streams in 1979 versus 2000. *Journal of Geophysical Research: Biogeosciences* 113(G3): GO3032.

Bernhardt, E. S., R. O. Hall Jr., and G. E. Likens. 2002. Whole-system estimates of nitrification and nitrate uptake in streams of the Hubbard Brook Experimental Forest. *Ecosystems* 5:419–430.

Bernhardt, E. S., G. E. Likens, D. C. Buso, and C. T. Driscoll. 2003. In-stream update dampens effects of major forest disturbance on watershed nitrogen export. *Proceedings of the National Academy of Sciences, USA* 100:10304–10308.

Bernhardt, E. S., G. E. Likens, R. O. Hall Jr., D. C. Buso, S. G. Fisher, T. M. Burton, J. L. Meyer, et al. 2005. Can't see the forest for the stream? In-stream processing and terrestrial nitrogen exports. *BioScience* 55:219–230.

Bidartondo, M. L., D. Redecker, I. Hijri, A. Wiemken, T. D. Bruns, L. Domínguez, A. Sérsic, J. R. Leake, and D. J. Read. 2002. Epiparasitic plants specialized on arbuscular mycorrhizal fungi. *Nature* 419:389–392.

Bilby, R. E. 1979. The function and distribution of organic debris dams in forest stream ecosystems. Ph.D. dissertation, Cornell University, Ithaca, N.Y. 143 pp.

Bilby, R. E., and G. E. Likens. 1980. Importance of organic debris dams in the structure and function of stream ecosystems. *Ecology* 61:1107–1113.

Bofinger, J. 2001. New Hampshire ice storm damage assessment and response. New York Society of American Foresters Ice Storm Symposium. USDA Forest Service, Cortland, N.Y. Publication NA-TP-03-01, pp. 73–76.

Bohlen, P. J., D. M. Pelletier, P. M. Groffman, T. J. Fahey, and M. C. Fisk. 2004. Influence of earthworm invasion on redistribution and retention of soil carbon and nitrogen in northern temperate forests. *Ecosystems* 7:13–27.

Bormann, F. H. 1985. Air pollution and forests: An ecosystem perspective. *BioScience* 35:434–441.

Bormann, F. H., and G. E. Likens. 1967. Nutrient cycling. *Science* 155:424–429.

Bormann, F. H., and G. E. Likens. 1977. The fresh air–clean water exchange. *Natural History* 86:62–71.

Bormann, F. H., and G. E. Likens. 1979. *Pattern and Process in a Forested Ecosystem: Disturbance, Development, and the Steady State Based on the Hubbard Brook Ecosystem Study.* Springer-Verlag, New York.

Bormann, F. H., G. E. Likens, and J. S. Eaton. 1969. Biotic regulation of particulate and solution losses from a forest ecosystem. *BioScience* 19:600–610.

Bormann F. H., T. G. Siccama, G. E. Likens, and R. H. Whittaker. 1970. The Hubbard Brook Ecosystem Study: Composition and dynamics of the tree stratum. *Ecological Monographs* 40:373–388.

Bormann, F. H., G. E. Likens, T. G. Siccama, R. S. Pierce, and J. S. Eaton. 1974. The export of nutrients and recovery of stable conditions following deforestation at Hubbard Brook. *Ecological Monographs* 44:255–277.

Bormann, R. E., F. H. Bormann, and G. E. Likens. 1985. Catastrophic disturbance and regional land use. In G. E. Likens, ed., *An Ecosystem Approach to Aquatic Ecology: Mirror Lake and Its Environment,* pp. 65–72. Springer-Verlag, New York.

Both, C., R. G. Bijlsma, and M. E. Visser. 2005. Climatic effects on timing of spring migration and breeding in a long-distance migrant, the pied flycatcher *Ficedula hypoleuca. Journal of Avian Biology* 36:368–373.

Brady, N. C., and R. R. Weil. 2007. *The Nature and Properties of Soils,* 14th ed. Prentice Hall, Upper Saddle River, N.J. 980 pp.

Burton, T. M., and G. E. Likens. 1975a. Salamander populations and biomass in the Hubbard Brook Experimental Forest, New Hampshire. *Copeia* 1975:541–546.

Burton, T. M., and G. E. Likens. 1975b. Energy flow and nutrient cycling in salamander populations in the Hubbard Brook Experimental Forest, New Hampshire. *Ecology* 56:1068–1080.

Buso, D. C., G. E. Likens, and J. S. Eaton. 2000. Chemistry of precipitation, stream water, and lake water from the Hubbard Brook Ecosystem Study: A record of sampling protocols and analytical procedures. General Technical Report NE-275. Newtown Square, Pa., USDA Forest Service, Northeastern Research Station. 52 pp.

Buso, D. C., G. E. Likens, J. W. LaBaugh, and D. Bade. 2009. Nutrient dynamics. In T. C. Winter and G. E. Likens, eds., *Mirror Lake: Interactions Among Air, Land, and Water,* pp. 69–203. University of California Press, Berkeley.

Butler, T. J., M. D. Cohen, F. M. Vermeylen, G. E. Likens, D. Schmeltz, and R. S. Artz. 2008. Regional precipitation mercury trends in the eastern USA, 1998–2005: Declines in the Northeast and Midwest, but no change in the Southeast. *Atmospheric Environment* 42:1582–1592.

Butler, T. J., F. M. Vermeylen, M. Rury, G. E. Likens, B. Lee, G. E. Bowker, and L. M. McCluney. 2011. Response of ozone and nitrate to stationary source NO$_x$ emission reductions in the eastern USA. *Atmospheric Environment* 45:1084–1094.

Campbell J. L, M. J. Mitchell, P. M. Groffman, L. M. Christensen, and J. P. Hardy. 2005. Winter in northeastern North America: A critical period for ecological processes. *Frontiers in Ecology and the Environment* 3:314–322.

Campbell, J. L., C. T. Driscoll, C. Eagar, G. E. Likens, T. G. Siccama, C. E. Johnson, T. J. Fahey, et al. 2007. Long-term trends from ecosystem research at the Hubbard Brook Experimental Forest. General Technical Report NRS-17. Newtown Square, Pa., USDA Forest Service, Northern Research Station. 41 pp.

Campbell, J. L., L. E. Rustad, E. W. Boyer, S. F. Christopher, C. T. Driscoll, I. J. Fernandez, P. M. Groffman, D. House, J. Kiekbusch, A. H. Magill, M. J. Mitchell, and S. V. Ollinger. 2009. Consequences of climate change for biogeochemical cycling in forests of northeastern North America. *Canadian Journal of Forest Research* 39:264–284.

Campbell, J. L., S. V. Ollinger, G. N. Flerchinger, H. Wicklein, K. Hayhoe, and A. S. Bailey. 2010. Past and projected future changes in snowpack and soil frost at the Hubbard Brook Experimental Forest, New Hampshire, USA. *Hydrological Processes* 24:2465–2480.

Campbell, J. L., A. S. Bailey, C. Eagar, M. B. Green, and J. J. Battles. 2013. Vegetation treatments and hydrologic responses at the Hubbard Brook Experimental Forest, New Hampshire. In A. E. Camp, L. C. Irland, and C. J. W. Carroll, eds., *Long-Term Silvicultural and Ecological Studies: Results for Science and Management.* 2:1–9. Yale University, Global Institute of Sustainable Forestry, Research Paper 013.

Campbell, J. L., A. M. Socci, and P. H. Templer. 2014. Increased nitrogen leaching following soil freezing is due to decreased root uptake in a northern hardwood forest. *Global Change Biology* 20:2663–2673.

Cary Institute of Ecosystem Studies. 2015. In a warmer world, ticks that spread disease are arriving earlier, expanding their ranges. *ScienceDaily,* February 18, 2015. www.sciencedaily.com/releases/2015/02/150218122947.htm.

Chittenden, A. K. 1905. Forest conditions of northern New Hampshire. *USDA Bureau of Forestry Bulletin* 55:1–100.

Cho, Y., C. T. Driscoll, C. E. Johnson, and T. G. Siccama. 2010. Chemical changes in soil and soil solution after calcium silicate addition to a northern hardwood forest. *Biogeochemistry* 100:3–20.

Cho, Y., C. T. Driscoll, C. E. Johnson, J. D. Blum, and T. J. Fahey. 2012. Watershed-level responses to calcium silicate treatment in a northern hardwood forest. *Ecosystems* 15:416–434.

Christenson, L. M., M. J. Mitchell, P. M. Groffman, and G. M. Lovett. 2010. Winter climate change implications for decomposition in northeastern forests: Comparisons of sugar maple litter to herbivore fecal inputs. *Global Change Biology* 16:2589–2601.

Christenson, L. M., M. J. Mitchell, P. M. Groffman, and G. M. Lovett. 2014. Cascading effects of climate change on forest ecosystems: Biogeochemical links between trees and moose in the northeast USA. *Ecosystems* 17:442–457.

Chuang, H. C., M. S. Webster, and R. T. Holmes. 1999. Extra-pair paternity and local synchrony in the black-throated blue warbler (*Dendroica caerulescens*). *Auk* 116:727–736.

Chuang-Dobbs, H. C., M. S. Webster, and R. T. Holmes.

2001. Paternity and parental care in the black-throated blue warbler (*Dendroica caerulescens*). *Animal Behavior* 62:83–92.

Cleavitt N. L, T. J. Fahey, P. M. Groffman, J. P. Hardy, K. S. Henry, and C. T. Driscoll. 2008a. Effects of soil freezing on fine roots in a northern hardwood forest. *Canadian Journal of Forest Research* 38:82–91.

Cleavitt, N. L., M. Fairbairn, and T. J. Fahey. 2008b. Growth and survivorship of American beech (*Fagus grandifolia* Ehrh.) seedlings in a northern hardwood forest following a mast event. *Journal of the Torrey Botanical Society* 135:328–345.

Cleavitt, N. L., T. J. Fahey, and J. J. Battles. 2011. Regeneration ecology of sugar maple (*Acer saccharum*): Seedling survival in relation to nutrition, site factors, and damage by insects and pathogens. *Canadian Journal of Forest Research* 41:235–244.

Cleavitt, N. L., J. J. Battles, T. J. Fahey, and J. D. Blum. 2014. Determinants of survival over seven years for a natural cohort of sugar maple seedlings in a northern hardwood forest. *Canadian Journal of Forest Research* 44:1112–1121.

Cogbill, C. V. 1989. Hubbard Brook revisited: Land use history of the Hubbard Brook valley. Unpublished manuscript.

Cogbill, C. V., and G. E. Likens. 1974. Acid precipitation in the northeastern United States. *Water Resources Research* 10:1133–1137.

Comerford D. P, P. G. Schaberg, P. H. Templer, A. M. Socci, J. L. Campbell, and K. F. Wallin. 2013. Influence of experimental snow removal on root and canopy physiology of sugar maple trees in a northern hardwood forest. *Oecologia* 171:261–269.

Conrod, C. A. 2008. Rodent population and habitat dynamics in a northern hardwood forest. Master of Science thesis, Plymouth State University, Plymouth, N.H.

Costello, C., M. Yamasaki, P. J. Pekins, W. B. Leak, and C. D. Neefus. 2000. Songbird response to group selection harvests and clearcuts in a New Hampshire northern hardwood forest. *Forest Ecology and Management* 127:41–54.

Davis, M. B. 1985. History of the vegetation on the Mirror Lake watershed. In G. E. Likens, ed., *An Ecosystem Approach to Aquatic Ecology: Mirror Lake and Its Environment,* pp. 53–65. Springer-Verlag, New York.

DeGraaf, R. M., and M. Yamasaki. 2003. Options for managing early-successional forest and shrubland bird habitats in the northeastern United States. *Forest Ecology and Management* 185:179–191.

Demers, J. D., C. T. Driscoll, and J. B. Shanley. 2010. Mercury mobilization and episodic stream acidification during snowmelt: Role of hydrologic flow paths, source areas, and supply of dissolved organic carbon. *Water Resources Research* 46, W01511. doi:10.1029/2008WR007021.

Dittman, J. A., J. B. Shanley, C. T. Driscoll, G. R. Aiken, A. T. Chalmers, J. E. Towse, and P. Selvendiran. 2010. Mercury dynamics in relation to dissolved organic carbon concentration and quality during high flow events in three northeastern U.S. streams. *Water Resources Research* 46, WO7522. doi:10.1029/2009WR008351.

Doran, P. J. 2003. Intraspecific spatial variation in bird abundance: Patterns and processes. Ph.D. dissertation, Dartmouth College, Hanover, N.H.

Doran, P. J., and R. T. Holmes. 2005. Habitat occupancy patterns of a forest-dwelling songbird: Causes and consequences. *Canadian Journal of Zoology* 83:1297–1305.

Draxler, R. R., and G. D. Hess. 1998. An overview of the Hysplit_4 modelling system for trajectories, dispersion, and deposition. *Australian Meteorological Magazine* 47:295–308.

Driscoll, C. T. 1984a. Aluminum chemistry and potential effects in acidic surface waters. In D. D. Hemphill, ed., *Trace Substances in Environmental Health* 17:215–229. Columbia, Mo.

Driscoll, C. T. 1984b. A procedure for the fractionation of aqueous aluminum in dilute acidic waters. *International Journal of Environmental and Analytical Chemistry* 16:267–283.

Driscoll, C. T., G. B. Lawrence, A. J. Bulger, T. J. Butler, C. S. Cronan, C. Eagar, K. Fallon Lambert, G. E. Likens, J. L. Stoddard, and K. C. Weathers. 2001. Acid rain revisited: Advances in scientific understanding since the passage of the 1970 and 1990 Clean Air Act Amendments. Hubbard Brook Research Foundation. *Science Links Publication* 1 (1):1–20.

Driscoll, C. T., D. Evers, K. F. Lambert, N. Kamman, T. Holsen, Young-Ji Han, C. Chen, W. Goodale, T. Butler, T. Clair, and R. Munson. 2007a. Mercury matters: Linking mercury science with public policy in the northeastern United States. Hubbard Brook Research Foundation, Science Links Publication 1(3):1–24.

Driscoll, C. T., Y.-J. Han, C. Chen, D. C. Evers, K. Fallon Lambert, T. M. Holsen, N. C. Kamman, and R. K. Munson. 2007b. Mercury contamination in forest and freshwater ecosystems in the northeastern United States. *BioScience* 57:17–28.

Driscoll, C. T., K. F. Lambert, and K. C. Weathers. 2011. Integrating science and policy: A case study of the Hubbard Brook Research Foundation's Science Links program. *BioScience* 61:791–801.

Dumanoski, D. 1983. Ecosystem data is unique: Hubbard

Brook's 20-year ecological records are a national treasure. *Boston Globe,* 19 September 1983.

Durán, J., J. L. Morse, P. M. Groffman, J. L. Campbell, L. M. Christenson, C. T. Driscoll, T. J. Fahey, M. C. Fisk, M. J. Mitchell, and P. H. Templer. 2014. Winter climate change affects growing-season soil microbial biomass and activity in northern hardwood forests. *Global Change Biology* 20:3568-3577.

Edwards, C. H., and P. J. Bohlen. 1996. *Biology and Ecology of Earthworms.* Chapman & Hall, London.

Edzwald, J. 2014. Acidity matters. *Interstate Water Report 2014.* September 2014.

Environmental Protection Agency. 2011. The benefits and costs of the Clear Air Act from 1990 to 2020. Final Report, Rev. A., April 2011. www.epa.gov/airsect812/prospective2 .html.

Faaborg, J., R. T. Holmes, A. D. Anders, K. L. Bildstein, K. M. Dugger, S. A. Gauthreaux Jr., P. Heglund, et al. 2010a. Recent advances in understanding migration systems of New World land birds. *Ecological Monographs* 80:3-48.

Faaborg, J., R. T. Holmes, A. D. Anders, K. L. Bildstein, K. M. Dugger, S. A. Gauthreaux Jr., P. Heglund, et al. 2010b. Conserving migratory landbirds in the New World: Do we know enough? *Ecological Applications* 20:398-418.

Fahey, T. J., T. G. Siccama, C. T. Driscoll, G. E. Likens, J. Campbell, C. E. Johnson, J. D. Aber, et al. 2005. The biogeochemistry of carbon at Hubbard Brook. *Biogeochemistry* 75:109-176.

Fahey, T. J., P. B. Woodbury, J. J. Battles, C. L. Goodale, S. P. Hamburg, S. V. Ollinger, and C. W. Woodall. 2010. Forest carbon storage: Ecology, management, and policy. *Frontiers in Ecology and the Environment* 8:245-252.

Fahey, T. J., P. H. Templer, B. T. Anderson, J. J. Battles, J. L. Campbell, C. T. Driscoll, A. R. Fusco, M. B. Green, K. K. Kassam, N. L. Rodenhouse, et al. 2015. The promise and peril of intensive-site based ecological research: Insights from the Hubbard Brook Ecosystem Study. *Ecology* 96:885-901.

Federer, C. A. 1969. New landmark in the White Mountains. *Appalachia* 36:589-594.

Federer, C. A., J. W. Hornbeck, L. M. Tritton, C. W. Martin, R. S. Pierce, and C. T. Smith. 1989. Long-term depletion of calcium and other nutrients in eastern U.S. forests. *Environmental Management* 13:593-601.

Fisher, S. G., and G. E. Likens. 1972. Stream ecosystem: Organic energy budget. *BioScience* 22:33-35.

Fisher, S. G., and G. E. Likens. 1973. Energy flow in Bear Brook, New Hampshire: An integrative approach to stream ecosystem metabolism. *Ecological Monographs* 43:421-439.

Fisk, M. C., T. J. Fahey, and P. M. Groffman. 2010. Carbon resources, soil organisms, and nitrogen availability: Landscape patterns in a northern hardwood forest. *Forest Ecology and Management* 260:1175-1183.

Fitzhugh, R. D., C. T. Driscoll, P. M. Groffman, G. L. Tierney, T. J. Fahey, and J. P. Hardy. 2001. Effects of soil freezing disturbance on soil solution nitrogen, phosphorus, and carbon chemistry in a northern hardwood ecosystem. *Biogeochemistry* 56:215-238.

Fitzhugh, R. D., G. E. Likens, C. T. Driscoll, M. J. Mitchell, P. M. Groffman, T. J. Fahey, and J. P. Hardy. 2003. Role of snow freezing events in interannual patterns of stream chemistry at the Hubbard Brook Experimental Forest, New Hampshire. *Environmental Science and Technology* 37:1575-1580.

Fiumara, C. 2006. The response of ground-dwelling beetle communities to calcium amendment and influence on litter decay and nutrient dynamics in a north-temperate forest. Master of Science thesis, Appalachian State University, Boone, N.C.

Forcier, L. K. 1975. Reproductive strategies and the co-occurrence of climax tree species. *Science* 189:808-810.

Franklin, J. F., and K. N. Johnson. 2014. Lessons in policy implementation from experiences with the Northwest Forest Plan, USA. *Biodiversity and Conservation* 23:3607-3613.

Friedland, A. J., R. Relyea, and D. Courard-Hauri. 2012. *Essentials of Environmental Science.* W. H. Freeman.

Galloway, J. N., and G. E. Likens. 1981. Acid precipitation: The importance of nitric acid. *Atmospheric Environment* 15:1081-1085.

Galloway, J. N., G. E. Likens, and E. S. Edgerton. 1976. Acid precipitation in the northeastern United States: pH and acidity. *Science* 194:722-724.

Galloway, J. N., G. E. Likens, W. C. Keene, and J. M. Miller. 1982. The composition of precipitation in remote areas of the world. *Journal of Geophysical Research* 87:8771-8786.

Gillen, C. P., S. W. Bailey, and K. J. McGuire. 2015. Mapping of hydropedologic spatial patterns in a steep headwater catchment. *Soil Science of America Journal* 79. doi:10.2136/sssaj2014.05.0189.

Goetz, S. J., D. Steinberg, M. G. Betts, R. T. Holmes, P. J. Doran, R. Dubayah, and M. Hofton. 2010. Lidar remote sensing variables predict breeding habitat of a Neotropical migrant bird. *Ecology* 91:1569-1576.

Goodale, C. L., J. D. Aber, and P. M. Vitousek. 2003. An unexpected nitrate decline in New Hampshire streams. *Ecosystems* 6:75-86.

Gosz, J. R., G. E. Likens, and F. H. Bormann. 1972. Nutrient content of litter fall on the Hubbard Brook Experimental Forest, New Hampshire. *Ecology* 53:769–784.

Gosz, J. R., R. T. Holmes, G. E. Likens, and F. H. Bormann. 1978. The flow of energy in a forest ecosystem. *Scientific American* 238:92–102.

Graveland, J., R. van der Wal, J. H. van Balen, and A. J. van Noordwijk. 1994. Poor reproduction in forest passerines from decline of snail abundance on acidified soils. *Nature* 368:446–448.

Green, M. B., A. S. Bailey, S. W. Bailey, J. J. Battles, J. L. Campbell, C. T. Driscoll, T. J. Fahey, et al. 2013. Decreased water flowing from a forest amended with calcium silicate. *Proceedings of the National Academy of Sciences* 110:5999–6003.

Greene, B. T., W. H. Lowe, and G. E. Likens. 2008. Forest succession and prey availability influence the strength and scale of terrestrial-aquatic linkages in a headwater salamander system. *Freshwater Biology* 53:2234–2243.

Groffman, P. M., J. P. Hardy, S. Nolan, R. D. Fitzhugh, C. T. Driscoll, and T. J. Fahey. 1999. Snow depth, soil frost, and nutrient loss in a northern hardwood forest. *Hydrological Processes* 13:2275–2286.

Groffman, P. M., C. T. Driscoll, T. J. Fahey, J. P. Hardy, R. D. Fitzhugh, and G. L. Tierney. 2001a. Colder soils in a warmer world: A snow manipulation study in a northern hardwoods forest ecosystem. *Biogeochemistry* 56:135–150.

Groffman, P. M., C. T. Driscoll, T. J. Fahey, J. P. Hardy, R. D. Fitzhugh, and G. L. Tierney. 2001b. Effects of mild winter freezing on soil nitrogen and carbon dynamics in a northern hardwood forest. *Biogeochemistry* 56:191–213.

Groffman, P. M., C. T. Driscoll, G. E. Likens, M. C. Fisk, T. J. Fahey, R. T. Holmes, G. E. Likens, and L. Pardo. 2004. Nor gloom of night: A new conceptual model for the Hubbard Brook Ecosystem Study. *BioScience* 54:139–148.

Groffman, P. M., L. E. Rustad, P. H. Templer, L. M. Christenson, N. K. Lany, A. M. Socci, et al. 2012. Long-term integrated studies show complex and surprising effects of climate change in the northern hardwood forest. *BioScience* 62:1056–1066.

Hall, R. J., and G. E. Likens. 1980. Ecological effects of experimental acidification on a stream ecosystem. In D. Drabløs and A. Tollan, eds., *Ecological Impact of Acid Precipitation,* pp. 375–376. Proceedings of the International Conference, Sandefjord, Norway. March 1980.

Hall, R. J., and G. E. Likens. 1981. Chemical flux in an acid-stressed stream. *Nature* 292:329–331.

Hall, R. J., G. E. Likens, S. B. Fiance, and G. R. Hendrey. 1980. Experimental acidification of a stream in the Hubbard Brook Experimental Forest, New Hampshire. *Ecology* 61:976–989.

Hall, R. J., J. M. Pratt, and G. E. Likens. 1982. Effects of experimental acidification on macroinvertebrate drift diversity in a mountain stream. *Journal of Water, Air, and Soil Pollution* 18:273–287.

Hall, R. J., C. T. Driscoll, G. E. Likens, and J. M. Pratt. 1985. Physical, chemical, and biological consequences of episodic aluminum additions to a stream. *Limnology and Oceanography* 30:212–220.

Hall, R. O., Jr. 2003. A stream's role in watershed nutrient export. *Proceedings of the National Academy of Sciences, USA* 100:10137–10138.

Hall, R. O., Jr., K. H. Macneale, E. S. Bernhardt, M. Field, and G. E. Likens. 2001. Biogeochemical responses of two forest streams to a two-month calcium addition. *Freshwater Biology* 46:291–302.

Hallett, R. A., S. W. Bailey, S. B. Horsley, and R. P. Long. 2006. Influence of nutrition and stress on sugar maple at a regional scale. *Canadian Journal of Forest Research* 36:2235–2246.

Halman, J. M., P. G. Schaberg, G. J. Hawley, C. F. Hansen, and T. J. Fahey. 2015. Differential impacts of calcium and aluminum on sugar maple and American beech growth dynamics. *Canadian Journal of Forest Research* 45:52–59.

Hamburg, S. P. 1984. Organic matter and nitrogen accumulation during 70 years of old-field succession in central New Hampshire. Ph.D. dissertation, Yale University, New Haven. 238 pp.

Hamburg, S. P., and C. V. Cogbill. 1988. Historical decline of red spruce populations and climatic warming. *Nature* 331:428–431.

Hamburg, S. P., M. A. Vadeboncoeur, A. D. Richardson, and A. S. Bailey. 2013. Climate change at the ecosystem scale: A 50-year record in New Hampshire. *Climate Change* 116:457–477.

Hawley, G. J., P. G. Schaberg, C. Eagar, and C. H. Borer. 2006. Calcium addition at the Hubbard Brook Experimental Forest reduced winter injury to red spruce in a high-injury year. *Canadian Journal of Forest Research* 36:2544–2549.

Hayhoe, K., C. P. Wake, T. G. Huntington, L. Luo, M. D. Schwartz, J. Sheffeild, E. Wood, B. Anderson, J. Bradbury, A. Degaetano, T. J. Troy, and D. Wolfe. 2007. Past and future changes in climate and hydrological indicators in the US Northeast. *Climate Dynamics* 28:381–407.

Hayhoe, K., C. Wake, B. Anderson, X.-Z. Liang, E. Maurer, J. Zhu, J. Bradbury, A. DeGaetano, A. M. Stoner, and D. Wuebbles. 2008. Regional climate change projections for the Northeast USA. *Mitigation and Adaptation Strategies for Global Change* 13:425–436.

Hedin, L. O., M. S. Mayer, and G. E. Likens. 1988. The effect of deforestation on organic debris dams. *Verhandlungen des Internationalen Verein Limnologie* 23:1135–1141.

Hendrey, G. R., R. J. Hall, and G. E. Likens. 1979. Experimental stream acidification: Effects on benthic algal productivity. Abstracts of Papers Submitted for the American Society of Limnology and Oceanography, 42nd Annual Meeting, Stony Brook, New York.

Hewlett, J. D. 1971. Comments on the catchment experiment to determine vegetal effects on water yield. *Water Resources Bulletin* 7:376–381.

Holmes, R. T. 1990. The structure of a temperate deciduous forest bird community: Variability in time and space. In A. Keast, ed., *Biogeography and Ecology of Forest Bird Communities,* pp. 121–139. SPB Academic Publications, The Hague.

Holmes, R. T. 2007. Understanding population change in migratory songbirds: Long-term and experimental studies of Neotropical migrants in breeding and wintering areas. *Ibis* 149 (Supplement 2):2–13.

Holmes, R. T. 2011. Avian population and community processes in forest ecosystems: Long-term research in the Hubbard Brook Experimental Forest. *Forest Ecology and Management* 262:20–32.

Holmes, R. T., and G. E. Likens. 1999. Organisms of the Hubbard Brook Valley, New Hampshire. USDA Forest Service, Northeastern Research Station, General Technical Report NE-257. 32 pp.

Holmes, R. T., and S. K. Robinson. 1981. Tree species preferences by foraging insectivorous birds in a northern hardwood forest. *Oecologia* 48:31–35.

Holmes, R. T., and J. C. Schultz. 1988. Food availability for forest birds: Effects of prey distribution and abundance on bird foraging. *Canadian Journal of Zoology* 66:720–728.

Holmes, R. T., and T. W. Sherry. 1992. Site fidelity of migratory warblers in temperate breeding and Neotropical wintering areas: Implications for population dynamics, habitat selection, and conservation. In J. M. Hagan and D. W. Johnston, eds., *Ecology and Conservation of Neotropical Migrant Landbirds,* pp. 563–575. Smithsonian Press, Washington, D.C.

Holmes, R. T., and T. W. Sherry. 2001. Thirty-year bird population trends in an unfragmented temperate deciduous forest: Importance of habitat change. *Auk* 118:589–610.

Holmes, R. T., and F. W. Sturges. 1973. Annual energy expenditure by the avifauna of a northern hardwood ecosystem. *Oikos* 24:24–29.

Holmes, R. T., and F. W. Sturges. 1975. Avian community dynamics and energetics in a northern hardwood ecosystem. *Journal of Animal Ecology* 44:527–531.

Holmes, R. T., R. E. Bonney Jr., and S. W. Pacala. 1979. The guild structure of the Hubbard Brook bird community: A multivariate approach. *Ecology* 60:512–520.

Holmes, R. T., T. W. Sherry, and F. W. Sturges. 1986. Bird community dynamics in a temperate deciduous forest: Long-term trends at Hubbard Brook. *Ecological Monographs* 56:201–220.

Holmes, R. T., T. W. Sherry, and L. R. Reitsma. 1989. Population structure, territoriality, and overwinter survival of two migrant warbler species in Jamaica. *Condor* 91:545–561.

Holmes, R. T., T. W. Sherry, P. P. Marra, and K. E. Petit. 1992. Multiple-brooding and annual productivity of a neotropical migrant, the black-throated blue warbler (*Dendroica caerulescens*), in an unfragmented temperate forest. *Auk* 109:321–333.

Holmes, R. T., N. L. Rodenhouse, and T. S. Sillett. 2005. Black-throated blue warbler (*Dendroica caerulescens*). *The Birds of North America Online,* ed. A. Poole, Cornell Laboratory of Ornithology. http://bna.birds.cornell.edu /BNA/account/Black-throated_Blue_Warbler.

Hong, B., D. P. Swaney, and D. A. Weinstein. 2006. Assessment of ozone effects on nitrate export from Hubbard Brook Watershed 6. *Environmental Pollution* 141:8–21.

Hoover, M. D. 1944. Effect of removal of forest vegetation upon water-yields. *Transactions of the American Geophysical Union,* Part 6, 969–977.

Hornbeck, J. 2001. Events leading to the establishment of the Hubbard Brook Experimental Forest. Unpublished manuscript, available at www.hubbardbrook.org.

Hornbeck, J. W., C. W. Martin, R. S. Pierce, F. H. Bormann, G. E. Likens, and J. S. Eaton. 1987. The northern hardwood forest ecosystem: Ten years of recovery from clearcutting. USDA Forest Service, Northeastern Forest Experiment Station, NE-RP-596. 30 pp.

Hornbeck, J. W., S. W. Bailey, D. C. Buso, and J. B. Shanley. 1997. Streamwater chemistry and nutrient budgets for forested watersheds in New England: Variability and management implications. *Forest Ecology and Management* 93:73–89.

Hornbeck, J. W., M. M. Alexander, C. Eagar, J. Y. Carlson, and R. B. Smith. 2001. Database for chemical contents of streams on the White Mountain National Forest. USDA

Forest Service, Northeastern Forest Research Station, General Technical Report NE-282. Newtown Square, Pa. 12 pp.

Horsley, S. B., R. P. Long, S. W. Bailey, R. A. Hallett, and P. M. Wargo. 2002. Health of eastern North American sugar maple forests and factors affecting decline. *Northern Journal of Applied Forestry* 19:34–44.

Houston, D. R. 1994. Major new tree disease epidemics: Beech bark disease. *Annual Review of Phytopathology* 32:75–87.

Huang, C. Y., P. F. Hendrix, T. J. Fahey, P. J. Bohlen, and P. M. Groffman. 2010. A simulation model to evaluate the impacts of invasive earthworms on soil carbon dynamics. *Ecological Modelling* 221:2447–2457.

Hunt, C. B. 1967. *Physiography of the United States.* Freeman and Company, San Francisco.

Hunt, P. D. 1996. Habitat selection by American redstarts along a successional gradient in northern hardwood forests: Evaluation of habitat quality. *Auk* 113:875–888.

International Panel on Climate Change (IPCC). 2012. Summary for Policymakers. In *Managing the Risks of Extreme Events and Disasters to Advance Climate Change Adaptation* [C. B. Field, V. Barros, T. F. Stocker, D. Qin, D. J. Dokken, K. L. Ebi, M. D. Mastrandrea, K. J. Mach, G.-K. Plattner, S. K. Allen, M. Tignor, and P. M. Midgley, eds.]. A Special Report of Working Groups I and II of the Intergovernmental Panel on Climate Change. Cambridge University Press, Cambridge, pp. 1–19.

International Panel on Climate Change (IPCC). 2014. Fifth Assessment Report, Contribution of Working Group II to the Fifth Assessment Report of the Intergovernmental Panel on Climate Change [C. B. Field, V. R. Barros, D. J. Dokken, K. J. Mach, M. D. Mastrandrea, T. E. Bilir, M. Chatterjee, K. L. Ebi, Y. O. Estrada, R. C. Genova, B. Girma, E. S. Kissel, A. N. Levy, S. MacCracken, P. R. Mastrandrea, and L. L. White, eds.]. Cambridge University Press, Cambridge, pp. 1–32.

Iverson, L. R., A. M. Prasad, S. N. Matthews, and M. Peters. 2008. Estimating potential habitat for 134 eastern U.S. tree species under six climate scenarios. *Forest Ecology and Management* 254:390–406.

Janzen, D. H. 1971. Seed predation by animals. *Annual Review of Ecology and Systematics* 2:465–492.

Johnson, C. E. 1989. The chemical and physical properties of a northern hardwood forest soil: Harvesting effects, soil-tree relations, and sample size determination. Ph.D. dissertation, University of Pennsylvania, Philadelphia. 221 pp.

Johnson, C. E., T. G. Siccama, C. T. Driscoll, G. E. Likens, and R. E. Moeller. 1995. Changes in lead biogeochemistry in response to decreasing atmospheric inputs. *Ecological Applications* 5:813–822.

Johnson, C. E., C. T. Driscoll, T. G. Siccama, and G. E. Likens. 2000. Element fluxes and landscape position in a northern hardwood forest watershed ecosystem. *Ecosystems* 3:159–184.

Johnson, C. E., T. G. Siccama, E. G. Denny, M. M. Koppers, and D. J. Vogt. 2014. In situ decomposition of northern hardwood tree boles: Decay rates and nutrient dynamics in wood and bark. *Canadian Journal of Forest Research* 44:1515–1524.

Johnson, M. D., T. W. Sherry, R. T. Holmes, and P. P. Marra. 2006. Assessing habitat quality for a migratory songbird wintering in natural and agricultural habitats. *Conservation Biology* 20:1433–1444.

Johnson, N. M., G. E. Likens, F. H. Bormann, D. W. Fisher, and R. S. Pierce. 1969. A working model for the variation in stream water chemistry at the Hubbard Brook Experimental Forest, New Hampshire. *Water Resources Research* 5:1353–1363.

Johnson, N. M., C. T. Driscoll, J. S. Eaton, G. E. Likens, and W. H. McDowell. 1981. Acid rain, dissolved aluminum, and chemical weathering at the Hubbard Brook Experimental Forest, New Hampshire. *Geochimica Cosmochimica Acta* 45:1421–1437.

Jones, J., P. J. Doran, and R. T. Holmes. 2003. Climate and food synchronize regional forest bird abundances. *Ecology* 84:3024–3032.

Judd, K. E., G. E. Likens, and P. M. Groffman. 2007. High nitrate retention during winter in soils of the Hubbard Brook Experimental Forest. *Ecosystems* 10:217–225.

Judd, K. E., G. E. Likens, D. C. Buso, and A. S. Bailey. 2011. Minimal response in watershed nitrate export to severe soil frost raises questions about nutrient dynamics in the Hubbard Brook Experimental Forest. *Biogeochemistry* 106:443–459.

Juice, S. M., T. J. Fahey, T. G. Siccama, C. T. Driscoll, E. G. Denny, C. Eagar, N. L. Cleavitt, R. Minocha, and A. D. Richardson. 2006. Response of sugar maple to calcium addition to northern hardwood forest. *Ecology* 87:1267–1280.

Kaiser, S. A., M. S. Webster, and T. S. Sillett. 2014. Phenotypic plasticity in hormonal and behavioural responses to changes in resource conditions in a migratory songbird. *Animal Behaviour* 96:19–29.

Kaiser, S. A., T. S. Sillett, B. B. Risk, and M. S. Webster. 2015. Experimental food manipulation experiment reveals habitat-dependent male reproductive investment in a migratory bird. *Proceedings of the Royal Society B* 282:20142523. http://dx.doi.org/10.1098/rspb.2014.2523.

Keene, W. C., J. N. Galloway, G. E. Likens, F. A. Deviney, K. N. Mikkelson, J. L. Moody, and J. R. Maben. 2015. Atmospheric wet deposition in remote regions: Benchmarks for environmental change. *Journal of the Atmospheric Sciences* 72:2947–2978.

Kelly, D. 1994. The evolutionary ecology of mast seeding. *Trends in Ecology and Evolution* 9:465–470.

Kosiba, A. M., P. G. Schaberg, G. J. Hawley, and C. F. Hansen. 2013. Quantifying the legacy of foliar winter injury on woody aboveground carbon sequestration of red spruce trees. *Forest Ecology and Management* 302:363–371.

Krebs, C. J., and B. Elwood. 2008. *The Ecological World View.* University of California Press, Berkeley.

Lany, N. K., M. P. Ayres, E. E. Stange, T. S. Sillett, N. L. Rodenhouse, and R. T. Holmes. 2015. Breeding timed to maximize reproductive success for a migratory songbird: The importance of phenological asynchrony. *Oikos.* doi:10.1111/oik.02412.

La Sorte, F. A., D. Fink, W. M. Hochachka, J. L. Aycrigg, K. V. Rosenberg, A. D. Rodewald, N. E. Bruns, A. Farnsworth, B. L. Sullivan, C. Wood, and S. Kelling. 2015. Documenting stewardship responsibilities across the annual cycle for birds on U.S. public lands. *Ecological Applications* 25:39–51.

Lawrence, G. B., M. B. David, and W. C. Shortle. 1995. A new mechanism for calcium loss in forest-floor soils. *Nature* 378:162–165.

Leopold, A. 1941. Lakes in relation to terrestrial life patterns. In *A Symposium on Hydrobiology,* pp. 17–22. University of Wisconsin Symposium, Volume on Hydrology, Madison.

Leopold, A. 1949. *A Sand County Almanac.* Oxford University Press, New York.

Likens, G. E. 1972. Mirror Lake: Its past, present, and future. *Appalachia* 39:23–41.

Likens, G. E. 1974. Acid in rain found up sharply in east; smoke curb. Cited by Boyce Rensberger, *New York Times,* June 13, 1974.

Likens, G. E. 1984a. Acid rain: The smokestack is the "smoking gun." *Garden* 8:12–18.

Likens, G. E. 1984b. We must take prompt action. *Wall Street Journal,* June 28, 1984.

Likens, G. E., ed. 1985a. *An Ecosystem Approach to Aquatic Ecology: Mirror Lake and Its Environment.* Springer-Verlag, New York. 516 pp.

Likens, G. E. 1985b. Mirror Lake: Cultural history. In G. E. Likens, ed., *An Ecosystem Approach to Aquatic Ecology: Mirror Lake and Its Environment,* pp. 72–83. Springer-Verlag, New York.

Likens, G. E. 1985c. An experimental approach for the study of ecosystems. *Journal of Ecology* 73:381–396.

Likens, G. E. 1987. Chemical wastes in our atmosphere—an ecological crisis. *Industrial Crisis Quarterly* 1:13–33.

Likens, G. E. 1989. Some aspects of air pollutant effects on terrestrial ecosystems and prospects for the future. *Ambio* 18:172–178.

Likens, G. E. 1992. *The Ecosystem Approach: Its Use and Abuse.* Excellence in Ecology, Book 3. The Ecology Institute, Oldendorf-Luhe, Germany. 166 pp.

Likens, G. E. 1998. Limitations to intellectual progress in ecosystem science. In M. L. Pace and P. M. Groffman, eds., *Successes, Limitations, and Frontiers in Ecosystem Science,* pp. 247–271. 7th Cary Conference, Institute of Ecosystem Studies, Millbrook, New York. Springer-Verlag, New York.

Likens, G. E. 1999. The science of nature, the nature of science: Long-term ecological studies at Hubbard Brook. *Proceedings of the American Philosophical Society* 143:558–572.

Likens, G. E. 2001. Biogeochemistry, the watershed approach: Some uses and limitations. *Marine and Freshwater Research* 52:5–12.

Likens, G. E. 2002. Biogeochemistry: Some opportunities and challenges for the future. In *Biogeomon 2002, 4th International Symposium on Ecosystem Behaviour.* The University of Reading, U.K., August 2002. p. 136.

Likens, G. E. 2004. Some perspectives on long-term biogeochemical research from the Hubbard Brook Ecosystem Study. *Ecology* 85:2355–2362.

Likens, G. E. 2010. The role of science in decision-making: Does evidence-based science drive environmental policy? *Frontiers in Ecology and the Environment* 8(6):e1–e8. doi:10.1890/090132.

Likens, G. E. 2011. Limnological measures related to climate change in the Hubbard Brook valley, USA. *Inland Waters* 1:93–99.

Likens, G. E. 2013. *Biogeochemistry of a Forested Ecosystem,* 3rd ed. Springer-Verlag, New York. 208 pp.

Likens, G. E., and S. W. Bailey. 2014. The discovery of acid rain at the Hubbard Brook Experimental Forest: A story of collaboration and long-term research. In D. C. Hayes, S. L. Stout, R. H. Crawford, and A. P. Hoover, eds., *USDA Forest Service Experimental Forests and Ranges,* pp. 463–482. Springer, New York.

Likens, G. E., and R. E. Bilby. 1982. Development, maintenance, and role of organic-debris dams in New England streams. In F. J. Swanson, R. J. Janda, T. Dunne, and D. W. Swanston, eds., *Sediment Budgets and Routing in Forested Drainage Basins,* pp. 122–128. USDA Forest Service, General Technical Report PNW-141.

Likens, G. E., and F. H. Bormann. 1972. Nutrient cycling in ecosystems. In J. Wiens, ed., *Ecosystem Structure and Function,* pp. 25–67. Oregon State University Press, Corvallis.

Likens, G. E., and F. H. Bormann. 1974a. Acid rain: A serious regional environmental problem. *Science* 184:1176–1179.

Likens, G. E., and F. H. Bormann. 1974b. Linkages between terrestrial and aquatic ecosystems. *BioScience* 24:447–456.

Likens, G. E., and F. H. Bormann, 1974c. Effects of forest clearing on the northern hardwood forest ecosystem and its biogeochemistry. In *Proceedings of the First International Congress of Ecology,* pp. 330–335. Wageningen, The Hague.

Likens, G. E., and D. C. Buso. 2006. Variation in streamwater chemistry throughout the Hubbard Brook valley. *Biogeochemistry* 78:1–30.

Likens, G. E., and D. C. Buso. 2010. Long-term changes in streamwater chemistry following disturbance in the Hubbard Brook Experimental Forest, USA. *Verhandlungen des Internationalen Verein Limnologie* 30:1577–1581.

Likens, G. E., and D. C. Buso. 2012. Dilution and the elusive baseline. *Environmental Science and Technology* 46:4382–4387.

Likens, G. E., and D. C. Buso. 2016. Decades of biogeochemical change at Hubbard Brook—Impacts on legacies. *Ecological Monographs.*

Likens, G. E., and T. J. Butler. 2014. Atmospheric acid deposition. *Encyclopedia of Natural Resources.* Taylor & Francis. doi:10.1081/E-ENRA-120047613.

Likens, G. E., and M. B. Davis. 1975. Post-glacial history of Mirror Lake and its watershed in New Hampshire, U.S.A.: An initial report. *Verhandlungen des Internationalen Verein Limnologie* 19:982–993.

Likens, G. E., and J. W. LaBaugh. 2009. Mirror Lake: Past, present, and future. In T. C. Winter and G. E. Likens, eds., *Mirror Lake: Interactions Among Air, Land, and Water,* pp. 301–328. University of California Press, Berkeley.

Likens, G. E., and K. Fallon Lambert. 1998. The importance of long-term data in addressing regional environmental issues. *Northeastern Naturalist* 5:127–136.

Likens, G. E., F. H. Bormann, N. M. Johnson, D. W. Fisher, and R. S. Pierce. 1970. Effects of forest cutting and herbicide treatment on nutrient budgets in the Hubbard Brook watershed-ecosystem. *Ecological Monograph* 40:23–47.

Likens, G. E., F. H. Bormann, and N. M. Johnson. 1972. Acid rain. *Environment* 14:33–40.

Likens, G. E., F. H. Bormann, R. S. Pierce, J. S. Eaton, and N. M. Johnson. 1977. *Biogeochemistry of a Forested Ecosystem.* Springer-Verlag, New York. 146 pp.

Likens, G. E., F. H. Bormann, R. S. Pierce, and W. A. Reiners. 1978. Recovery of a deforested ecosystem. *Science* 199:492–496.

Likens, G. E., R. F. Wright, J. N. Galloway, and T. J. Butler. 1979. Acid rain. *Scientific American* 241:43–51.

Likens, G. E., E. S. Edgerton, and J. N. Galloway. 1983. The composition and deposition of organic carbon in precipitation. *Tellus* 35B:16–24.

Likens, G. E., W. C. Keene, J. M. Miller, and J. N. Galloway. 1987. Chemistry of precipitation from a remote, terrestrial site in Australia. *Journal of Geophysical Research* 92:13299–13314.

Likens, G. E., C. T. Driscoll, D. C. Buso, T. G. Siccama, C. E. Johnson, D. F. Ryan, G. M. Lovett, T. Fahey, and W. A. Reiners. 1994. The biogeochemistry of potassium at Hubbard Brook. *Biogeochemistry* 25:61–125.

Likens, G. E., C. T. Driscoll, and D. C. Buso. 1996. Long-term effects of acid rain: Response and recovery of a forest ecosystem. *Science* 272:244–246.

Likens, G. E., C. T. Driscoll, D. C. Buso, T. G. Siccama, C. E. Johnson, G. M. Lovett, T. J. Fahey, et al. 1998. The biogeochemistry of calcium at Hubbard Brook. *Biogeochemistry* 41:89–173.

Likens, G. E., C. T. Driscoll, D. C. Buso, M. J. Mitchell, G. M. Lovett, S. W. Bailey, T. G. Siccama, et al. 2002. The biogeochemistry of sulfur at Hubbard Brook. *Biogeochemistry* 60:235–316.

Likens, G. E., D. C. Buso, B. K. Dresser, E. S. Bernhardt, R. O. Hall Jr., K. H. Macneale, and S. W. Bailey. 2004a. Buffering an acidic stream in New Hampshire with a silicate mineral. *Restoration Ecology* 12:419–428.

Likens, G. E., B. K. Dresser, and D. C. Buso. 2004b. Short-term, temperature response in forest floor and soil to ice storm disturbance in a northern hardwood forest. *Northern Journal of Applied Forestry* 21:209–219.

Likens, G. E., D. C. Buso, and T. J. Butler. 2005. Long-term relationships between SO_2 and NO_x emissions and SO_4^{2-} and NO_3^- concentration in bulk deposition at the Hubbard Brook Experimental Forest, New Hampshire. *Journal of Environmental Monitoring* 7:964–968.

Likens, G. E., T. J. Butler, and M. A. Rury. 2009. Acid rain. In H. Anheier and M. Juergensmeyer, eds., *Encyclopedia of Global Studies.* Sage Publications, Los Angeles.

Lindenmayer, D. B., and G. E. Likens. 2009. Adaptive

monitoring: A new paradigm for long-term research and monitoring. *Trends in Ecology and Evolution* 24:482–486.

Lindenmayer, D. B., and G. E. Likens. 2010. *Effective Ecological Monitoring.* CSIRO Publishing and Earthscan, London. 170 pp.

Lindenmayer, D. B., G. E. Likens, and J. F. Franklin. 2010a. Rapid responses to facilitate ecological discoveries from major disturbances. *Frontiers in Ecology and the Environment* 8:527–532.

Lindenmayer, D. B., G. E. Likens, C. J. Krebs, and R. J. Hobbs. 2010b. Improved probability of detection of ecological "surprises." *Proceedings of the National Academy of Sciences, USA* 107:21957–21962.

Long, R. P., S. B. Horsley, R. A. Hallett, and S. W. Bailey. 2009. Sugar maple growth in relation to nutrition and stress in the northeastern United States. *Ecological Applications* 19:1454–1456.

Lovett, G. M., and M. J. Mitchell. 2004. Sugar maple and nitrogen cycling in the forests of eastern North America. *Frontiers in Ecology and the Environment* 2:81–88.

Lovett, G. M., G. E. Likens, and S. S. Nolan. 1992. Dry deposition of sulfur to the Hubbard Brook Experimental Forest: A preliminary comparison of methods. In S. E. Schwartz and W. G. N. Slinn, coordinators, Fifth International Conference on Precipitation Scavenging and Atmosphere-Surface Exchange, pp. 1391–1401. Vol. 3, The Summers Volume: Applications and Appraisals. Hemisphere Publishing.

Lovett, G. M., K. Weathers, and M. Arthur. 2002. Control of nitrogen loss from forested watersheds by soil carbon: Nitrogen ratio and tree species composition. *Ecosystems* 5:712–718.

Lovett, G. M., G. E. Likens, D. C. Buso, C. T. Driscoll, and S. W. Bailey. 2005. The biogeochemistry of chlorine at Hubbard Brook, New Hampshire, USA. *Biogeochemistry* 72:191–232.

Lovett, G. M., C. D. Canham, M. A. Arthur, K. C. Weathers, and R. D. Fitzhugh. 2006. Forest ecosystem responses to exotic pests and pathogens in eastern North America. *BioScience* 56:395–405.

Lovett, G. M., D. A. Burns, C. T. Driscoll, J. C. Jenkins, M. J. Mitchell, L. Rustad, J. B. Shanley, G. E. Likens, and R. Haeuber. 2007. Who needs environmental monitoring? *Frontiers in Ecology and the Environment* 5:253–260.

Lovette, J. I., and R. T. Holmes. 1995. Foraging behavior of American redstarts (*Setophaga ruticilla*) in breeding and winter habitats: Implications for relative food availability. *Condor* 97:782–791.

Lowe, W. H., G. E. Likens, and M. Power. 2006. Linking scales in stream ecology. *BioScience* 56:591–597.

Lowe, W. H., M. A. McPeek, G. E. Likens, and B. J. Cosentino. 2008. Linking movement behaviour to dispersal and divergence in plethodontid salamanders. *Molecular Ecology* 17:4459–4469.

Macneale, K. H., B. L. Peckarsky, and G. E. Likens. 2005. Stable isotopes identify dispersal patterns of stonefly populations living along stream corridors. *Freshwater Biology* 50:1117–1130.

Mann, C. C. 2005. *1491: New Revelations of the Americas Before Columbus.* Alfred A. Knopf, New York.

Marks, P. L. 1974. The role of pin cherry (*Prunus pensylvanica* L.) in the maintenance of stability in northern hardwood ecosystems. *Ecological Monographs* 44:73–88.

Marks, P. L., and F. H. Bormann. 1972. Revegetation following forest cutting: Mechanisms for return to steady-state nutrient cycling. *Science* 176:914–915.

Marra, P. P. 2000. The role of behavioral dominance in structuring habitat occupancy of a migrant bird during the non-breeding season. *Behavioral Ecology* 11:299–308.

Marra, P. P., and R. T. Holmes. 2001. Consequences of dominance-mediated habitat segregation in a migrant passerine bird during the non-breeding season. *Auk* 118:94–106.

Marra, P. P., K. A. Hobson, and R. T. Holmes. 1998. Linking winter and summer events in a migratory bird by using stable-carbon isotopes. *Science* 282:1884–1886.

Marra, P. P., C. E. Studds, S. Wilson, T. S. Sillett, T. W. Sherry, and R. T. Holmes. 2015. Non-breeding season habitat quality mediates the strength of density-dependence for a migratory bird. *Proceedings of the Royal Society Series B* 282: 20150624. http://dx.doi.org/10.1098/rspb.2015.0624.

Martin, C. W., and J. W. Hornbeck. 1989. Revegetation after strip cutting and block clearcutting in northern hardwoods: A 10-year history. USDA Forest Service, Northeastern Forest Experiment Station, Research Paper NE-625. 17 pp.

Martin, C. W., and J. W. Hornbeck. 1990. Regeneration after strip cutting and block clearcutting in northern hardwoods. *Northern Journal of Applied Forestry* 7:65–68.

Martin, C. W., and R. S. Pierce. 1980. Clearcutting patterns affect nitrate and calcium in New Hampshire streams. *Journal of Forestry* 78:268–272.

Martin, C. W., R. S. Pierce, G. E. Likens, and F. H. Bormann. 1986. Clearcutting affects stream chemistry in the White Mountains of New Hampshire. USDA Forest Service Research Paper NE-579. 12 pp.

Matthews, S. N., L. R. Iverson, A. M. Prasad, and M. P. Peters. 2011. Changes in potential habitat of 147 North American breeding bird species in response to redistribution of trees and climate following predicted climate change. *Ecography* 34:1–13.

Mayer, M. S., and G. E. Likens. 1987. The importance of algae in a shaded headwater stream as food for an abundant caddisfly (Trichoptera). *Journal of the North American Benthological Society* 6:262–269.

McCutchan, J. H., Jr., and G. E. Likens. 2002. Relative importance of carbon sources for macroinvertebrates in three streams at the Hubbard Brook Experimental Forest. Abstract for North American Benthological Society. Pittsburgh, Pa., June 2002.

McDowell, W. H. 1982. Mechanisms controlling the organic chemistry of Bear Brook, New Hampshire. Ph.D. dissertation, Cornell University, Ithaca, N.Y. 152 pp.

McDowell, W. H., and G. E. Likens. 1988. Origin, composition, and flux of dissolved organic carbon in the Hubbard Brook valley. *Ecological Monographs* 58:177–195.

McGuire, K. J., and G. E. Likens. 2011. Historical roots of forest hydrology and biogeochemistry. Chapter 1 in D. F. Levia, D. Carlyle-Moses, and T. Tanaka, eds., *Forest Hydrology and Biogeochemistry: Synthesis of Past Research and Future Directions.* Ecological Studies Series, No. 216, pp. 3–26. Springer-Verlag, Heidelberg, Germany.

McGuire, K. J., C. E. Torgersen, G. E. Likens, D. C. Buso, W. H. Lowe, and S. W. Bailey. 2014. Network analysis reveals multiscale controls on streamwater chemistry. *Proceedings of the National Academy of Sciences* 111:7030–7035.

McNeil, B. E., and K. L. Culcasi. 2014. Maps on acid: Cartographically constructing the acid rain environmental issue, 1972–1980. *The Professional Geographer.* doi:10.1080/00330124.2014.922016.

McPeek, M. A., N. L. Rodenhouse, T. W. Sherry, and R. T. Holmes. 2001. Site dependent population regulation: Population-level regulation without individual-level interactions. *Oikos* 94:417–424.

Merrens, E. J., and D. R. Peart. 1992. Effects of hurricane damage on individual growth and stand structure in a hardwood forest in New Hampshire, USA. *Journal of Ecology* 80:787–795.

Meyer, J. L. 1979. The role of sediments and bryophytes in phosphorus dynamics in a headwater stream. *Limnology and Oceanography* 24:365–375.

Meyer, J. L., and G. E. Likens. 1979. Transport and transformation of phosphorus in a forest stream ecosystem. *Ecology* 60:1255–1269.

Meyer, J. L., G. E. Likens, and J. Sloane. 1981. Phosphorus, nitrogen, and organic carbon flux in a headwater stream. *Archiv für Hydrobiologie* 91:28–44.

Mitchell, M. J., and G. E. Likens. 2011. Watershed sulfur biogeochemistry: Shift from atmospheric deposition dominance to climatic regulation. *Environmental Science and Technology* 45:5267–5271.

Mitchell, M. J., G. M. Lovett, S. W. Bailey, F. Beall, D. Burns, D. C. Buso, T. Clair, et al. 2011. Comparisons of watershed sulfur budgets in southeast Canada and northeast U.S.: New approaches and implications. *Biogeochemistry* 103:181–207.

Molina, M., J. McCarthy, D. Wall, R. Alley, K. Cobb, J. Cole, et al. 2014. What we know: The reality, risks, and response to climate change. American Association for the Advancement of Science Climate Science Panel (http://whatweknow.aaas.org/wp-content/uploads/2014/07/whatweknow).

Moore, J.-D., R. Ouimet, and L. Duchesne. 2012. Soil and sugar maple response 15 years after dolomitic lime application. *Forest Ecology and Management* 281:130–139.

Moreira, X., L. Abdala-Roberts, Y. B. Linhart, and K. A. Mooney. 2014. Masting promotes individual- and population-level reproduction by increasing pollination efficiency. *Ecology* 95:801–807.

Morse, S. C. 2015. Declining moose populations: What does the future hold? *Northern Woodlands* 22:34–41.

Mou, P. U., T. J. Fahey, and J. W. Hughes. 1993. Effects of soil disturbance on vegetation recovery and nutrient accumulation following whole-tree harvest of a northern hardwoods ecosystem. *Journal of Applied Ecology* 30:661–675.

Muller, R. N. 1978. The phenology, growth, and ecosystem dynamics of *Erythronium americanum* in the northern hardwood forest. *Ecological Monographs* 48:1–20.

Muller, R. N., and F. H. Bormann 1976. Role of *Erythronium americanum* Ker. in energy flow and nutrient dynamics of a northern hardwood forest ecosystem. *Science* 193:1126–1128.

Nagy, L. R., and R. T. Holmes. 2005. Food limits annual fecundity of a migratory songbird: An experimental study. *Ecology* 86:675–681.

Norris, D. R., P. P. Marra, G. J. Bowen, L. M. Ratcliffe, J. A. Royle, and T. K. Kyser. 2006. Migratory connectivity of a widely distributed songbird, the American redstart, *Setophaga ruticilla. Ornithological Monographs* 61:14–28.

North American Bird Conservation Initiative, U.S. Committee. 2014. The state of the birds report, 2014. Available at www.stateofthebirds.org.

Odum, E. P. 1959. *Fundamentals of Ecology,* 2nd ed. W. B. Saunders, Philadelphia. 546 pp.

Ollinger, S. V., J. D. Aber, P. B. Reich, and R. J. Freuder. 2002a. Interactive effects of nitrogen deposition, tropospheric ozone, elevated CO_2, and land use history on the carbon dynamics of northern hardwood forests. *Global Change Biology* 8:545-562.

Ollinger, S. V., M. L. Smith, M. E. Martin, R. A. Hallett, C. L. Goodale, and J. D. Aber. 2002b. Regional variation in foliar chemistry and soil nitrogen status among forests of diverse history and composition. *Ecology* 83:339-355.

Ollinger, S. V., A. D. Richardson, M. E. Martin, D. Y. Hollinger, S. E. Frolking, P. B. Reich, L. C. Plourde, et al. 2008. Canopy nitrogen, carbon assimilation, and albedo in temperate and boreal forests: Functional relations and potential climate feedbacks. *Proceedings of the National Academy of Sciences, USA* 105:19335-19340.

Pabian, S. E., and M. C. Brittingham. 2007. Terrestrial liming benefits birds in an acidified forest in the Northeast. *Ecological Applications* 17:2184-2194.

Pabian, S. E., and M. C. Brittingham. 2011. Soil calcium availability limits forest songbird productivity and density. *Auk* 128:441-447.

Palmer, S. M., B. I. Wellington, C. E. Johnson, and C. T. Driscoll. 2005. Landscape influences on aluminum and dissolved organic carbon in streams draining the Hubbard Brook valley, New Hampshire, USA. *Hydrological Processes* 19:1751-1769.

Pardo, L. H., H. F. Hemond, J. P. Montoya, T. J. Fahey, and T. G. Siccama. 2002. Response of the natural abundance of ^{15}N in forest soils and foliage to high nitrate loss following clear-cutting. *Canadian Journal of Forest Research* 32:1126-1136.

Peart, D. R., C. V. Cogbill, and P. A. Palmiotto. 1992. Effects of logging history and hurricane damage on canopy structure in a northern hardwood forest. *Bulletin of the Torrey Botanical Club* 119:29-38.

Phillips, R. P., and T. J. Fahey. 2005. Patterns of rhizosphere carbon flux in sugar maple (*Acer saccharum*) and yellow birch (*Betula alleghaniensis*) saplings. *Global Change Biology* 11:983-995.

Pierce, R. S., C. W. Martin, C. C. Reeves, G. E. Likens, and F. H. Bormann. 1972. Nutrient loss from clearcuttings in New Hampshire. In *Symposium on Watersheds in Transition*, pp. 285-295. Fort Collins, Colorado.

Pletscher, D. H., F. H. Bormann, and R. S. Miller. 1989. Importance of deer compared to other vertebrates in nutrient cycling and energy flow in a northern hardwood ecosystem. *American Midland Naturalist* 121:302-311.

Potter, G. L. 1974. Population dynamics, energy flow, and nutrient cycling of mice in the Hubbard Brook Experimental Forest, New Hampshire. Ph.D. dissertation, Dartmouth College, Hanover, N.H.

Rastetter, E. B., R. D. Yanai, R. Q. Thomas, M. A. Vadeboncoeur, T. J. Fahey, M. C. Fisk, B. L. Kwiatkowski,

and S. P. Hamburg. 2013. Recovery from disturbance requires resynchronization of ecosystem nutrient cycles. *Ecological Applications* 23:621-642. doi:10.1890/12-0751.1.

Raven, P. H., G. B. Johnson, K. A. Mason, J. B. Losos, and S. S. Singer. 2014. *Biology*, 10th edition. McGraw-Hill, New York.

Reece, J. B., L. A. Urry, M. L. Cain, S. A. Wasserman, and P. V. Minorsky. 2013. Campbell Biology, 10th edition. Benjamin Cummings, San Francisco.

Reiners, W. A. 1992. Twenty years of ecosystem reorganization following experimental deforestation and regrowth suppression. *Ecological Monographs* 62:503-523.

Reiners, W. A., K. L. Driese, T. J. Fahey, and K. G. Gerow. 2012. Effects of three years of regrowth inhabitation on the resilience of a clear-cut northern hardwood forest. *Ecosystems* 15:1351-1362.

Reitsma, L. R., R. T. Holmes, and T. W. Sherry. 1990. Effects of removal of red squirrels (*Tamiasciurus hudsonicus*) and eastern chipmunks (*Tamias striatus*) on nest predation in a northern hardwood forest: An experiment with artificial nests. *Oikos* 57:375-380.

Reudink, M. W., P. P. Marra, T. K. Kyser, P. T. Boag, K. M. Langin, and L. M. Radcliffe. 2009. Non-breeding season events influence sexual selection in a long-distance migratory bird. *Proceedings of the Royal Society of London, Series B* 276:1619-1626.

Reynolds, L. V., M. P. Ayres, T. G. Siccama, and R. T. Holmes. 2007. Climatic effects on caterpillar fluctuations in northern hardwood forests. *Canadian Journal of Zoology* 37:481-491.

Rhoads, A. G., S. P. Hamburg, T. J. Fahey, T. G. Siccama, E. N. Hane, J. J. Battles, C. Cogbill, J. Randall, and G. Wilson. 2002. Effects of an intense ice storm on the structure of a northern hardwood forest. *Canadian Journal of Forest Research* 32:1763-1775.

Richardson, A. D., A. S. Bailey, E. G. Denny, C. W. Martin, and J. O'Keefe. 2006. Phenology of a northern hardwood forest canopy. *Global Change Biology* 12:1174-1188.

Richey, J. S., W. H. McDowell, and G. E. Likens. 1985. Nitrogen transformation in a small mountain stream. *Hydrobiologia* 124:129-139.

Robbins, C. S., J. R. Sauer, R. S. Greenberg, and S. Droege. 1989. Population declines in North American birds that migrate to the tropics. *Proceedings of the National Academy of Sciences, USA* 86:7658-7662.

Robinson, S. K., and R. T. Holmes. 1982. Foraging behavior of forest birds: The relationships among search tactics, diet, and habitat structure. *Ecology* 63:1918-1931.

Robinson, S. K., and R. T. Holmes. 1984. Effects of plant

species and foliage structure on the foraging behavior of forest birds. *Auk* 101:672–684.

Rodenhouse, N. L., and R. T. Holmes. 1992. Effects of experimental and natural food reductions for breeding black-throated blue warblers. *Ecology* 73:357–372.

Rodenhouse, N. L., T. S. Sillett, P. J. Doran, and R. T. Holmes. 2003. Multiple density-dependent mechanisms regulate a migratory bird population during the breeding season. *Proceedings of the Royal Society of London, Series B* 270:2105–2110.

Rodenhouse, N. L., S. N. Matthews, K. P. McFarland, J. D. Lambert, L. R. Iverson, A. Prasad, T. S. Sillett, and R. T. Holmes. 2008. Potential effects of climate change on birds of the Northeast. *Mitigation and Adaptation Strategies for Global Change* 13:517–540.

Rodenhouse, N. L., L. M. Christenson, D. Parry, and L. E. Green. 2009. Climate change effects on native fauna of northeastern forests. *Canadian Journal of Forest Research* 39:249–263.

Rubenstein, D. R., C. P. Chamberlain R. T. Holmes, M. P. Ayres, J. R. Waldbauer, G. R. Graves, and N. C. Tuross. 2002. Linking breeding and wintering ranges of a Neotropical songbird using stable isotopes. *Science* 295:591–593.

Rustad, L., J. Campbell, J. S. Dukes, T. Huntington, K. F. Lambert, J. Mohan, and N. Rodenhouse. 2012. Changing climate, changing forests: The impacts of climate change on forests of the northeastern United States and Canada. U.S. Forest Service, General Technical Report NRS-99. pp. 1–48.

Samuel, W. M. 2007. Factors affecting epizootics of winter ticks and mortality of moose. *Alces* 43:39–48.

Schwarz, P. A., T. J. Fahey, and C. E. McCulloch. 2003. Factors controlling spatial variation of tree species abundance in a forested landscape. *Ecology* 84:1862–1878.

Seuss, Dr. 1971. *The Lorax.* Random House, New York.

Sherry, T. W., and R. T. Holmes. 1992. Population fluctuations in a long-distance neotropical migrant: Demographic evidence for the importance of breeding season events in the American redstart. In J. M. Hagan and D. W. Johnston, eds., *Ecology and Conservation of Neotropical Migrant Landbirds,* pp. 431–442. Smithsonian Press, Washington, D.C.

Sherry, T. W., and R. T. Holmes. 1995. Summer versus winter limitation of Neotropical migrant land bird populations: Conceptual issues and evidence? In T. E. Martin and D. Finch, eds., *Ecology and Management of Neotropical Migratory Birds: A Synthesis and Review of the Critical Issues,* pp. 85–120. Oxford University Press, New York.

Sherry, T. W., and R. T. Holmes. 1996. Winter habitat quality, population limitation, and conservation of Neotropical-Nearctic migrant birds. *Ecology* 77:36–48.

Sherry, T. W., and R. T. Holmes. 1997. American redstart (*Setophaga ruticilla*). In A. Poole and F. Gill, eds., *The Birds of North America,* no. 277, pp. 1–32. The Academy of Natural Sciences, Philadelphia, and The American Ornithologists' Union, Washington D.C.

Sherry, T. W., S. Wilson, S. Hunter, and R. T. Holmes. 2015. Multiple direct and indirect factors impact reproductive success and population limitation in a long-distance migratory songbird. *Journal of Avian Biology* 46:1–11.

Siccama, T. G., and W. H. Smith. 1978. Lead accumulation in a northern hardwood forest. *Environmental Science and Technology* 12:593–594.

Siccama, T. G., F. H. Bormann, and G. E. Likens. 1970. The Hubbard Brook Ecosystem Study: Productivity, nutrients, and phytosociology of the herbaceous layer. *Ecological Monographs* 40:389–402.

Siccama, T. G., W. H. Smith, and D. L. Mader. 1980. Changes in lead, zinc, copper, dry weight, and organic matter content of the forest floor of white pine stands in central Massachusetts over 16 years. *Environmental Science and Technology* 14:54–56.

Siccama, T. G., T. J. Fahey, C. E. Johnson, T. Sherry, E. G. Denny, E. B. Girdler, G. E. Likens, and P. Schwarz. 2007. Population and biomass dynamics of trees in a northern hardwood forest at Hubbard Brook. *Canadian Journal of Forest Research* 37:737–749.

Sillett, T. S., and R. T. Holmes. 2002. Variation in survivorship of a migratory songbird throughout its annual cycle. *Journal of Animal Ecology* 71:296–308.

Sillett, T. S., and R. T. Holmes. 2005. Long-term demographic trends, limiting factors, and the strength of density dependence in a breeding population of a migratory songbird. In R. Greenberg and P. P. Marra, eds., *Birds of Two Worlds: Advances in the Ecology and Evolution of Temperate-Tropical Migration Systems,* pp. 426–436. Johns Hopkins University Press, Baltimore.

Sillett, T. S., R. T. Holmes, and T. W. Sherry. 2000. Impacts of a global climate cycle on the population dynamics of a migratory songbird. *Science* 288:2040–2042.

Sillett, T. S., N. L. Rodenhouse, and R. T. Holmes. 2004. Experimentally reducing neighbor density affects reproduction and behavior of a migratory songbird. *Ecology* 85:2467–2477.

Skeldon, M. A., M. A. Vadeboncoeur, S. P. Hamburg, and J. D. Blum. 2007. Terrestrial gastropod responses to an ecosystem-level calcium manipulation in a northern hardwood forest. *Canadian Journal of Zoology* 85:994–1007.

Sloan, S. S., R. T. Holmes, and T. W. Sherry. 1998. Depredation rates and predators at artificial bird nests in an unfragmented northern hardwood forest. *Journal of Wildlife Management* 62:529–539.

Sloane, J. 1979. Nitrogen flux in small mountain streams in New Hampshire. Master of Science thesis, Cornell University, Ithaca, N.Y. 131 pp.

Smith, W. H., and T. G. Siccama. 1981. The Hubbard Brook Ecosystem Study: Biogeochemistry of lead in the northern hardwood forest. *Journal of Environmental Quality* 10:323–333.

Sopper, W. E., and H. W. Lull. 1965. The representativeness of small forested experimental watersheds in northeastern United States. *Bulletin of the International Society of Hydrology* 66:441–446.

Sopper, W. E., and H. W. Lull. 1970. *Streamflow Characteristics of the Northeastern United States.* Bulletin 766. The Pennsylvania State University, University Park. 129 pp.

Sridevi, G., R. Minocha, S. A. Turlapati, K. C. Goldfarb, E. L. Brodie, L. S. Tisa, and S. C. Minocha. 2012. Soil bacterial communities of a calcium-supplemented and a reference watershed at the Hubbard Brook Experimental Forest (HBEF), New Hampshire, USA. *FEMS Microbial Ecology* 79:728–740.

Stange, E. E., M. P. Ayres, and J. E. Bess. 2011. Concordant population dynamics of Lepidoptera herbivores in a forest ecosystem. *Ecography* 34:772–779.

Steinhart, G. S., G. E. Likens, and P. M. Groffman. 2000. Denitrification in stream sediments in five northeastern (USA) streams. *Verhandlungen des Internationalen Verein Limnologie* 27:1331–1336.

Stelzer, R. S., J. Heffernan, and G. E. Likens. 2003. The influence of dissolved nutrients and particulate organic matter quality on microbial respiration and biomass in a forest stream. *Freshwater Biology* 48:1925–1937.

Stunder, B. J. B. 1996. An assessment of the quality of forecast trajectories. *Journal of Applied Meteorology* 35:1319–1331.

Sturges, F. W., R. T. Holmes, and G. E. Likens. 1974. The role of birds in nutrient cycling in a northern hardwoods ecosystem. *Ecology* 55:149–155.

Svensson, T., G. M. Lovett, and G. E. Likens. 2012. Is chloride a conservative ion in forest ecosystems? *Biogeochemistry* 107:125–134.

Swatantran, A., R. Dubayah, S. Goetz, M. Hofton, M. G. Betts, M. Sun, M. Simard, and R. T. Holmes. 2012. Mapping migratory bird prevalence using remote sensing data fusion. *PLoS One* 7(1):e28922. doi:10.1371/journal.pone .0028922.

Taliaferro, E. H., R. T. Holmes, and J. D. Blum. 2001. Eggshell characteristics and calcium demands of a migratory bird breeding in two New England forests. *Wilson Bulletin* 113:94–100.

Tank, J. L., P. Mulholland, J. L. Meyer, W. B. Bowden, J. R. Webster, B. J. Peterson, and D. Sanzone. 1998. Nitrogen cycling in grazing vs. detrital pathway in three temperate, forested streams. Abstract for 46th Annual Meeting, Prince Edward Island, Canada. *Bulletin of the North American Benthological Society* 15:128.

Taylor, D. L., T. N. Hollingsworth, J. W. McFarland, N. J. Lennon, C. Nusbaum, and R. W. Ruess. 2014. A first comprehensive census of fungi in soils reveals both hyperdiversity and fine-scale niche partitioning. *Ecological Monographs* 84:3–20.

Templer, P. H., A. F. Schiller, N. W. Fuller, A. M. Socci, J. L. Campbell, J. E. Drake, and T. H. Kunz. 2012. Impact of a reduced winter snowpack on litter arthropod abundance and diversity in a northern hardwood forest. *Biology and Fertility of Soils* 48:413–424.

Terborgh, J. W. 1989. *Where Have All the Birds Gone? Essays on the Biology and Conservation of Birds That Migrate to the American Tropics.* Princeton University Press, Princeton.

Tewksbury, J. J., J. G. T. Anderson, J. D. Bakker, et al. 2014. Natural history's place in science and society. *BioScience* 64:300–310.

Thornton, G. A. 1974. Some effects of deforestation on stream macroinvertebrates. Master of Science thesis, Cornell University, Ithaca, N.Y. 84 pp.

Tierney, G. L., and T. J. Fahey. 1998. Soil seed bank dynamics of pin cherry in a northern hardwood forest, New Hampshire. *Canadian Journal of Forest Research* 28:1471–1480.

Tierney, G. L., T. J. Fahey, P. M. Groffman, J. P. Hardy, R. D. Fitzhugh, and C. T. Driscoll. 2001. Soil freezing alters fine root dynamics in a northern hardwood forest. *Biogeochemistry* 56:175–190.

Tingley, M. W., D. A. Orwig, and R. Field. 2002. Avian response to removal of a forest dominant: Consequences of hemlock woolly adelgid infestations. *Journal of Biogeography* 29:1505–1516.

Townsend, A. K., T. S. Sillett, N. K. Lany, S. A. Kaiser, N. L. Rodenhouse, M. S. Webster, and R. T. Holmes. 2013. Warm springs, early lay dates, and double-brooding in a North American migratory songbird, the black-throated blue warbler. *PLoS ONE* 8(4): e59467. doi:10.1371/journal.pone .0059467.

Townsend, A. K., E. G. Cooch, T. S. Sillett, N. L. Rodenhouse,

R. T. Holmes, and M. S. Webster. 2015. The interacting effects of food, spring temperature, and global climate cycles on population dynamics of a migratory songbird. *Global Climate Change.* doi:10.1111/gcb.13053.

Union of Concerned Scientists. 2006. *The Changing Northeast Climate: Our Choices, Our Legacy.* Summary Report. Two Battle Square, Cambridge, Mass.

Vadeboncoeur, M. A., S. P. Hamburg, C. V. Cogbill, and W. Y. Sugimura. 2012. A comparison of presettlement and modern forest composition along an elevation gradient in central New Hampshire. *Canadian Journal of Forest Research* 42:190–202.

Vadeboncoeur, M. A., S. P. Hamburg, R. D. Yanai, and J. D. Blum. 2014. Rates of sustainable forest harvest depend on rotation length and weathering of soil minerals. *Forest Ecology and Management* 318:194–205.

Van Doorn, N. S. 2014. Patterns and process of forest growth: The role of neighborhood dynamics and tree demography in a northern hardwood forest. Ph.D. dissertation, University of California, Berkeley.

Van Doorn, N. S., J. J. Battles, T. J. Fahey, T. G. Siccama, and P. A. Schwarz. 2011. Links between biomass and tree demography in a northern hardwood forest: A decade of stability and change in Hubbard Brook valley, New Hampshire. *Canadian Journal of Forest Research* 41:1369–1379.

Visser, M. E., L. J. M. Holleman, and P. Gienapp. 2006. Shifts in caterpillar biomass phenology due to climate change and its impact on the breeding biology of an insectivorous bird. *Oecologia* 147:164–172.

Warren, D. R., E. S. Bernhardt, R. O. Hall Jr., and G. E. Likens. 2007. Forest age, wood and nutrient dynamics in headwater streams of the Hubbard Brook Experimental Forest, N.H. *Earth Surface Processes and Landforms* 32:1154–1163.

Warren, D. R., G. E. Likens, D. C. Buso, and C. E. Kraft. 2008. Status and distribution of fish in an acid-impacted watershed in the northeastern United States. *Northeastern Naturalist* 15:375–390.

Warren, D. R., C. E. Kraft, W. S. Keeton, J. S. Nunery, and G. E. Likens. 2009. Dynamics of wood recruitment in streams of the northeastern U.S. *Forest Ecology and Management* 258:804–813.

Warren, D. R., K. E. Judd, D. L. Bade, G. E. Likens, and C. E. Kraft. 2013. Effects of wood removal on stream habitat and nitrate uptake in two northeastern U.S. headwater streams. *Hydrobiologia* 717:119–131.

Weathers, K. C., G. E. Likens, F. H. Bormann, J. S. Eaton, W. B. Bowden, J. Andersen, D. A. Cass, J. N. Galloway, W. C. Keene, K. D. Kimball, et al. 1986. A regional acidic cloud/fog water event in the eastern United States. *Nature* 319:657–659.

Weathers, K. C., G. E. Likens, F. H. Bormann, S. H. Bicknell, B. T. Bormann, B. C. Daube Jr., J. S. Eaton, J. N. Galloway, W. C. Keene, K. D. Kimball, et al. 1988a. Cloudwater chemistry from ten sites in North America. *Environmental Science and Technology* 22:1018–1026.

Weathers, K. C., G. E. Likens, F. H. Bormann, J. S. Eaton, K. D. Kimball, J. N. Galloway, T. G. Siccama, and D. Smiley. 1988b. Chemical concentrations in cloud water from four sites in the eastern United States. In M. H. Unsworth and D. Fowler, eds., *Acid Deposition at High Elevation Sites,* pp. 345–357. Kluwer Academic Publishers.

Weathers, K. C., G. E. Likens, and T. J. Butler. 2006. Acid rain. In W. N. Rom, ed., *Environmental and Occupational Medicine,* 4th ed., pp. 1507–1520. Lippincott, Williams & Wilson, Philadelphia.

Webster, M. S., H. C. Chuang-Dobbs, and R. T. Holmes. 2001. Microsatellite identification of extra-pair sires in a socially monogamous warbler. *Behavioral Ecology* 12:439–446.

Weeks, B. C., S. P. Hamburg, and M. A. Vadeboncoeur. 2009. Ice storm effects on the canopy structure of a northern hardwood forest after 8 years. *Canadian Journal of Forest Research* 39:1475–1483.

Wexler, S. K., C. L. Goodale, K. J. McGuire, S. W. Bailey, and P. M. Groffman. 2014. Isotopic signals of summer denitrification in a northern hardwood forested catchment. *Proceedings of the National Academy of Sciences, USA* 111:16413–16418.

White, E. H. 1974. Whole-tree harvest depletes soil nutrients. *Canadian Journal of Forest Research* 4:530–535.

Whitehurst, A. S., A. Swatantran, J. B. Blair, M. A. Hofton, and R. Dubayah. 2013. Characterization of canopy layering in forested ecosystems using full waveform Lidar. *Remote Sensing* 5:2014–2036.

Whitlaw, H. A., and M. W. Lankster. 1994. The co-occurrence of moose, white-tailed deer, and *Paralaphostrongylus tenuis* in Ontario. *Canadian Journal of Zoology* 72:819–825.

Whittaker, R. H., F. H. Bormann, G. E. Likens, and T. G. Siccama. 1974. The Hubbard Brook Ecosystem Study: Forest biomass and production. *Ecological Monographs* 44:233–254.

Wilson, S., S. L. LaDeau, A. P. Tøttrup, and P. P. Marra. 2011. Range-wide effects of breeding and non-breeding season climate on the abundance of a Neotropical migrant songbird. *Ecology* 92:1789–1798.

Winter, T. C., and G. E. Likens, eds. 2009. *Mirror Lake: Interactions Among Air, Land, and Water.* University of California Press, Berkeley. 361 pp.

Yanai, R. D., N. Tokuchi, J. L. Campbell, M. B. Green, E. Matsuzaki, S. N. Laseter, C. L. Brown, et al. 2014. Sources of uncertainty in estimating stream solute export from headwater catchments at three sites. *Hydrological Processes.* doi:10.1002/hyp.10265.

Zak, D. R., P. M. Groffman, K. S. Pregitzer, S. Christensen, and J. M. Tiedje. 1990. The vernal dam: Plant microbe competition for nitrogen in northern hardwood forest. *Ecology* 71:651–656.

INDEX

Page numbers in italic type refer to illustrations

abiotic factors, 13, *14*

abscission cells, 2

Acari, 59

acid-neutralizing capacity (ANC), *121*

acid rain, viii, 15, 16, 40, 62, 65–66, 77, 95, 103, 114–123, 157, 199, 218, 226–227, 228; aquatic ecosystems and, 136–137; calcium depleted by, 81, 105, 112–113, 121, 122, 129, 130, 131, 135, 136; consequences of, 127–137; forest affected by, 129–133; policy response to, 215–216; soil development affected by, 128–129; soil nutrients depleted by, 132–135; testing for, xii, 22, 116

acid salts, 104

Advancing Informal STEM Learning (AISL), 221

air pollution, 6, 10, 11, 21, 113, 185, 215; effects of, 96, 127; outlook for, 226; ozone, 128, 129; regulation of, 95, 121, 127, 216, 218; sources of, 31, 116–117, 118

alder, 35, *36*, 55, 59

algae, 30, 49, 62, 77, 86, 89, 155, *157*, 166

alkalinization, 157, 158, 159–160

aluminum, 81; in precipitation, 98, 127, 227; in soil, 28, 122, 127, 131, 132, 142; in streams, 99, 113, 132, 137, 142, *144*, 159, 198

American beech, 3, 48, 57, 132, 149, *150*, 228; beech bark disease and, 1, 45, *46*, 53; caterpillars and, 63; dams formed by, 163; ice storm's damage to, 41; masting by, 68, 194; prevalence of, 33, 36, 37, 46, 49, 50, 187; reproduction of, 52–53

American marten, 72, 178

American redstart, *168*, 170–175, 178, 179–182, 183, *195*

American toad, 64, *65*

ammonium, 111, 117, *144*, 145

amphibians, 4, 41, 50, 62, 64–66, 76, 136, 229

anions, 97, 98, 108, 113, *121*, 230

anthocyanin, 2

ants, 87

aphids, 77

Arachnids, 59

Araneae, 64

Arctiidae, 59

arthropods, 59, 67

asexual reproduction, 52–53

ash, 35, 50, 154, 228, 233; white, 46, 48, 50

aspen, 35, 154; big-tooth, 50; quaking, 50, 149

asters, 50, 57, 148

automobiles, 98, 115

autotrophs, 76, 156

balsam fir, 72, 149, 190; climate change and, 206, 207, 228, 229; logging of, 38, 39; prevalence of, 33, 35, 37, 46, 49, 187; snowfall and, 4–5

barred owl, 5, 68, 76, 236

bats, 68, 70, 77, 234

bears, 45, 108; black, 3–4, 5, *71*, 72, 77, 78–79

beaver ponds, 25, 29, 55, 64, 66

bedrock, 1, 17, *18*, 21, 28, 29, 33, 92, 105, 198

beech, 1, 3, 35, 45–46, 48, 173, 233. *See also* American beech

beech bark disease, 1, 41, 43, 45, 48, 53, 129, 173, 228

bees, 50, 77

beetles, *45*, 50, 59, 62, 64, 76, 77, 87

Bicknell's thrush, 229

big-tooth aspen, 50

biodiversity, 13, *14*, 15, 190

biogeochemistry, 12, 95–113, 148

biomass, 57; of birds, 67; calculating, *62*, 63, 65, 75, 76; dead vs. living, 89; defined, 76; after disturbances, 148, 150; of plants, 75, 79–83; productivity of, 76, 79; of salamanders, 82–83; spatial patterns of, 188–190; storage in, 106–112; of trees, 129, 130, 133

biotic factors, 13, *14*

birch, 35, *36*, 50; paper, 33, 37, 41, 46, 48, 49, 149, 187. *See also* yellow birch

birds, xi, 12, 50, 57, 77, 83, 200; aquatic invertebrates consumed by, 62; calcium in, 107; caterpillars and, 86; climate change and, 209–212, 229; counting, 169, 187, 194; energy consumption by, 87; foraging by, 45, 69, 76; migratory, 4, 6, 66–67, 167–185, 211, 218; population change among, 15, 16, 167–185; population trends, 172; as predators, 89; predators on, 178; prevalence of, 192, 194; as prey, 68, 76, 194; reproduction of, 136, 218; sulfur in, 110;

tree preferences for foraging, 69. *See also* songbirds

black bear, 3–4, 5, *71*, 72, 77, 78–79

blackberry, 152

blackburnian warbler, *67*, 172, 229

black-capped chickadee, 179, 229

black flies, 2, 6, 50, 61, 62, 64, 77, 159, 234

blacknose dace, 64, 137, 197

blackpoll warbler, *195*, 229

black-throated blue warbler, 171–174, *176*, 178, 194, *195*; climate change and, 208–212, 229; mating behavior and paternity, 177; migration of, 179–181, 183–184; nesting by, 55; predators on, 178; prevalence of, 170, 209; reproduction of, 136, 175, 210; winter and summer distribution, 180

black-throated green warbler, 170, *171*, 172, 173, 229

blue-bead lily, 1, *58*

blueberry, 55

blue-headed vireo, 172, 229

blue jays, 4, 6, 66, 178

bobbins, 38

bobcat, 68, 72

Boletus, *78*

boreal forest, 211

Bormann, F. Herbert, 13, 17, *217*, 231

broad-winged hawk, 68

bromacil, 141, 151

bromeliads, 49

brook trout, 64, 137, 197

brown creeper, 68

brown-headed cowbird, 175

butterflies, 50, 59–60

caddisflies, 50, 61, 62, *63*, 77, 233

cadmium, 97

calcium, 95, 98, 128, 152; acid rain and, 81, 105, 112–113, 121, 122, 129, 130, 131, 135, 136; in animal shells, 62, 136; climate change and, 204–205; declining levels of, 226; flux and cycling of, 12, 18, 21, 105–108; in foliage, 205; leaching of, 112, 131, 228; mass balance, 109; in precipitation, 97, 102–103, 135; in rocks, 35; scarcity of, 153; in streams, 29, 97, 99, 102, *103*, 104, 142, *144*, 145, 159–160, 198 calcium bicarbonate, 226

calcium carbonate, 131, 132

calcium chloride, 160

calcium silicate, xii, 133, 134, 160-161

Canada warbler, *67*

Canadian tiger swallowtail, 60

canids, 72

carbohydrates, 57

carbon, 10, 18, 76, 95, 103, 106, 179-180; dissolved organic, 101, 105, 158, 162, 198, 208

carbon capture, 13

carbon dioxide, 89, 114, 218, 228, 229

carbonic acid, 114

carbon sequestration, 100, 226

cardinals, 229, 234

carnivores, 72, 75

Carolina wren, 229, 234

carotenoid, 2

caterpillars, 1, 53, 57, 76, 77, 83, 86, 87, 89, 175-177, 178, 234; feeding habits of, 60, 78; outbreak of (1969-1971), xi, 41, 42, 43, *44*; prevalence of, 61, 63, 192; as prey, 209

cations, 97, 98, 104, 113, 129, 131; base, 121, 122, 127, 135, 142, 204-205, 230

cedar, 35

centipedes, 49, 59, 64, 65, 76, 77

charcoal, 36, 38

chemical weathering, 104-105, 108, 110, 227

cherries, 50. *See also* pin cherry

chickadees, 4, 5, 169, 179, 229, 234, 236

Chilopoda, 64

chipmunk, eastern, 4, 5, 52, 68, *71*, 77, 87, *178*; food sources, 70; as nest predator, 70, 178, 194; population density, 52, 194-195, 196

Chironomidae, 62

chloride, 95, 104, 110, 112, 113, *129*, 160, 198, 226

chlorine, 18, 103, 106, 110

chlorophyll, 2, 57

Cladocera, 62

clay, 145

Clean Air Act (1970), 103, 118, *125*; amendments to (1990), 98, 118, 121, 122, 128, 216

Clean Air Status and Trends Network (CASTNet), xii, 20

clear-cutting, xi, 16, 22, 27, 151, *153*, 163-164; decline in, 147, 154; experimental, 139-140, 143-148, 190; limiting impacts of, 154, 217. *See also* deforestation; logging

Clemensia albata (lichen moth), 59, 60

climate change, viii, 6, 11, 13, *15*, 31, 36, 48, 76, 182, 188, 233; balsam fir affected by, 206,

207, 228, 229; birds affected by, 208-212, 229; calcium levels affected by, 204-205; deciduous trees affected by, 208; flux and cycling affected by, 201, 203, 206; magnesium levels affected by, 205; migration affected by, 203, 208; nitrate levels affected by, 204, 207, 230; nitrogen levels affected by, 205, 207, 208; precipitation affected by, 202, 228; rain affected by, 204; red spruce affected by, 228, 229; soil affected by, 205-206; streams affected by, 208; sugar maple affected by, 207, 208, *210*, 228-229

clouds, 31, 115; water chemistry in, 117

clubmosses, 50, 57, *58*

coal, 98, 110, 114, 115, 116, 226

Coleoptera, 62, 64

Collembola, 59, 62

colluvial action, 18

combustion engines, 98, 115

commensalism, sapsucker and hummingbird, 70

conifers, *34, 35,* 49, 50, 66, 70, 194, 208; acid rain's effect on, 131; caterpillars and, 60; foliar nitrogen in, 190; ice storm and, 41; snowfall and, 4

Copepoda, 62

copper, 97, 159

cordwood, 38, 39

corrosion, 127

Coweeta Hydrologic Laboratory, 17

coyotes, 68, *71,* 72, 76, 77, 108

crane flies, 61-62, 64, 76

Craterellus, 78

crossbills, 68

crows, 234, 236

Cryptococcus fagisuga (woolly beech scale), 43, 45

cyanobacteria, 60

cycling. *See* nutrient cycling and flux

damselflies, 64

dark-eyed juncos, 169

daylight, 3

debris dams, 30, 86, 101, 137, 157, 158, 161-164

deciduous trees, 2, 5, 6, *34, 51,* 170, 190, 194; acid rain's effect on, 131; climate change and, 208; ice storm's damage to, 41; moth larvae and, 60; prevalence of, 1, 49; transpiration and, 29, 91

decomposition, 11, 57, 59, 148, 152, 206; of nitrogen, 144-145; of plants, 10, 35, 41, 144, 203, 208; of rocks, 9

deer, 57, 68, 72, 76, 77, 79, 87, 229, 233

deer mouse, 70-71

defoliation, xi, 15, 41-43, *44*

deforestation, xi, *26*, 138, 145; of bird habitats, 179, 182; experimental, 139-144, 148, 150, 151, 153, 190. *See also* clear-cutting; logging

denitrification, 30, 87, 111

detritivores, 59, 60; aquatic invertebrate, 62, *63*; in "brown" food web, 75, 76-77, 79, 86, 86, *88,* 89; rodent, 68, 78

diameter at breast height (DBH), 80

Diplopoda, 64

Diptera, 61-62, 64

dissolved inorganic nitrogen (DIN), 109, 111, 204

dissolved organic carbon (DOC), 101, 105, 158, 162, 198, 208

dissolved organic nitrogen, 111

disturbance events, 39, 41, 149

dormancy, 3, 6

downy woodpecker, 172, 229

dragonflies, 64

drought, xi, 29, 40, 41, 90, 138, 228, 229, 230, 234

Dryas integrifolia (mountain avens), 35, *36*

dry deposition, 20, *21*, 103-104, 108, 110, *112*

Dumanoski, Diane, vii

dusky salamander, 65

dust, 18

Dytiscus, 62

earthworms, 64, 76, 234

eastern chipmunk. *See* chipmunk, eastern

eastern hemlock, 46, 50, 187-188

ecosystem boundary, *12, 19,* 20

ecosystem services, 14, 94, 137, 211, 231

elderberry, 55

electrical conductivity, 135, *136*

electrical power, 114, 117, 121

elm, *35*

El Niño/La Niña Southern Oscillation (ENSO), 177

emerald ash borer, 225, 229

energy flow and transfer, 19, 22, 82-83, 84-89; in streams, 155-156

Environmental Literacy Program (ELP), 222

Environmental Protection Agency, 98, 127, 216

ephemeral stream, 29

Ephemeroptera, 61

epiphytes, 49

Erebidae, *61*

erosion, 35; of debris dams, 162, 163; from forest harvesting, 145, 148, 154; measur-

ing, 17, 22; of stuctures, 127; transpiration vs., 94, 101, 148, 152
eutrophication, 111
evaporation, 6, 17, *18*, 22, 90, 92–94
evapotranspiration, *18*, 22, 90, 92–94, 105, 133, 228
extrapair paternity (EPP), 177

Falls Brook, viii
fecundity, 173, 174, *175*, 176, 178, 212
ferns, 1, *34*, 49, 50, 57, *58*, 148, 149, 152, 236
finches, 5, 45, 66, 68, 236
Finger Lakes, 115–116
fir, *28*, *35*, 50. *See also* balsam fir
fish, 50, 62, 64, 127, 136, 197, *198*
fishers, 5, 68, *71*, 72, 76, 77, 178
flooding, 86, 138, 230
flycatchers, 169–170
flying squirrel, 6, 70, 178
fog, 115, 117
food web and feeding relations, 77
foraging, 45, 68
forest development: following glaciers, 36; following disturbance, 149
forest harvesting. *See* logging
Forest Science Dialogues, 221–222
fossils, 33
foxes, 68, *71*, 72, 77
frass, 1, 43, 87, 89, 234
freezing, xii, 3, 31; of conifers, 131, *133;* of soil, xii, 5, 29, 41, *43*, 100, 145, 195, 197, 203, 204, 205, 207, 228
frogs, 5, 64, 77, 87, 195
fungi, *46*, 68, *71*, 76, 77, *78*, 83, 87, 89, 156; parasitizing of, 6, 57; pathogenic, 53, 59; taxonomy of, 49, 50

Gale River, 143–144
garlic mustard, 233
garter snake, 62, 66
gasoline, lead in, 98
gastropods, 62
Geometridae, 59, 60, *61*
germination, 45, 52, 64
glacial till, 27–28, 33
glaciers, xi, 27, 33, *36*, 64, 129
global warming. *See* climate change
goldenrods, 57
goldfinch, 5, 68, 236
Gomphus, 78
granite, 27, 28
grasses, 34, *36*, 50, 57, 149
grazing food web, 77
green frog, 64, *65*

gristmills, 38, 39
grosbeaks, 5, 67, 68, 236
grouse, 68
growing season, 31, 203, 208
guilds, 68
gypsy moth, 225

hail, xi, 40, 41, 103, 115
hairy woodpecker, 172, 229
hardpan, 28–29
hare, 57, *71*, 72, 77
hardwoods, xi, *28*, 38, 39, 40, 41, 131–132, 190, 211
hawk moths, 59
hawks, 68, 76, 77, 178
hay-scented fern, *58*
heath, xi, 50
heather, 35
Hemiptera, 62, 64
hemlock, eastern, xi, 35, 36, 37, 38, 39, 46, 48, 50, 187–188, 190, 228, 233
hemlock woolly adelgid, 225, 228, 229
herbicides, 139, 140
herbivores, 59, 60, *61*, *62*, 68, 76, 78, 79, 206; mammalian, 72; as prey, 75; plants consumed by, 86, 87, 89; seeds consumed by, 53, 87
herbs, 6, *35*, 36, 48, 49, *58*, 149
hermit thrush, 6, 169, 172
heterotrophs, 76, 79, 81, 83, 85, 86, 87, 156
Hexatoma, 62
hibernation, 4, 6, 70, 71, 203, 229, 234
hobblebush, 1, *5*, *34*, 207, 229, 233; distribution of, 48, 54–55; moose consumption of, 72, *206*
Holmes, Richard T., *142*, *217*
honeysuckle, 50, 55
hornbeam, *35*
Hubbard Brook Cooperators' Meeting, 223
Hubbard Brook Research Foundation, viii, xii, 218, 219
Hubbard Brook Roundtable, 221
humidity, 31
hummingbirds, 68, *70*, 77, 78
humus, 28, 59, 79, 129, 228
hurricanes, xi, 40–41, 148, 151, 173, 183, 236
Hydracarina, 62
hydrochloric acid, 110
hydrogen, 95, 179–180
hydrogen ion, 104, 114, *115*, *131;* acid rain and, 96–97; declining levels of, 118, 135, 142; in precipitation, 98, 102, 112, *125;* in streams, 99, 113, 142, 145, 198, 228
hydrology, xi, 12, 90–94, 191, 228

hymenoptera, 59, 206
hyperparasites, 59

ice, 5, 203; melting of, 6, *204*
ice storms, xii, 15, 31, 40, 150, 203; climate change and, 230; of 1998, 41, *42*, 72; nitrate concentration and, 145
idia moth (*Idia americalis*), 60
inchworm moths, 59
indian pipe, 6, 57, *58*
Industrial Revolution, 96, 116
insectivores, 68, 69, 70–71, 86
insects, 1, 5, 36, 40, *45*, 57, 65, 71, 76, 77; amphibians' consumption of, 78; birds' consumption of, 67–69, 70, 173, 175, 176, 182, 183, 209, 210; defoliating, 15, 43; taxonomy of, 29, 49, 50, 59
Irene, Tropical Storm, xii, 41
iron, 28, 100

Jamaica, 179–181, 184
Johnson, Noye M., 13
juncos, 68, 169
juniper, 35

land use, 15, 215
La Niña, 177
larch, *35*
larvae, 60, *61*, *63*, 77, 87, 206
Lasiocampidae, *61*
lead, 21, 97–98, 100
leaf fall, 3, 31
least flycatcher, 172, 173
leather, 38, 39
Leopold, Aldo, 9, 108, 236
Lepidoptera, 59, 60–61, 76, 87, 175–176, 229
lianas, 49
lichen moth (*Clemensia albata*), 59, 60
lichens, 49, 50, 60
LIDAR (light detecting and ranging), 190–191, 192, *193*
lightning, 31
Likens, Gene E., 13, 17, 115, *216*, *217*, 231
lilies, 50, 57, *58;* blue-bead, 1; trout, 1, 6, 55, *56*, 209, 233
little white lichen moth, *60*
liverworts, 49, 50
logging, xi, 11, 38, *40*, 138–147, 153, 154, 230–231; salvage, 27, 39, 40, 41. *See also* clear-cutting; deforestation
long-tailed weasel, 72
Long-Term Ecological Research (LTER), xii, 13
long-term monitoring, 15

Long-Term Research in Environmental Biology (LTREB), xii, 13
luna moths, 59
Lyme disease, 225, 230

macrobenthos, 61
magnesium: climate change and, 205; leaching of, 129, 131, *132;* in precipitation, 97, 98, 102–103; in rocks, 35; in streams, 29, 97, 99, 102, *103,* 104, 113, 121, 142, *144,* 145
maize, 38
manganese, 159
mangroves, 181–182
maple, *36,* 37; mountain, 50; red, *35,* 46, 49, 149; striped, 50. *See also* sugar maple
maple syrup, 6, 39, 234, 236
marsh violet, *56*
marten, American, 72, 178
mast, 5, 15; animal consumption of, 4, 45, 68, 70, 72, 194; production of, 50–52, 70
mayflies, 50, 61, 62, 77, 137, 159
meadowlark, 170
Mecoptera, 64
mercury, 97, 98–99, 218
Merrimack River, 25, 38, 39, 230
metabolism, 83
mice, 5, 68, 70–71, 77, 78, 178, 229, 236
microbursts, xii, 41, 203
midges, 62, 159
migration, 4, 6, 66–67, 167–185, 211, 218; climate change and, 203, 208
millipedes, 49, 64, 76, 89
mink, 72
Mirror Lake, *10, 26,* 226; development near, 25, 27, 39; freezing of surface, 6, 203–204; pollutants in, 198; research on, 22; sediment core of, 33, 35; vegetation near, 33–38; wildlife near, 64, 66
mites, 59, 62, 77
moles, 68, 70
moose, xii, 2, *4,* 5, 13, 68, 76, *206,* 230; calcium in, 107, 108; feeding habits of, 55, 57; fluctuating population of, 72, 89; in grazing food web, 77, 83, 87
mosquitoes, 2, 6, 234
mosses, 1, 49, 50, 60, 77, 86, 156
moths, 42, 50, 59–60, *61,* 208, 225
mountain avens (*Dryas integrifolia*), 35, 36
mountain maple, 50
Mount Cushman, *26*
Mount Kineo, 25, *26, 47*
mourning cloak, 60
multiflora rose, 234
mushrooms, 3, 57, *78*

mustelids, 72
mutualism, 57
Mycena, 78
mycoheterotroph, 57
mycorrhizae, 57, 71

National Acid Deposition Program/National Trends Network (NADP/NTN), xii
National Atmospheric Deposition Program, 103
National Center for Atmospheric Research (NCAR), 118
National Dry Deposition Network (NDDN), xii
National Science Foundation, ix, xi, xi, 13, 130, 221, 222
nectar, 57, 68
nematode worms, 59, 62
Neotropical migrants. *See* American redstart; black-throated blue warbler; songbirds
net ecosystem flux (NEF), 107–108
net hydrologic flux (NHF), 110
net primary productivity, 79, 81, 148, 188–190, 207
net soil release, 103, 104–105, 108, 109, 110, *112*
New Hampshire Education and Environment Team (NHEET), 222
newts, 66
nitrate, 18, 100, *102,* 104, 111, 113, 117, *144,* 217; climate change and, 204, 207, 230; declining levels of, 226, 228; in precipitation, 121–122, *129;* in streams, 112, 121–122, 133, 135, 142, *143,* 145, 147, 197, 198, 226
nitric acid, 115, 122, 127
nitrogen, 2–3, 18, 21, 36, 57, 59, 103, 108, 121, 152, 158, 218; anaerobic transformation of, 30; climate change and, 205, 207, 208; decomposition of, 144–145; dissolved inorganic, 109, 111, 204; dissolved organic, 111; flux and cycling of, 11, 16, 34–35, 43, 109, 110–111; marking with isotope of, 164–165; mass balance of, 112; nitrification of, 145, 229; remote sensing of, 190; scarcity of, 153; in streams, 41, 43, 111, 112, 204
nitrogen oxide, 226; as acid rain precursor, 114–115, *116,* 122, 123, 218; increases in, 127; regulation of, 111, 121, 128, 226; toxicity of, 99
Noctuidae, 59, *61*
nor'easters, 31
northern cardinal, 229

northern hardwood forests, 33, *34,* 46, 50, *51,* 55, 89, 132, 147, 149, 153–154, 186
Notodontidae, 59, 60, *61*
nuthatches, 68, 169, 179, 229
nutrient cycling and flux, 10, *12,* 17, *19,* 20, 22, 27, 57, 59, 64, 95–113, 147, 207, 208; acid rain and, 129; of calcium, 12, 18, 21, 105–108; of chlorine, 110; climate change and, 201, 203, 206; of nitrogen, 11, 16, 34–35, 43, 109, 110–111; of sulfur, 108–110
Nymphaliidae, 60

oaks, 35, *36,* 37, 50, 188, 208, 228, 234
Odonata, 64
Odum, E. P., 75
Oligochaeta, 62
orchids, 2, 49, 50, 57
ovenbird, *67,* 68, 172, 178
owlet moths, 59, 60
owls, 6, 77; barred, 5, 68, 76, 236; saw-whet, 68
oxygen, 29
ozone, 99–100, *101,* 127, 128, 129; depletion of, *15*

painted trillium, *56*
paper birch, 33, 37, 41, 46, 48, 49, 149, 187
Papilioniidae, 60
parasites, 6, 57, 59, 175
particulates, 110, 158; calcium, 108; in emissions, 102, 117, 135; estimates of, 20; forest disturbances and, 145; storms and, 100–101; waterborne, 18, 29, 30, 86, 100, 103, 111, 159
pathogens, 43–45, 188, 228
Pawtucket Confederation, xi
Pemigewasset River, 11, 25, *26,* 27, 38, 39, 64, 91, 198, 230
perennial stream, 29
pests, 43–45, 228, 233
petioles, 2, 69
phenology, 207–208, 211
Philadelphia vireo, 172
phosphate, 226
phosphorus, 35, 57, 100, 108, 153, 158, 159, 204
photosynthesis, 41, 57, 75, 83, 85; carbon dioxide and, 89; defoliation and, 43, 45; herbivory and, 87; by mosses and algae, 77; nitrogen and, 190; during vernal transition, 207
phreatic boundary, 11
Pierce, Robert S., xi, *149, 217*
Pierce Ecosystem Laboratory, U.S. Forest Service, 10, 31, 223

pileated woodpecker, *45*

pin cherry, 141, 148, 234; acid rain's effect on, 129, *131;* dams formed by, 163; after disturbances, 40–41, 89, 149, *150,* 151–153, 154

pine, 33, 38, 208; white, 35, 36, 37, 188, 228

pink lady slipper, 57

pink wood sorrel, *58*

pit-and-mound topography, 27

Platyhelminthes, 62

Plectoptera, 61

poison ivy, 234, 236

pollen, 33–34, 35, 36, 50; tree, xi, 51

polyphemus moths, 59

porcupines, 5, *71,* 72, 77, 87, 108

potassium, 106, 113, 129; in precipitation, *97,* 98, 102; in rocks, 35; in streams, 29, *97,* 98, 103, 121, 142, *144,* 145; uptake of, 57, 152, 208

power plants, 114, 117

precipitation, 3, 5–6, 19, 20, 85–86; chemistry of, 12, 97–98, 102–103, 106, 108, 110, 135–136, 216, 226; climate change and, 202, 228; collectors, *21, 96;* declining solutes in, 135; dissolved organic carbon in, 105; evapotranspiration and, 22; frequency of, 31, 32; infiltration of, 29; interception of, 101; lead in, 100; nitrates in, 121–122, 129; nitrogen in, 35, 111, *112,* 127; stream networks fed by, 17, *18,* 22, 90–92, 94, 96–97; sulfates in, 15, 112, 121–122, 129. *See also* acid rain; hail; rain; snow

predators, 11, 66, 75, 79, 86; birds as, 76, 77, 87, 89; of birds, 173, 176, 178, 185; eggs consumed by, 65, 70, 72; invertebrate, 59, 62, 64; mammalian, 76, 77, 87; seeds consumed by, 50, 52

primary consumers, 76, 78–79

procyonid, 72

productivity, 76

prominent moths, 59

protozoa, 49

pseudoscorpions, 64, 206

quaking aspen, 50, 149

raccoons, 62, 72

radiocarbon isotope dating, 34

railroads, 38, 39

rain, 3, 85, 86, 90, 101, 103, 153, 183, 207, 227; climate change and, 204; dissolved organic carbon in, 105; soil chemistry affected by, 11. *See also* acid rain

rainstorms, 91, 119

raptors, 68

raspberry, 55, 148, 149, *150,* 152, 154, 234

Reagan, Ronald, 216

red-backed salamander, 65–66, 82, *83*

red-backed vole, 53, 70, *71*

red-bellied woodpecker, 229, 234

red eft, 2, *66*

red-eyed vireo, 1, *67,* 170, *171, 172*

red fox, *71,* 72, 76

red maple, *35,* 46, 49, 149

red oak, 188, 228

red-spotted newt, 66

red spruce, *47,* 112, *132, 133, 135,* 149; acid rain's effect on, 131; climate change and, 228, 229; prevalence of, 33, 36, 37, 46, 49, 187; snowfall and, 4–5

red squirrel, 5, 52, 68–70, *71,* 236; food sources, 70; as nest predator, 6, 70, 178, 194; population density, 52, 194–195, 196

redstart, American, *168,* 170–175, 178, 179–182, 183, *195*

red-tailed hawk, 68

red trillium, 1, 55, *56,* 233

regeneration, 45, 52

remote sensing, 190–191

reptiles, 50, 66

Research Experience for Teachers (RET), 222

Research Experience for Undergraduates (REU), 222–223

respiration: of animals, 87; of humans, 127; of plants, 10, 86

Rhizoctonia spp., 53

Rhyacophilia, 62

rime ice, 115

road salt, 226

robins, 236

rodents, 53, 57, 70, 71, 77, 78; masting and, 45, 52, 72; prevalence of, 194–195; as prey, 5, 68, 76; seeds consumed by, 152

root exudates, 107, 110

rose-breasted grosbeak, *67,* 172

rose-twisted stalk, *58*

Rothamsted Experimental Station, 14

round-leaved orchid, 57, 59

round worms, 62

ruby-throated hummingbird. *See* hummingbirds

ruffed grouse, 68

running pine, *58*

rushes, 50, 57

Russula, 78

saddled prominent caterpillar, 41, 42, *44,* 61, 62

salamanders, 2, 31, 64, 66, 77, 157, 195, 229; biomass of, 82–83; calcium in, 107; dispersal of, 164, 165–166; energy consumption by, 87; hibernation of, 4, 5; population decline among, 89; as predators, 62, 65, 78; in streams, 30, 65

salmon, 230

salvage logging, 27, 39, 40, 41

samaras, 52

sampling, 187

sap, 6, 68, 77

sapsuckers, 68, *70,* 77, 78

Saturnidae, 59, *61*

sawmills, xi, 25, 38, 39, *40*

saw-whet owl, 68

scarlet tanager, 1, *67,* 68, 172

Science Links, 218, 220–221

scorpionflies, 64

sculpin, 64, 137

Scutellinia, 78

sedges, 34, *36,* 50, 57

sedimentary rock, 28

seeds, seedlings, 4, 64, 148; of beech trees, 45, 48, 52, *53;* dispersal of, 52, 53, 89, 152; on forest floor, 1; germination of, 45, 62, 68, 89; of pin cherries, 40–41, 89, 141, 152; predators and, 62, 66, 70, 71, 76, 77, 78, 152; production of, 50–52, 70; of sugar maple, 48, 52, *53,* 57, 58; survival, 53; uprooting and, 27; of yellow birch, 53–54. *See also* masting

seepage, 21–22, 100

seeps, 57

shade tolerance, 52, 150

sharp-shinned hawk, 178

shelf fungi, 57

short-tailed shrew, 70

short-tailed weasel, 72

shrews, 68, 70, 76, 77, 87

shrubs, xi, 48, 49, 54–55, 68, 70, 170

silica, 198

silicate, 132

Simuliidae, 61

siskin, 5, 68, 236

skunk, 72

slash, 38

sleet, 103, 115, 203, 204

slimy sculpin, 64, 197

slugs, 50, 57, 62, 77

snails, 50, 62, 65, 76, 77, 136

snakes, 62, 66, 77

snow, 39, 71, 90, 103, 153, 154, 207; acidity of, 114, 115, 120; depth of, 4, 29, 31, 202–203, 204, 206; dissolved organic carbon in,

snow (continued)
105; erosion and, 101; as insulation, 5, 29, 41; melting of, 5–6, 27, 29, 32, 55, 91, 99, 100, 137, 208, 211; organic matter in, 85

snowshoe hare, 57, *71*, 72, 87, 236

snowstorms, 119

Society for the Protection of New Hampshire Forests, 39

sodium, 102, 113, 121, 129, 142, *144*, 145, 198

sodium bicarbonate, 160

sodium chloride, 198

sodium sulfates, 104, 226

softwoods, xi, 39

soil, 10, 13, *28*, 191; acid rain and, 132–135; buffering capacity of, 122, 127, 129, 131, 132–135, 226, 230; climate change and, 205–206; fertility of, 35; freezing of, xii, 5, 29, 41, *43*, 100, 145, 195, 197, 203, 204, 205, 207, 228; infiltration of, 29, 91, 100, 101; nutrients in, 9, *12*; organisms in, 59, 83, 89; rainwater's effect on, 11; sampling of, 80; temperature of, 31, 151, 202, 208; transpiration from, 92, 94; warming of, 31–32. *See also* net soil release

solar radiation, 31, 32, 55, 75, 84–86, 155, 156, 207, 231

solutes, 135

songbirds, 60, *67*, 70, 87, 167, 170, 174, 175, 178–182; effects of breeding season on, 170–179; impact of winter period on, 179–185; mortality of, 183–184

soot, 128

sorrels, 57, *58*

sparrows, 170

Sphingidae, 59

spiders, 49, 59, 64, 65, 68, 76, 77, 206

spodosol, 28

spring beauty, 6, 55, *56*, 233

spring ephemerals, *56*, 57, 208, *209*, 233

spring peeper, 64, *65*

spring salamander, Northern, 65, *166*

springtails, 5, 59, 62, 206

spruce, xi, *28*, *35*, 36, 37, 38, 39, 50, 229

squash, 38

squirrels, 50, 68, 77, 108; flying, 6, 70, 178; red, 5, 6, 52, 70, *71*, 178, 194, *196*, 236

state factors, *14*

stochastic factors, 13, *14*

stomata, 92

stoneflies, 5, 50, 61, 62, 77, 87, 157, 164–165

storms, 29, 30, 100–101, 183, 233; bird mortality after, 179; hail, xi, 40, 41, 103, 115;

ice, xii, 15, 31, 40, 41, *42*, 72, 145, 150, 203, 230; rain, 91, 119; snow, 119; thunder, 31; tropical, xii, 41; wind, 3, 6, 27, 31, 40, 85, 117, 150, 230

stream order classification, Strahler-Horton system, *29*

streams, *73*, *92*, *93*; acidity of, 96, 133, 197, 198, 226–227; chemistry of, 11, 15, 96, 97, 98, 99, 100, 102, 106, 108, 110, 133, 135–136, 157–159, 187, 188, 197–198, *199*, 200, 216, 226–228; climate change and, 208; declining solutes in, 135; dissolved organic carbon in, 105; as ecosystems, 155–166; energy budget and, 85–86; ephemeral, 29; experimental manipulation of, 157–161; fish species in, 64, 197, *198*; hierarchy of, 29; invertebrates in, 61–62; nitrates in, 112, 121–122, 133, 135, 142, *143*, 145, 147, 197, 198, 226; nitrogen in, 41, 43, 111, 112, 204; perennial, 29; precipitation and, 17, *18*, 22, 90–92, 94, 96–97; salamanders and, 30, 65; in springtime, 6; sulfates in, 112, 121–122, 135, 198; suspended materials in, 30, 86; watersheds drained by, 11–12, 17, *18*, 86

striped maple, 50

Sturges, Frank W., *44*

sublimation, 5

subsistence farming, 38

sugar maple, 1, *27*, 48, 129, 131, *132*, 133, 134, 149, *150*, 190, 217, 233, 234; climate change and, 206, 207, 208, *210*, 228–229; dams formed by, 163; leaves of, 69; masting by, 68, 194; northward migration of, 35, 36, 208; prevalence of, 33, 37, 41, 46, 49, 50, 187; reproduction of, 52, *53*, 57, 59; responses to calcium addition, 134; sap of, 6, 39

sulfate, 102, 109, 110, 113, 117, 120, *121*, 216, 226, 230; in precipitation, 15, 112, 121–122, *129*; in streams, 112, 121–122, 135, 198; sulfur dioxide emissions and, 15

sulfur, 18, 106, 120, 122, 128; dry deposition of, 103–104; flux and cycling of, 108, 110, *111*; mass balance of, 111; as nutrient, 127; as pollutant, 95, 103, 117–118, 121, 123

sulfuric acid, 108, 112, 115, 127, 158

sulfur oxides, 104, 122; as acid rain precursor, 114–115, 118, 218; regulation of, 112, 121, 127, 128, 226; sources of, 114, *116*, 117, 120, 123; sulfates and, 15

sunlight, 1, 155, 173

Susquehanna River, 30

Swainson's thrush, 172

swallowtails, 60

tannins, 38, 39

temperature: of air, 3; decomposition and, 148; of soil, xii, 5, 29, 31–32, 41, *43*, 100, 145, 151, 195, 197, 202, 203, 204, 205, 207, 208, 228; of water, 29. *See also* climate change

termites, 87

Tewksbury, Joshua J., 49

thrushes, 6, 68, 169–170, 172, 173, 178

thunderstorms, 31

ticks, 72, 225, 229–230, 233

tiger moths, 59

Tipulidae, 62

titmice, 229, 234

toads, 5, 64, *65*, 77

topographic boundary, 11, *18*, 21

toxic metals, 21

transpiration, 90, 92, 101, 152; deciduous trees and, 29, 91; deforestation and, 141, 148; importance of, 94; light's effect on, 45; solar radiation and, 85; water loss from, 6, 91. *See also* evapotranspiration

tree life history traits, 52

tree pollen, xi, 51

tree rings, 80

Trichoptera, 61

trillium, 1, 55, *56*, 233

Tropical Storm Irene, xii, 41

trout lily, 1, 6, 55, *56*, *209*, 233

tufted titmouse, 229

tundra, xi, 34, 36

turkey, 68

turtles, 66

two-lined salamander, 65, *166*

ultraviolet radiation, 99

UNESCO, xii

U.S. Forest Service, xi, 13, 31, 39–40, 90, 147, 201–202, 222

U.S. Weather Bureau, 90

uprooting, 1, 27, 41

vascular plants, 49

veery, 172

vernal transition, 207–208

vernal window, 155

viceroy, 60

vines, 234, 236

violets, 6, 50, 55, *56*

vireos, 1, *67*, 68

Virginia creeper, 234
voles, 5, 53, 68, 71, 77, 78

warblers, 55, 68, 174–177, 179–181, 234; black-
    burnian, *67,* 172, 229; blackpoll, *195,* 229;
    black-throated green, 170, *171,* 172, 173,
    229; Canada, *67;* yellow-rumped, 170,
    172, 194, *195. See also* black-throated blue
    warbler
wasps, 59, 77
water cycle, 90
waterfalls, 137, 197
water fleas, 62
water quality, 14, 207, 215, 230
watershed-ecosystem, definition, 11, 18, 19
water striders, 62
water supply, 9, 14
weasels, 5, 68, 72, 76, 77
Weeks Act (1911), 39
weevils, 77
weirs, 20–21

West Nile virus, 234
white admiral, 60
white ash, 46, 48, 50. *See also* ash
white-breasted nuthatch, 229
white-footed mouse, 178
White Mountains, *26,* 35, 38, 114, 115–116, 139,
    143–144, 154, 211
white oak, 228
white pine, 35, 36, 37, 188, 228
wild boar, 236
wild grape, 234
wild turkey, 68
willow, xi, 35, 55
wind, 3, 6, 27, 31, 85, 117, 150, 230; hurricanes,
    xi, 40–41, 148, 151, 173, 183, 236; pollen
    dispersed by, 50, 51, 52
windstorms, 6, 40
winter wren, 172
witness trees, 37
Wollastonite, 132–133, *134, 135,* 161
woodbine, 234, 236

wood frog, 64–65
woodland jumping mouse, 71, 78, 229
woodpeckers, 5, *45,* 66, *70,* 169, 172, 179, 229,
    234, 236
wood thrush, 169–170, 172, 173
wool, 38
woolly beech scale (*Cryptococcus fagisuga*),
    43, 45
worms, 62, 64
wrens, 6, 172

yellow-bellied sapsucker, *70,* 172
yellow birch, 1, 27, 52, 149, *150,* 217, 228, 234;
    dams formed by, 163; ice storm's damage
    to, 41; prevalence of, 33, 46, 49, 50, 187;
    seeds of, 53–54
yellow-rumped warbler, 170, 172, 194, *195*
yellow violet, 55, *56*

Zig-Zag Brook, viii
zinc, 97